CONTROL AND DYNAMIC SYSTEMS

Advances in Theory and Applications

Volume 39

CONTRIBUTORS TO THIS VOLUME

J. K. AGGARWAL
A. K. BEJCZY
GEORGE A. BEKEY
YILONG CHEN
S. GANGULY
WITOLD JACAK
JUNG-HA KIM
VIJAY R. KUMAR
SUKHAN LEE
Z. LI
NEIL D. MCKAY
ZVI S. ROTH
KANG G. SHIN
T. J. TARN
TAKESHI TSUJIMURA
KENNETH J. WALDRON
Y. F. WANG
TETSURO YABUTA
HANQI ZHUANG

CONTROL AND DYNAMIC SYSTEMS

ADVANCES IN THEORY AND APPLICATIONS

Edited by
C. T. LEONDES

Department of Electrical Engineering
University of Washington
Seattle, Washington

VOLUME 39: ADVANCES IN ROBOTIC SYSTEMS
Part 1 of 2

ACADEMIC PRESS, INC.
Harcourt Brace Jovanovich, Publishers
San Diego   New York   Boston
London   Sydney   Tokyo   Toronto

Academic Press, Inc.
San Diego, California 92101

United Kingdom Edition published by
ACADEMIC PRESS LIMITED
24-28 Oval Road, London NW1 7DX

Library of Congress Catalog Card Number: 64-8027

ISBN   0-12-012739-3   (alk. paper)

PRINTED IN THE UNITED STATES OF AMERICA
91 92 93 94   9 8 7 6 5 4 3 2 1

CONTENTS

CONTRIBUTORS

Numbers in parentheses indicate the pages on which the authors' contributions begin.

J. K. Aggarwal (435), *Computer and Vision Research Center, The University of Texas, Austin, Texas 78712*

A. K. Bejczy (129, 177), *Jet Propulsion Laboratory, California Institute of Technology, Pasadena, California 91109*

George A. Bekey (1), *Computer Science Department, University of Southern California, Los Angeles, California 90007*

Yilong Chen (317), *General Motors Research Laboratories, Warren, Michigan 48090*

S. Ganguly (129), *Department of Systems Science and Mathematics, Washington University, St. Louis, Missouri 63130*

Witold Jacak (249), *Institute of Technical Cybernetics, Technical University of Wroclaw, Wroclaw, Poland*

Jung-Ha Kim (289), *Department of Mechanical Engineering and Applied Mechanics, University of Pennsylvania, Philadelphia, Pennsylvania 19104*

Vijay R. Kumar (289), *Department of Mechanical Engineering and Applied Mechanics, University of Pennsylvania, Philadelphia, Pennsylvania 19104*

Sukhan Lee (1), *Jet Propulsion Laboratory, California Institute of Technology, Pasadena, California 91109, and Department of Electrical Eng.-Systems, University of Southern California, Los Angeles, California 90007*

Z. Li (177), *Mallinckrodt Institute of Radiology, Washington University School of Medicine, St. Louis, Missouri 63110*

Neil D. McKay (345), *Computer Science Department, General Motors Research Laboratories, Warren, Michigan 48090*

Zvi S. Roth (71), *Robotics Center, College of Engineering, Florida Atlantic University, Boca Raton, Florida 33431*

Kang G. Shin *(345), Department of Electrical and Computer Engineering, The University of Michigan, Ann Arbor, Michigan 48109*

T. J. Tarn (129, 177), *Department of Systems Science and Mathematics, Washington University, St. Louis, Missouri 63130*

Takeshi Tsujimura (405), *NTT Transmission Systems Laboratories, Tokai, Ibaraki-ken, 319-11, Japan*

Kenneth J. Waldron (289), *Department of Mechanical Engineering, The Ohio State University, Columbus, Ohio 43210*

Y. F. Wang (435), *Department of Computer Science, University of California, Santa Barbara, California 93106*

Tetsuro Yabuta (405), *NTT Transmission Systems Laboratories, Tokai, Ibaraki-ken, 319-11, Japan*

Hanqi Zhuang (71), *Robotics Center, College of Engineering, Florida Atlantic University, Boca Raton, Florida 33431*

PREFACE

Research and development in robotic systems has been an area of interest for decades. However, because of increasingly powerful advances in technology, the activity in robotic systems has increased significantly over the past decade. Major centers of research and development in robotic systems were established on the international scene, and these became focal points for the brilliant research efforts of many academicians and industrial professionals. As a result, this is a particularly appropriate time to treat the issue of robotic systems in this international series. Thus this volume and Volume 40 in this series are devoted to the timely theme of "Advances in Robotic Systems Dynamics and Control."

The first contribution to this volume, "Applications of Neural Networks to Robotics," by Sukhan Lee and George A. Bekey, is an excellent example of the impact of powerful advances in technology on advances in robotic systems. Specifically, while neural network theory has been pursued for decades, it is only now, with the tremendous advances in integrated electronics, that it is possible to reduce neural network techniques to practice and in particular, to do so in the case of robotic systems. The control of robot manipulators involves three fundamental problems: task planning, trajectory planning, and motion control. To date, most of the useful work in robotics has been in the areas of trajectory planning and motion control. It is these areas to which neural network techniques are presented in this first contribution.

The next contribution, "A Unified Approach to Kinematic Modeling, Identification, and Compensation for Robot Calibration," by Hanqi Zhuang and Zvi S. Roth, is a rather comprehensive treatment of the robot calibration problem, which is the process of enhancing the accuracy of a robot manipulator through modification of the robot control software. Three distinct actions are required in this process of robot calibration, namely, measurement, identification, and modification. The need for robot calibration arises in many applications, and its importance is further manifested by the growing number of publications in this area in recent years. In addition to critically examining the status of this area of major importance to robotic systems, this contribution presents a unified approach to all phases of model-based calibration of robotic manipulators. As such, this is an important element of these two volumes.

The systems aspects of robotics, in general, and of robot control, in particular, are manifested through a number of technical facts. These are their degrees of freedom, the system dynamics descriptions and sensors involved in task performance, modern robotics computer control implementations, coordination in multiple robotic elements systems, and the development of planning and task decomposition programs. The next contribution, "Nonlinear Control Algorithms in Robotic Systems," by T.J. Tarn, S. Ganguly, and A.K. Bejczy, focuses on robot control algorithms and their real-time implementation. It presents a rather comprehensive treatment of the issues involved in addition to powerful techniques for this problem.

The next contribution, "Kinematic and Dynamic Task Space Motion Planning for Robot Control," by Z. Li, T.J. Tarn, and A.K. Bejczy, presents techniques for an integrated treatment of robotic motion planning and control, which traditionally have been treated as separate issues. A rather comprehensive analysis of the literature on robot motion planning is presented. The techniques presented in this contribution provide a framework within which intelligent robot motion planners can be designed; as such, this contribution is an essential element of these two volumes on advances in robotic systems dynamics and control.

The next contribution, "Discrete Kinematic Modeling Techniques in Cartesian Space for Robotic System," by Witold Jacak, presents rather powerful techniques for kinematic modeling in robotic systems. The techniques for modeling robotic kinematics presented in this contribution have the essential features of convenience in computer simulation of robotic motion, facility in analysis of obstacle avoidance, and functional simplicity (i.e., computational complexity is kept to a minimum). Because of the fundamental importance of the issues treated in this contribution, it is an essential element of these two volumes.

Dexterous, multifingered grippers have been the subject of considerable research in robotic systems. The kinematic and force control problems engendered by these devices have been analyzed in depth in the literature. In the next contribution, "Force Distribution Algorithms for Multifingered Grippers," by Jung-Ha Kim, Vijay R. Kumar, and Kenneth J. Waldron, highly effective techniques are presented for computing finger forces for multifingered grippers through the means of decomposition of the finger forces field into equilibrating forces and interacting forces. The techniques presented are optimal for two- and three-fingered grippers and suboptimal for more complicated grippers.

Because of their simplicity, PD (or PID) controllers are widely used with various robot arm control methods. Other methods utilized include approximate linearization techniques, the computed torque method, hierarchical control techniques, the feedforward compensation method, and adaptive control techniques. In the next contribution, "Frequency Analysis for a Discrete-Time Robot System," by Yilong Chen, it is shown that lag-lead compensation techniques are substantially more effective than PID controllers with respect to static accuracy, better stability, reduced sensitivity to system model uncertainty, and less sensitivity to noise. In other words, they are more robust.

The goal of automation is to produce goods at as low a cost as possible. In practice, costs may be divided into two groups: fixed and variable. Variable costs depend upon details of the manufacturing process and include, in the cases where robots are used, that part of the cost of driving a robot, which varies with robot motion, and some maintenance costs. Fixed costs include taxes, heating costs, building maintenance, and, in the case of a robot, robot operating costs. If one assumes that the fixed costs dominate, then cost per item produced will be proportional to the time taken to produce the item. In other words, minimum production cost is closely related to minimum production time. The next contribution, "Minimum Cost Trajectory Planning for Industrial Robots," by Kang G. Shin and Neil D. McKay, presents an in-depth treatment of this significant issue of minimum cost utilization of robots in industrial production and techniques for accomplishing this.

An essential issue in many robotic systems is the detection of shapes of objects with which a robotic system is to interact. The next contribution, "Tactile Sensing Techniques in Robotic Systems," by Takeshi Tsujimura and Tetsuro Yabuta, presents techniques for dealing with this major issue through the utilization of force/torque sensors and probes. Computer vision is one of the means examined frequently in the literature with respect to this issue of environment recognition, but it is not without significant computational limitations. As a result, this contribution is significant in that it presents techniques for an important alternative where there are, indeed, few alternatives.

In robotic systems many types of sensors may be used to gather information on the surrounding environment. Different sensors possess distinct characteristics, which are designed based on differing physical principles, operate in a wide range of the electromagnetic spectrum, and are geared toward a variety of applications. A single sensor operating alone provides a limited sensing range and can be inherently unreliable due to possible operational errors. However, a synergistic operation of many sensors provides a rich body of information on the sensed environment from a wide range of the electromagnetic spectrum. In the next contribution, "Sensor Data Fusion in Robotic Systems," by J.K. Aggarwal and Y.F. Wang, an in-depth treatment is presented of techniques and systems for data fusion, once again a major issue in many robotics applications.

This volume is a particularly appropriate one as the first of a companion set of two volumes on advances in robotic systems dynamics and control. The authors are all to be commended for their superb contributions, which will provide a significant reference source for workers on the international scene for years to come.

APPLICATIONS OF NEURAL NETWORKS

TO ROBOTICS

SUKHAN LEE   and   GEORGE A. BEKEY

Jet Propulsion Laboratory   Computer Science Department
California Institute of Technology   University of Southern California
and
Dept. of Electrical Eng.-Systems
University of Southern California

I.   Introduction

The field of robotics concerns the design and application of articulated mechanical systems to manipulate and transfer objects, to perform
mechanical tasks with versatility approaching that of human arms and
to provide mobility. A variety of autonomous and semi-autonomous systems are termed "robots" if they involve processing of sensory inputs
from the environment and some mechanical interaction with it.

In view of the fact that robot manipulators (or legs) are open-chain
kinematic mechanisms, their control is difficult. There is clearly coupling
between motions of individual segments. Furthermore, the parameters
of a manipulator depend upon its configuration and the governing equations are highly nonlinear. The control of robots is particularly difficult
since the desired trajectory of the end-point of the arms (or legs) is
normally specified in Cartesian space, while motions are actually obtained from actuators located at the joints. The transformation from
Cartesian to joint coordinates is a computationally intensive problem,

CONTROL AND DYNAMIC SYSTEMS, VOL. 39
1

the accurate solution of which depends both on the algorithms used and on precise knowledge of robot parameters. Living organisms with articulated extremities perform the transformation from goal space to actuator (muscle) coordinates whenever they move. While some aspects of this transformation appear to be pre-programmed in the genes (thus enabling animals to move almost immediately after birth), other aspects appear to be learned from experience. This aspect of motion control in biological systems has provided a model for the application of connectionist approaches to robot control, since neural networks can, in principle, be trained to approximate relations between variables regardless of their analytical dependency [35]. Hence, it is appealing to attempt to solve various aspects of the robot control problem without accurate knowledge of the governing equations or parameters, by using neural networks trained by a sufficiently large number of examples.

The control of a robot manipulator involves three fundamental problems: task planning, trajectory planning and motion control. In the task planning phase, high level planners manage and coordinate information concerning the job to be performed. Trajectory planning involves finding the sequence of points through which the manipulator end-point must pass, given initial and goal coordinates, intermediate (or via) points and appropriate constraints.

Such constraints may include limits on velocity and acceleration or the need to avoid obstacles. Given such a trajectory, the motion control problem consists of finding the joint torques which will cause the arm to follow it while satisfying the constraints. Artificial neural networks find applications in all three of the problem areas indicated above. However, since most of the useful work to date has been done in trajectory planning and motion control, we shall discuss these two areas first. There are two approaches to the trajectory planning problem, which are referred to as *joint space planning* and *Cartesian space planning* respectively. Since trajectory constraints are generally specified in Cartesian space, planning a trajectory in joint space requires that the location of the end points and via points be transformed to their corresponding joint space coordinates. A smooth trajectory can then be obtained (say by fitting a polynomial to these points) [7]. Alternatively, the path planning can be done in Cartesian coordinates and then each path point converted into its corresponding joint space values for control. Clearly, the key to Cartesian space trajectory planning is the transformation of information from Cartesian to joint coordinates, known as a robot arm inverse kinematic problem, which we consider next.

II. Robot Arm Inverse Kinematics

In a robot arm, the joint coordinates θ are related to the Cartesian coordinates $\mathbf{x}$ by the kinematic equation

$$\mathbf{x} = \mathbf{f}(\theta) \tag{1}$$

For a six-degree of freedom arm, both θ and $\mathbf{x}$ are six dimensional vectors. When path planning is done in Cartesian coordinates, the required trajectory is obtained by the planning algorithm and then transformed to joint space by solving eq. (1). Since this solution requires inverting eq. (1), this approach is termed *position-based inverse kinematic control*. In many cases, the trajectory specification includes velocity constraints, in which case the forward kinematic equation is obtained by differentiating eq. (1):

$$\dot{\mathbf{x}} = \mathbf{J}(\theta)\dot{\theta} \tag{2}$$

where the elements of the Jacobian matrix J are the partial derivatives

$$\frac{\partial x_i}{\partial \theta_j} \quad \forall\, i, j$$

Solution of eq. (2) yields the inverse relation

$$\dot{\theta} = \mathbf{J}^{-1}(\theta)\dot{\mathbf{x}} \tag{3}$$

At any joint position θ the planner now computes the velocity $\dot{\mathbf{x}}$ which causes the manipulator end point to move toward the next via point or end point. Thus, trajectory planning based on (3) is referred to as velocity- based inverse kinematic control (or inverse Jacobian control). Clearly, one must assume that the Jacobian is invertible at each point for this method to be feasible. In practice, the Jacobian matrix is well behaved, except near singularity points [7]. Since efficient inversion of the Jacobian is evidently the key to successful application of this method, a number of algorithms have been proposed, e.g. [6, 23].

A manipulator having more degrees of freedom than required by the given task is called a *redundant manipulator*, e.g., a manipulator working in the 6 dimensional Cartesian space with more than 6 joints. The forward kinematics equations, (1) and (2), of a redundant manipulator represent underdetermined set of equations, and the corresponding inverse kinematic solutions yield solution manifolds instead of a unique solution. In this case, the inverse kinematic problem concerns about an optimal solution based on additional constraints or performance indices such as *manipulability*.

It should be noted that success of the inverse kinematic method depends not only on efficient inversion of the Jacobian, but on accurate knowledge of the robot kinematic parameters. In the absence of such knowledge, it may be necessary to use system identification techniques to obtain parameter estimates before trajectory planning can begin. Since neural network approaches do not depend on accurate a-priori knowledge, they are attractive alternatives to the inverse Jacobian method.

One of the earliest connectionist approaches to robot control is due to Albus [1]. His "Cerebellar Model Articulation Controller" (CMAC) uses a three-layer network, the first set of connections being random while the second uses adjustable weights. The network has no advance knowledge of the structure of the system being controlled and thus can be trained to accomplish the robot control task, provided there are sufficient adjustable and random connections. The basic idea of CMAC is to compute control commands by look-up tables rather than by solving control equations analytically. The table is organized in a distributed fashion, so that the function value for any point in the input space is derived by summing the contents over a number of memory locations. While the work of Albus pioneered the application of neural networks to robotics, he did not attempt to model the structural characteristics of networks of neurons as did later investigators.

Kuperstein [22] concerned himself with models of visual motor co-ordination in robots. While he did not explicitly address the inverse kinematics problem, his work did in fact use neural networks to obtain the transformation needed to convert desired hand coordinates in Cartesian space into the appropriate joint coordinates. The work is based on that of Grossberg and Kuperstein on adaptive sensory-motor control [11]. The system was designed to teach a three-joint robot arm to move to a point in three-dimensional Cartesian space as located by a vision system. No kinematic relationships nor the calibration of joint angles to actuator signals were known *a priori*. The architecture of the system consisted of an input layer fed by a stereo camera, whose outputs are connected to three arrays which convert the visual inputs into distributions in terms of camera orientations and their disparity. These distributions are connected to a target map with adjustable weights. The strategy followed by Kuperstein is the following. First, a random generator activates the target map which orients the robot arm into random positions. These positions are sensed by means of the camera and registered on the input map. The outputs of this map are then correlated with the desired or target locations. At the same time the network receives the visual activation corresponding to the end of the arm and determines an activation pattern which is compared with the actual pattern. Errors are used to adjust weights in the network by means of Hebb's rule [35]. Basically, this is a *circular sensory-motor reaction* [11] in which a spatial representation is formed based on signals used to orient and move in the space.

Ritter, Martinez and Schulten [34, 29] presented a different approach to the above visuomotor coordination problem dealt by Kuperstein. They applied the Kohonen's self-organizing feature mapping algorithm [20] for the construction of topology conserving mappings between the camera "retinas" and the (two or three dimensional) neural lattice, as well as between the neural lattice and the joint space. Each neuron in the neural lattice is associated with a pair of camera retinal coordinates, a joint angle vector, and Taylor series coefficient matrix to be used for proper interpolation. An object is presented randomly to the system to induce camera retina coordinates and the corresponding neuron in the neural lattice. The joint angle vector associated with the neuron is applied to the robot arm, so as to compute the error between the camera retinal coordinates of the presented object and that of the robot end effector. Learning is based on the Widrow-Hoff type error correction rule [38].

Inverse kinematic control has also been studied by a number of other investigator, e.g., Elsley [8], Guez and Ahmad [12], Josin et al. [15] and Psaltis et al [33], using the capability of multilayer feedforward network in generalizing a mapping and the backpropagation algorithm for learning. Bahren et al. [2] used a heteroassociative memory as a content addressable memory, to store and retrieve the solutions of inverse kinematics based on the dynamics of a network formed by the bidirectional connection between neurons, and the concept of terminal attractors [42].

However, there still remains a number of problems to be resolved to make the neural network approach to inverse kinematic problem a viable alternative to the conventional approach based on numerical computation. First, all the papers cited above have demonstrated the algorithms on robots with either 2 or 3 degrees of freedom (DOF). Attempts to apply backpropagation directly to systems with more DOF's have not been very successful, since these systems typically exhibit high-order nonlinearities and hence very slow learning rates. In order to achieve reasonable convergence time in learning, it appears to be necessary to decompose the systems into smaller subsystems which have better scale-up properties. One approach to the first problem is discussed below in Section II, A. Second, we need to ensure the required control accuracy uniformly throughout the workspace with a limited number of neural elements. Furthermore, we need to handle "one-to-many" inverse mapping required for redundant arm kinematic control, which has not been addressed by previous works. One approach to the second problem is discussed in Section II, B.

A. Context-Sensitive Networks

The usual way to represent a general computation unit by means of neural networks is shown in Fig. 1a, where the box labeled "Network" contains the required number of input and output units and at least two hidden layers. For the inverse kinematic problem, θ and $\dot{x}$ from Eq. (3) are used as the input, while $\dot{\theta}$ is the output of the network. A 6-DOF robot requires 12 input and 6 output units. The mapping is highly nonlinear since the transformation depends on the location of the system in the coordinate frame. It is well-known that learning in highly-nonlinear mappings may be very time-consuming. As shown by Yeung [40, 41], the set of input variables can be partitioned into two groups (Fig. 1b). One set is used as the input to the network which approximates the basic mathematical operations being represented (the function network), while the second set determines the setting or context within which the function is determined. In the robotics case, the context is the spatial location of the manipulator.

1. Architecture And Properties of Context Networks

We now consider context-sensitive networks of the feedforward type, such as commonly used with with the back-propagation algorithm. Suppose w_{ij} is a weight in either the function or the context network, connecting unit i in one layer with unit j in the next layer. Let the inputs be denoted by x and the outputs by y as usual. Then the total input to unit j can be written as

$$x_j = \sum_i w_{ji} y_i + b_j \tag{4}$$

where b_j is the bias term associated with unit j. The output y_j is related to the input through a sigmoid activation function $f_i(\cdot)$. Since the output units of the context network are used to set up the weights in the function

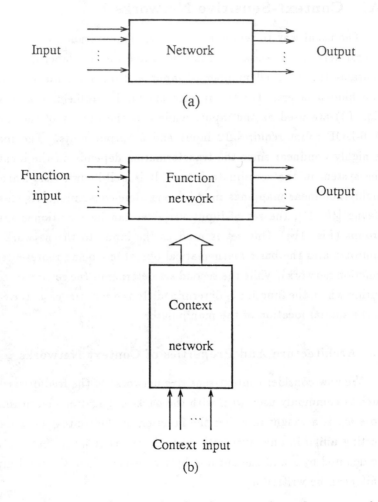

Figure 1: Structure of conventional and context-dependent networks.

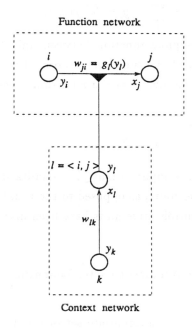

Figure 2: Coupling from context to function network.

network, the context network has as many output units as there are weights in the function network. Since this number can be very large, it is desirable that the function network be as simple as possible; ideally, it should be linear. The discussion which follows will concentrate on linear function networks; clearly, other choices are possible.

Learning in feedforward context-sensitive networks can be accomplished by an extension of the back-propagation algorithm [40]. Consider Fig. 2 which shows unit k in the last hidden layer of the context network being connected to output unit l, which sets up the weight w_{ji} in the function network. It can be shown that the required gradient components can be computed using back propagation in the function network.

It is interesting to note the nature of the coupling. Consider the weight w_{ji} in Fig. 2. The total input to unit j can be expressed as

$$x_j = \sum_i w_{ji} y_i = \sum_i g_l(y_l) y_i \tag{5}$$

where g($\cdot$) is the coupling function between unit l of the context network and weight w_{ji} of the function network. For the special case when g is a linear function, i.e., $g(x) = a_l x$ for some constant $a_l \neq 0$, we have

$$x_j = \sum_i a_l y_l y_i \tag{6}$$

Thus, in this case the total input to unit j is a quadratic function of the activation values of other units, as opposed to the usual form of being a linear function. Such mutiplicative units have been discussed by Hinton [14].

2. Context Sensitive Architecture for Learning Inverse Models

The issue of choosing an appropriate set of input variables for the context input depends on the specific problem. If possible, the context should be chosen in such a way that the function network now represents a linear function, or at least one of reduced complexity. In the case of robot kinematics, it is natural to choose the configuration of the arm (as given by the joint state vector θ) as the ontext input and $\mathbf{x}$ as the function input. For a given context θ, the joint velocity vector $\dot{\theta}$ and the Cartesian coordinates $\mathbf{x}$ are linearly related. Thus the function network only needs to represent linear functions with no constant terms. An n-DOF arm requires a function network with n input and n output units, as shown in Fig. 3. The output units have no bias terms and their activation functions are just the identity function $f(x) = x$. The network has n^2 weights, which correspond to the entries in the inverse Jacobian matrix $\mathbf{J}^{-1}(\theta)$ evaluated at θ. The context network consists of n^2 decoupled learning subnetworks (Fig. 3), each of which is responsible for learning a single scalar function corresponding one entry in the matrix. The n input units are common to all the subnetworks, each of which has two hidden

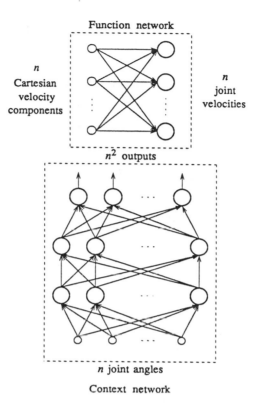

Figure 3: Context network for robot inverse kinematics.

layers. Generally, both the hidden and output units will be chosen with sigmoid activation functions. Note that decoupling an n-to-n^2 mapping into n^2 n-to-1 functions simplifies the learning problem, since the role of each hidden unit is much clearer. Furthermore, the functions can be learned in parallel. Increasing the number of degrees of freedom n only increases the number of functions to be learned.

3. Simulation of A 2-DOF Planar Arm

A 2-degree of freedom arm is shown in Fig. 4. The two links are of length d_1 and d_2 respectively. The Cartesian coordinates of the end

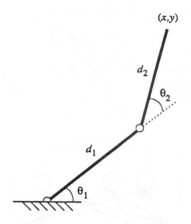

Figure 4: Two degree-of-freedom arm.

effector are related to the joint angles by the kinematic equations

$$\begin{bmatrix} x \\ y \end{bmatrix} = \begin{bmatrix} d_1 cos\theta_1 + d_2 cos(\theta_1 + \theta_2) \\ d_1 sin\theta_1 + d_2 sin(\theta_1 + \theta_2) \end{bmatrix} \tag{7}$$

The corresponding equations for velocity-based control are obtained by differentiating (7) with respect to time to obtain

$$\begin{bmatrix} \dot{x} \\ \dot{y} \end{bmatrix} = \begin{bmatrix} -d_1 sin\theta_1 - d_2 sin(\theta_1 + \theta_2) & -d_2 sin(\theta_1 + \theta_2) \\ d_1 cos\theta_1 + d_2 cos(\theta_1 + \theta_2) & d_2 cos(\theta_1 + \theta_2) \end{bmatrix} \begin{bmatrix} \dot{\theta}_1 \\ \dot{\theta}_2 \end{bmatrix} \tag{8}$$

In this simple case, the inverse kinematic equations can be obtained analytically:

$$\begin{bmatrix} \dot{\theta}_1 \\ \dot{\theta}_2 \end{bmatrix} = \begin{bmatrix} \frac{cos(\theta_1 + \theta_2)}{d_1 sin\theta_2} & \frac{sin(\theta_1 + \theta_2)}{d_1 sin\theta_2} \\ -\frac{cos\theta_1}{d_2 sin\theta_2} - \frac{cos(\theta_1 - \theta_2)}{d_1 sin\theta_2} & -\frac{sin\theta_1}{d_2 sin\theta_2} - \frac{sin(\theta_1 - \theta_2)}{d_1 sin\theta_2} \end{bmatrix} \begin{bmatrix} \dot{x} \\ \dot{y} \end{bmatrix} \tag{9}$$

During off-line learning of the inverse model, the training set consisted of examples generated by solving the above equations for 400 random

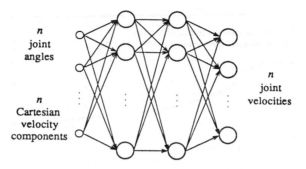

n joint angles

n Cartesian velocity components

n joint velocities

Figure 5: "Naive" neural network architecture for inverse kinematics.

contexts $(\theta_1, \theta_2)^T$. For each context, four input-output pairs representing the Cartesian and joint velocity components were generated for the function network. The context and function network input variables were linearly scaled from their actual values to the range [-1,1]. The segment lengths were both set to 0.5. The context network we used had 20 units in the first hidden layer and 10 units in the second layer (selected by experimentation). The simulation was compared to a standard, "naive" neural network architecture as shown in Figure 5, both being run with identical numbers of hidden units, learning rate and momentum factor. The mean-squared output error, averaged over 100 trials, for both the standard and the context-sensitive networks, is shown in Fig. 6. An epoch corresponds to cycling through the entire set of training examples once. It is evident that the context-sensitive architecture exhibits dramatically faster convergence than the standard network, for this simple inverse kinematics problem.

4. Simulation of 3-DOF PUMA Kinematics

The example above, while encouraging, was restricted to 2 degrees of freedom. In order to validate the context network techniques, it was necessary to apply the method to a more realistic problem. We have

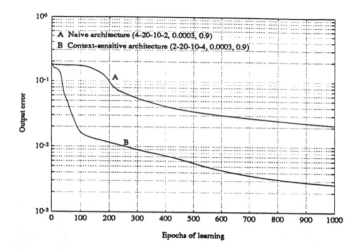

Figure 6: Learning curves for standard and context-sensitive networks.

applied context- sensitive networks to inverse kinematics computation in both 3-DOF and 6-DOF problems using models of the Unimation PUMA robot, a standard industrial manipulator. The PUMA 560 arm has 6 DOF. However, by careful choice of coordinate frame assignments (as in Fig. 7) it is possible to reduce some of the elements of the Jacobian to zero, thus producing a system with only 3 DOF [7].

The training set used in off-line learning of the feedforward model for the 3-DOF PUMA arm consisted of examples corresponding to 400 randomly generated contexts or joint configurations. The details of the simulation are given by Yeung [41]. For simplicity, the context network was implemented as a single multi-layer network. Using decoupled subnetworks, as in Fig. 3, would be expected to further improve performance. The context network had 20 units in the first hidden layer and 15 units in the second hidden layer. As before, the learning rate was compared with that of a standard network using the same parameters (same number of hidden layers, number of units in each layer, learning rate

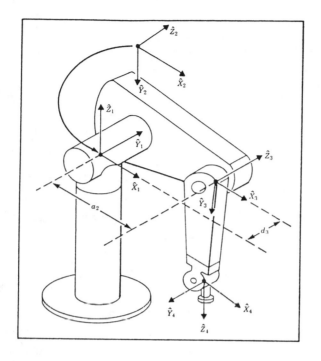

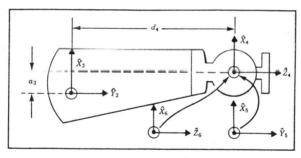

Figure 7: Some kinematic parameters and frame assignments for PUMA 560.

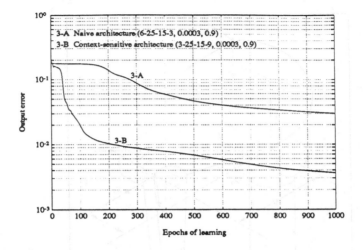

Figure 8: Learning curves for 3 DOF robot arm.

and momentum factor). The corresponding learning curves are shown in Figure 8. It can be seen from the results that the performance of the context-sensitive network is significantly better than that of a standard architecture network. After learning for 1000 epochs, the rms errors of the standard and the context-sensitive network decreased to 0.0301 and 0.0036 respectively. The rate of improvement is particularly dramatic during the first 200 training epochs.

5.   Inverse Jacobian Control

Having described the network architecture and the learning proce-
dure, we now consider the use of the neural network in a feedback control system for arm control. Fig. 9 shows the block diagram of such a system, which is sometimes referred to as "inverse Jacobian control". The box labeled "Neural network controller" solves the inverse kinematics problem, by computing the joint velocity $\dot{\theta}$ from the context input θ and the Cartesian velocity $\dot{x}$. The arm is then controlled with the given joint

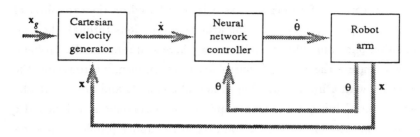

Figure 9: Inverse Jacobian control for robot manipulators.

velocity. Sensors on the arm provide the feedback signals θ and $\mathbf{x}$. The current position $\mathbf{x}$ and the next target location $\mathbf{x}_g$ are used to estimate the velocity $\dot{\mathbf{x}}$. It is important to note that the control scheme of Fig. 9 does not take system dynamics into account. Furthermore, there are other ways in which feedforward models and feedback information can be combined for control purposes. One such approach, due to Kawato et al [17] will be discussed in Section III in conjunction with dynamics problems.

B. Robot Kinematic Control Based on
A Bidirectional Mapping Neural Network

Robot kinematic control is based on the computation of forward and inverse mapping between joint space and Cartesian space. Robot kinematic control is computationally expensive and requires frequent calibration to maintain accuracy. Especially, the advent of multiple and redundant robot arms makes robot kinematic control based on numerical computation extremely difficult, if not impossible, not only due to the intensive computational complexity involved but also due to the difficulty of obtaining inverse kinematic solutions. Therefore, it is quite attractive to develop a neural network which automatically generates robot arm forward and inverse kinematic solutions based on the input-output associations stored in the network. Attempts have been made to build neural networks performing such associations based on the capability of a neural network in approximating arbitrary functions through the generalization of sample patterns. However, there still remains the problem of how to ensure the required control accuracy uniformly throughout the workspace with a limited number of neural elements and how to handle "one-to-many" inverse mapping required for redundant arm kinematic control. This section presents a solution to the above problem based on a *Bidirectional Mapping Neural Network* (BMNN) [25, 26]. The BMNN constructed here is composed of a multilayer feedforward network with hidden units having sinusoidal activation functions, or simply sinusoidal hidden units, and a feedback network connecting from the output to the input of the forward network to form a recurrent loop. The forward network exactly represents robot forward kinematics equations, allowing accurate computation of kinematic solutions with simple training. The feedback network generates input updates based on a Liapunov function to modify the current input in such a way that the output of the feedforward network moves toward the desired output. The parameters of the feedback network can be set in terms of the parameters of the forward network and thus, no additional training is required. Robot kinematic control based on the BMNN offers advantages over conventional approaches: it provides an accurate computation of both forward and inverse kinematic solutions, it requires simple training, and it enables to handle "one-to-many" inverse mapping for redundant arm kinematic control. Furthermore, it allows the control of convergence trajectory based on a Liapunov function, which can be directly used for arm trajectory generation.

1.   The Model of Forward Kinematics

The forward kinematic equation of a robot arm with n revolute joints can be represented by the weighted sum of sinusoidal functions in the form of $\prod_{i=1}^{n} g(\theta_i)$ where $g(\theta_i)$ is either $sin\theta_i, cos\theta_i$ or 1. Since $\prod_{i=1}^{n} g(\theta_i)$ can be derived from the set $\{g(\sum_{j=1}^{n} w_j \theta_j)| \ w_j \in (-1, 0, 1), \ for \ j = 1, \cdots, n\}$ through trigonometric identities, the forward kinematics of a robot arm with n revolute joints can be represented by

$$f_k(\boldsymbol{\theta}) = \sum_{i=1}^{m} l_i^k sin[(\mathbf{w}_i^k)^t\boldsymbol{\theta}], \quad k = 1, \cdots, l \tag{10}$$

where f_k indicates the kth output representing the kth Cartesian variable (as an element of a position vector or a 3×3 orientation matrix of the homogeneous transformation), $\boldsymbol{\theta}$, $\boldsymbol{\theta} \equiv [\frac{\pi}{2}, \theta_1, \theta_2, \cdots, \theta_n]^t$, represents an augmented joint vector, and $\mathbf{w}_i^k$, $\mathbf{w}_i^k \equiv [w_{i0}^k, w_{i1}^k, w_{i2}^k, \cdots, w_{in}^k]^t$, represents a weight vector of the ith sinusoidal function for the kth output with $w_{ij} \in \{-1, 0, 1\}$.

It should be noted that (10) represents the input-output function of a multilayer feedforward network with sinusoidal hidden units. Figure 10 illustrates a schematic diagram of the network implementing a forward kinematic equations represented by (10). The network with sinusoidal hidden units as shown in Figure 10, can be trained for accurate computation of forward kinematic solutions, whereas the conventional approaches using sigmoidal hidden units can only provide approximate solutions.

The maximum number of sinusoidal hidden units necessary for the implementation of (10) are 3^n.[1] Note that, in case the ith joint of a robot arm is prismatic, θ_i becomes a constant and some of l_i^ks needs to be treated as variables.

The training of the feedforward network shown in Figure 10 simply requires the estimation of the parameters, l_i^ks, since w_i^ks can be preset according to (10). l_i^ks define the linear relationship between the network output, f_k, and the outputs of sinusoidal units, therefore, l_i^ss can be trained by the sequential parameter estimation based on *Least Mean Square* (LMS) algorithm [39]:

$$l_i^{k \, (new)} = l_i^{k \, (old)} + \eta(y_k^d - f_k(\boldsymbol{\theta}^d))sin[(\mathbf{w}_i^k)^t\boldsymbol{\theta}^d] \tag{11}$$

where y_k^d and $\boldsymbol{\theta}^d$ represent respectively the kth desired output value in Cartesian coordinate and the desired input vector in joint coordinate, and η represents a positive constant called the *learning rate*.

---

[1] The maximum possible combinations of sinusoidal units for an n revolute joint robot arm is $\sum_{i=0}^{n} nCi2^i = (2+1)^n = 3^n$.

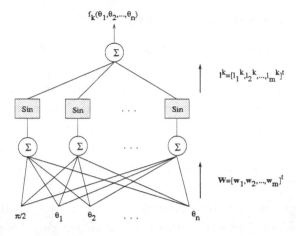

Figure 10: The Model of Forward Kinematics for a Robot Arm.

The training of the network with the maximum number of sinusoidal units may produce a large number of l_i^ks with zero values. It is possible to use less than the maximum number of sinusoidal hidden units with their corresponding w_i^ks undetermined and train both l_i^ks and w_i^ks based on the *Backpropagation algorithm* [35] or its varieties. However, this may require a large number of learning cycles or handling of flat surfaces and local minima incurred by sinusoidal hidden units involved in training. Alternatively, we can build a training algorithm which is able to automatically identify and recruit sinusoidal units corresponding to only nonzero l_i^ks. Lee and Kil developed a *Hierarchically Self-Organizing Learning* (HSOL) algorithm [24] which automatically and incrementally recruits a required number of sinusoidal hidden units by monitoring network performance. The recruitment of sinusoidal hidden units in HSOL is based on the decisions whether a new teaching pattern can be accommodated by the sinusoidal hidden units previously generated or a new sinusoidal hidden unit should be recruited. This decision is based on the accommodation boundaries of individual sinusoidal hidden units, defined in the frequency domain as the range of the amplitudes individual sinusoidal hidden units can accommodate to adapt to the change of frequency response incurred by a new sample. The gradual reduction of the range of accommodation boundaries ensures not only error convergence but also fast training.

2. The Inverse Kinematic Solutions Based on Liapunov Function

Inverse kinematic solutions can be obtained by attaching a feedback network around a forward network to form a recurrent loop, such that, given a desired output (Cartesian position) of a feedforward network, the feedback network iteratively generates input joint angle correction terms to move the output of feedforward network toward the desired output. The inverse kinematic solution is obtained by taking the input joint angle at the time the output of forward network converges to the desired output. To handle this concept with mathematical formalism, let us define the inverse mapping problem as follows: Given the output pattern, $\mathbf{y}^d$, generate an input pattern, $\mathbf{x}$ which satisfies the forward mapping $\mathbf{y}^d = \mathbf{y}(\mathbf{x})$, while optimizing the given performance index, $P(\mathbf{x})$. An example of $P(\mathbf{x})$ is $(\mathbf{x}^d - \mathbf{x})$, where $\mathbf{x}^d$ represents a particular input pattern to which we want to locate $\mathbf{x}$ as closely as possible. We adopt the following approach to the solution of the above inverse mapping problem: Generate $\dot{\mathbf{x}}$ in the direction of decreasing the Liapunov function (defined in terms of the error, $\mathbf{y}^d - \mathbf{y}(\mathbf{x})$, and $P(\mathbf{x})$) to update $\mathbf{x}$ iteratively until $\mathbf{y}(\mathbf{x})$ converges to $\mathbf{y}^d$.

Let us first select the *Liapunov function candidate*, V, as follows:

$$V = \frac{1}{2}\tilde{\mathbf{y}}^t\tilde{\mathbf{y}} + \frac{1}{2}\tilde{\mathbf{x}}^t\tilde{\mathbf{x}} \tag{12}$$

where $\tilde{\mathbf{y}} = \mathbf{y}^d - \mathbf{y}(\mathbf{x})$: an $M \times 1$ output error vector, $\tilde{\mathbf{x}} = \mathbf{x}^d - \mathbf{x}$: an $N \times 1$ input error vector; $\mathbf{y}^d$ = the given output vector; $\mathbf{y}(\mathbf{x})$ = the actual output vector for the given input pattern, $\mathbf{x}$, and $\mathbf{x}^d$ = the input constraint vector.

The time derivative of the Liapunov function is given by

$$\dot{V} = (\frac{\partial V}{\partial \mathbf{x}})^t\dot{\mathbf{x}} = -(\tilde{\mathbf{y}}^t\frac{\partial \mathbf{y}}{\partial \mathbf{x}} + \tilde{\mathbf{x}}^t)\dot{\mathbf{x}} = -(\tilde{\mathbf{y}}^t\mathbf{J} + \tilde{\mathbf{x}}^t)\dot{\mathbf{x}} \tag{13}$$

where $\mathbf{J} \equiv \frac{\partial \mathbf{y}}{\partial \mathbf{x}}$: the $M \times N$ Jacobian matrix.

Let us investigate the *input pattern update rule* based on the Liapunov function technique. First, let us consider the following system equation:

$$\dot{\mathbf{x}} = \mathbf{F}(\mathbf{x}) \ \ and \ \ \mathbf{F}(\mathbf{x}) = \frac{1}{2} \frac{\| \tilde{\mathbf{y}} \|^2}{\| \mathbf{J}^t \tilde{\mathbf{y}} + \tilde{\mathbf{x}} \|^2} (\mathbf{J}^t \tilde{\mathbf{y}} + \tilde{\mathbf{x}}). \tag{14}$$

For the equilibrium points of (14), the following lemma is suggested:

Lemma 1 $\mathbf{x}^*$ *is an equilibrium point of (14) if and only if* $\mathbf{x}^*$ *is a solution vector of* $\tilde{\mathbf{y}}$, *i.e.,*

$$\dot{\mathbf{x}} = \mathbf{F}(\mathbf{x}^*) = \mathbf{0} \ \ iff \ \ \tilde{\mathbf{y}}(\mathbf{x}^*) = \mathbf{0}. \tag{15}$$

For the proof of lemma, see Appendix A.

The following theorem is introduced to compute $\dot{\mathbf{x}}$ based on the data from forward mapping while preserving the convergence of the input pattern update rule:

Theorem 1 *If an arbitrarily selected initial input pattern,* $\mathbf{x}(0)$ *is updated by*

$$\mathbf{x}(t^{'}) = \mathbf{x}(0) + \int_0^{t^{'}} \dot{\mathbf{x}} dt \tag{16}$$

where $\dot{\mathbf{x}}$ *is given by*

$$\dot{\mathbf{x}} = \frac{1}{2} \frac{\| \tilde{\mathbf{y}} \|^2}{\| \mathbf{J}^t \tilde{\mathbf{y}} + \tilde{\mathbf{x}} \|^2} (\mathbf{J}^t \tilde{\mathbf{y}} + \tilde{\mathbf{x}}), \tag{17}$$

then $\tilde{\mathbf{y}}$ *converges to* $\mathbf{0}$ *under the condition that* $\dot{\mathbf{x}}$ *exists along the convergence trajectory.*

For the proof of theorem, see Appendix B.

The update of $\mathbf{x}$ based on $\dot{\mathbf{x}}$ determined by (17) guarantees the convergence as long as $\dot{\mathbf{x}}$ exists. Note however that the Liapunov function V defined by (12) is a nonlinear function of $\mathbf{x}$ and may have local minima where the following equation holds:

$$(\frac{\partial V}{\partial \mathbf{x}})^t = -(\tilde{\mathbf{y}}^t \mathbf{J} + \tilde{\mathbf{x}}^t) = \mathbf{0} \tag{18}$$

If $\tilde{\mathbf{y}} \neq \mathbf{0}$, (17) results in a large norm of $\dot{\mathbf{x}}$. This explosion of $\dot{\mathbf{x}}$ near a local minimum is referred to as a *jumping* phenomenon. The jumping is a useful phenomenon for escaping from a local minimum but it should be controlled to curtail jumping which goes beyond the domain of the input space and to avoid jumping which wanders around the same local minimum. See the next subsection on jumping control for more details on this subject. Theorem 1 presents the following features:

- It allows the direct calculation of $\dot{\mathbf{x}}$ based on the quantities generated by the forward mapping.

- If $\| \tilde{\mathbf{y}} \| \gg 0$, $\dot{\mathbf{x}}$ moves fast when $\| \frac{\partial V}{\partial \mathbf{x}} \|$ is small and vice versa.

- At the local minima or maxima of V, $\dot{\mathbf{x}}$ is subject to jumping ($\| \dot{\mathbf{x}} \|$ explodes to infinity). Note that the jumping direction is dependent upon the direction of the gradient of V.

Note that the Jacobian $\mathbf{J} = \frac{\partial \mathbf{y}}{\partial \mathbf{x}}$ can be obtained directly from the output of sinusoidal units as illustrated in Figure 11. The Model of inverse kinematics for a robot arm is illustrated in Figure 12.

One problem in (17) is that near the solution of $\tilde{\mathbf{y}}$, $\mathbf{x}$ may not directly converge to the solution because $\dot{\mathbf{x}}$ follows the gradient of (12), not the gradient of (V.). To overcome this problem, we include the Lagrangian multiplier λ in the Liapunov function:

$$V(\lambda, \mathbf{x}) = \frac{\lambda}{2}\tilde{\mathbf{y}}^t\tilde{\mathbf{y}} + \frac{1}{2}\tilde{\mathbf{x}}^t\tilde{\mathbf{x}} \qquad (19)$$

The purpose of the Lagrangian multiplier is to force $\tilde{\mathbf{y}}$ to converge to zero by increasing λ exponentially when $\mathbf{x}$ is near the solution of $\tilde{\mathbf{y}}$. Based on (19), the following theorem of input pattern update rule is introduced:

Theorem 2 *If an arbitrarily selected initial input pattern, $\mathbf{x}(0)$ and an arbitrarily selected positive Lagrangian multiplier, $\lambda(0)$ are updated by*

$$\mathbf{x}(t') = \mathbf{x}(0) + \int_0^{t'} \dot{\mathbf{x}}dt \ \ and \ \ \lambda(t') = \lambda(0) + \int_0^{t'} \dot{\lambda}dt \ \ respectively \qquad (20)$$

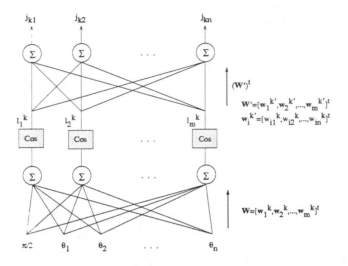

Figure 11: The Calculation of Jacobian, $\mathbf{J} = [j_{kl}]$ Based on the Model of Forward Kinematics.

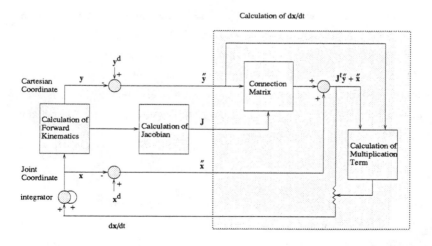

Figure 12: The Model of Inverse Kinematics for a Robot Arm: the calculation of $\dot{\mathbf{x}}$ is done based on the following assumptions: the connection matrix is changed according to the element values of the jacobian matrix, $\mathbf{J}$, and the output of the calculation of the multiplication term generates $\frac{1}{2} \frac{\|\dot{\mathbf{y}}\|^2}{\|\mathbf{J}^t \dot{\mathbf{y}} + \ddot{\mathbf{x}}\|^2}$.

where $\dot{\mathbf{x}}$ and $\dot{\lambda}$ are given by

$$\dot{\mathbf{x}} = \frac{\frac{\lambda}{2}\parallel\tilde{\mathbf{y}}\parallel^2 + \frac{1}{2}\parallel\tilde{\mathbf{y}}\parallel\parallel\tilde{\mathbf{x}}\parallel}{\parallel\lambda\mathbf{J}^t\tilde{\mathbf{y}}+\tilde{\mathbf{x}}\parallel^2}(\lambda\mathbf{J}^t\tilde{\mathbf{y}}+\tilde{\mathbf{x}}) \text{ and } \dot{\lambda} = \frac{\parallel\tilde{\mathbf{x}}\parallel}{\parallel\tilde{\mathbf{y}}\parallel} \qquad (21)$$

then $\tilde{\mathbf{y}}$ converges to $\mathbf{0}$ under the condition that $\dot{\mathbf{x}}$ exists along the convergence trajectory.

For the proof of theorem, see Appendix C.

Note that $\dot{\lambda}$ goes to infinity when $\tilde{\mathbf{y}}$ approaches zero. This forces the input pattern to converge to the solution when $\tilde{\mathbf{y}}$ approaches zero. In this input pattern update rule, the property of convergence can be controlled by changing $\dot{\mathbf{x}}$ in (21) to

$$\dot{\mathbf{x}} = \frac{\lambda(\frac{1}{2}\parallel\tilde{\mathbf{y}}\parallel^2)^\alpha + \frac{1}{2}\parallel\tilde{\mathbf{y}}\parallel\parallel\tilde{\mathbf{x}}\parallel}{\parallel\lambda\mathbf{J}^t\tilde{\mathbf{y}}+\tilde{\mathbf{x}}\parallel^2}(\lambda\mathbf{J}^t\tilde{\mathbf{y}}+\tilde{\mathbf{x}}). \qquad (22)$$

To see the effect of α, let us investigate the convergence time, t_c, of the input pattern update rule when the input pattern is near the solution vector, $\mathbf{x}^*$. The convergence time of the input pattern update rule for the given initial position, $\mathbf{x}(1)$ near $\mathbf{x}^*$ is determined as follows:

$$t_c = \int_{V(\mathbf{x}(1))}^{V(\mathbf{x}^*)}\frac{dV}{V} \approx -\int_{V(\mathbf{x}(1))}^{V(\mathbf{x}^*)}V^{-\alpha}dV = \begin{cases} \infty & \text{if } \alpha \geq 1 \\ \frac{V(\mathbf{x}(1))^{1-\alpha}}{1-\alpha} & \text{if } \frac{1}{2} < \alpha < 1 \end{cases} \qquad (23)$$

where $V = \frac{1}{2}\parallel\tilde{\mathbf{y}}\parallel^2$.

This result implies the followings:

- If $\alpha \geq 1$, $\mathbf{x}$ converges to $\mathbf{x}^*$ asymptotically, i.e., $\mathbf{x}$ approaches $\mathbf{x}^*$ as t goes to infinity. Note that $V(t) = V(\mathbf{x}(1))e^{-t}$ when $\alpha = 1$.
- If $\frac{1}{2} < \alpha < 1$, $\mathbf{x}$ converges to $\mathbf{x}^*$ within a finite time. This type of stable point, $\mathbf{x}^*$, is called a *terminal attractor* [42].

For the simulation on a digital computer, the following input pattern update rule for discrete time is introduced:

$$\mathbf{x}(n+1) = \mathbf{x}(n) + \eta\Delta\mathbf{x}(n) \qquad (24)$$

where n represents the *number of iterations*, and η represents a small positive constant called the *update rate*.

$\Delta \mathbf{x}$ is calculated from $\dot{\mathbf{x}}$ in (21). Since it is desirable to prevent jumping beyond the input space domain, $\Delta \mathbf{x}$ is determined as follows:

$$\Delta x_i = \begin{cases} x_L + \gamma & \text{if } \Delta x_i > x_L, \\ \Delta x_i & \text{if } -x_L \leq \Delta x_i \leq x_L, \\ -x_L + \gamma & \text{if } \Delta x_i < -x_L, \end{cases} \tag{25}$$

where Δx_i represents the ith component of $\Delta \mathbf{x}$, x_L represents a saturation constant, and γ is a random variable having a small variance to avoid the possible cyclic trajectory of $\mathbf{x}$.

When we use the input pattern update rule for discrete time, the convergence to the global minima depends upon the value of η and x_L. If ηx_L is large enough to jump out of the largest region of attraction of local minima, this will guarantee an asymptotic global convergence. But η should be small enough to guarantee that the input pattern update rule will converge - especially, near the solution. For the condition of η which guarantees the convergence of the input pattern update rule for discrete time, it can be shown that the following condition is sufficient:

$$0 < \eta < 4(\frac{\lambda_{min}}{\lambda_{max}})^2 \tag{26}$$

where λ_{min} and λ_{max} respectively represent the minimum and the maximum eigenvalues of $\frac{\partial^2 V}{\partial \mathbf{x}^2}(\mathbf{x}^*)$ in which $V = \frac{1}{2} \parallel \tilde{\mathbf{y}} \parallel^2$.

3. Jumping Control

In the input pattern update rule, $\mathbf{x}$ is updated by $\Delta \mathbf{x}$, the local information related to the gradient of the Liapunov function, V. When $\mathbf{x}$ reaches the local minima or maxima of V, jumping occurs in the perturbed direction of the gradient of V. Jumping control is intended to guide the trajectory of $\mathbf{x}$ to the desired solution based on the global information of the given function defined in the forward mapping. This is done by adding a new term, $\frac{1}{2}\tilde{\mathbf{x}}_c{}^t\tilde{\mathbf{x}}_c$, to the Liapunov function:

$$V = \frac{1}{2}\tilde{\mathbf{y}}^t\tilde{\mathbf{y}} + \frac{1}{2}\tilde{\mathbf{x}}_c{}^t\tilde{\mathbf{x}}_c \tag{27}$$

where $\tilde{\mathbf{x}}_c$ is a *control term* and is defined as $\tilde{\mathbf{x}}_c = c_f(\mathbf{m} - \mathbf{x})$ with c_f representing a positive constant called the *control factor* and $\mathbf{m}$ representing a selected vector from the set, $\{m^k| \ (\mathbf{m}^k, \mathbf{y}(\mathbf{m}^k)), \ for \ k = 1, \cdots, l\}$, which represents the predefined global information of forward mapping.

The net-effect of adding a control term to the convergence behavior is that the convergence trajectory of $\mathbf{x}$ becomes more concentrated around $\mathbf{m}$. The selection of $\mathbf{m}$ is done as follows:

1. Test whether there exists a solution between $\mathbf{m}^k$ and the current input pattern, $\mathbf{x}$ for all k. A solution can exist between $\mathbf{m}^k$ and $\mathbf{x}$, if the following condition holds:

$$\mathbf{f}(\mathbf{y}^d - \mathbf{y}(\mathbf{m}^k)) \cdot \mathbf{f}(\tilde{\mathbf{y}}) = -M \qquad (28)$$

   where $\mathbf{f}(\mathbf{x}) = [f(x_1), f(x_2), \cdots, f(x_M)]^t$, $f(x) = \begin{cases} 1 & if \ x > 0 \\ -1 & otherwise \end{cases}$

   and M is the dimension of output, $\mathbf{y}$.

2. $\mathbf{m}$ is selected from S^k, a set of $\mathbf{m}^k$s satisfying (28):

$$\mathbf{m} = \begin{cases} \mathbf{m}^j | \ \| \mathbf{x}^d - \mathbf{m}^j \| = min_{\mathbf{m}^k \in S^k} \| \mathbf{x}^d - \mathbf{m}^k \| & if \ S^k \neq \phi, \\ \mathbf{0} & otherwise. \end{cases} \qquad (29)$$

If there is no $\mathbf{x}^d$, we can choose $\mathbf{m}$ arbitrarily from S^k. In the general case of multiple output functions, (28) does not guarantee the existence of a solution between $\mathbf{m}^k$ and $\mathbf{x}$. However, (28) selects $\mathbf{m}$ correctly if the output is one dimensional or binary. If the selected vector, $\mathbf{m}$, is near the solution, a fast and stable convergence can be achieved, regardless of the initial $\mathbf{x}$. In the case that $\mathbf{m}$ is not selected correctly, it has high possibility of the reoccurrence of jumping around the local minima.

4.   Simulation

We applied the inverse mapping algorithm based on a BMNN to a 6 degree of freedom planar robot arm. In the simulation, a constraint term is added to the Liapunov function to get the inverse solution representing

a desirable joint configuration. Two types of constraints are considered:
1) the minimization of the difference between the initial and final angles
of the first joint, 2) the closeness of individual angles to the middle of
their joint-limits. The parameters involved in the simulation of inverse
kinematics are listed as follows:

- update rate, $\eta = 0.002$
- initial Lagrangian multiplier, $\lambda(0) = 0.01$

The simulation results indicate that the end point of the manipulator
with the given initial joint configuration, converges accurately to the
desired position, while satisfying the given constraint. This is illustrated
in Figure 13, where the initial and the final configurations as well as
the intermediate convergence history are described in the 2 dimensional
Cartesian coordinates. Note that in most cases, the input pattern update
rule worked well even without jumping control. This is due to fact that
the input space (joint space) is defined as a compact set and the solution
space is defined as a manifold.

5. Discussion

It is shown that robot arm forward kinematics can be exactly represented by a multilayer feedforward network with sinusoidal hidden units,
and that such a network can be trained to accurately represent robot
arm forward kinematic solutions. Moreover, it is demonstrated that
robot arm kinematic solutions can be accurately obtained by the feedback network iteratively generating joint angle updates based on a Liapunov function until the output of the feedforward network converges to
the desired Cartesian position. The robot kinematic control based on the
presented BMNN provides a number of advantages over conventional approaches with respect to the control accuracy, the capability of handling
"one-to-many" inverse mapping for redundant arm kinematic solutions,
the ease of training, and the application to automatic generation of arm
trajectory.

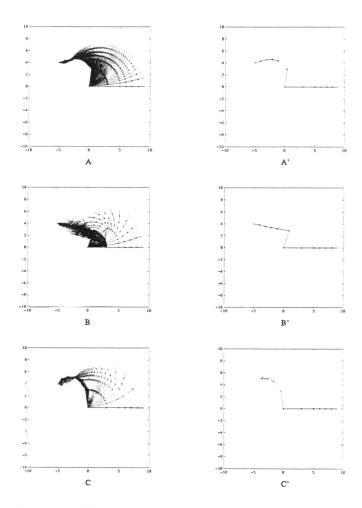

Figure 13: Inverse Kinematics of 6-DOF Planar Robot Arm: The initial configuration for all cases is set as (9,0) while the desired points in the Cartesian coordinates is given by (-5,4). In A, no constraint term is added to Liapunov function. In B, a constraint term minimizing the angle difference of the first joint is added to Liapunov function. In C, a constraint term locating the joints in the middle of joint-limits is considered; the limit of the first joint is given by $[0,\pi]$ while the limit of other joints are given by $[0,\frac{\pi}{2}]$.

The training of a forward network by incrementally recruiting sinu-soidal hidden units is an attractive feature, which needs further attention in future research. An interesting observation concerning with the speed of convergence of the input pattern update rule is as follows: it is possible to achieve faster convergence by controlling the update rate in such a way that the update rate is initially assigned a large value but reduced grad-ually according to the progress of convergence. Furthermore, this paper showed that, by setting α between $\frac{1}{2}$ and 1, we can turn the desired solu-tions into terminal attractors which provides even more speed-up. Note, however, that the update rate should be carefully controlled when the input pattern is near the desired solution.

III.   Inverse Dynamic Solutions

The discussion in Section II above has been focused on the problem of transforming the representation of manipulator position from Cartesian to joint coordinates. (Such a transformation probably also occurs in the position control of human arms and the legs or wings of animals, since they must also transform points in the external world to their equivalent skeletal joint angles). Given the desired joint configuration, it is now necessary to compute the torques to be applied at the joints to drive the arms, legs or robot manipulators to the desired orientation. Gener-ally, the desired motions will be subject to constraints or performance criteria, such minimum overshoot, minimum energy or minimum time. The computation of the necessary torques requires consideration of such parameters as inertia and damping. In view of the fact that the arm seg-ments are coupled, the inertia matrix is not diagonal and the dynamical equations are highly nonlinear. Gravitational, Coriolis and centripetal forces must also be considered.

The general form of the dynamics equation can be expressed a cite-crai:

$$\tau = \mathbf{M}(\theta)\ddot{\theta} + \tau_v(\theta, \dot{\theta}) + \tau_g(\theta) + \tau_f(\theta, \dot{\theta}) \qquad (30)$$

where τ is the torque vector, $M(\theta)$ is the mass matrix, $\tau_v(\theta, \dot{\theta})$ is a vector of centrifugal and Coriolis terms, $\tau_g(\theta)$ is a vector of gravity terms and $\tau_f(\theta, \dot{\theta})$ is a vector of friction terms. The acceleration is often specified in Cartesian coordinates. In order to relate the Cartesian acceleration to the joint acceleration, we return to the kinematic relationship

$$x = f(\theta)$$

and differentiate it twice with respect to time, obtaining

$$\ddot{x} = J(\theta)\ddot{\theta} + H(\theta)\dot{\theta} \tag{31}$$

where $H(\theta)$, the matrix of second derivative is known as the Hessian. Solving this expression for the joint acceleration and substituting in (30), we obtain

$$\tau = M(\theta)J^{-1}(\theta)\ddot{x} - M(\theta)J^{-1}(\theta)H(\theta)\dot{\theta} + \tau_v(\theta, \dot{\theta}) + \tau_g(\theta) + \tau_f(\theta, \dot{\theta}) \tag{32}$$

For small values of angular velocity and neglecting the effects of gravity, the torque can be approximated by the first term on the right hand side of equation (32), i.e.,

$$\tau_x = M(\theta)J^{-1}(\theta)\ddot{x} \tag{33}$$

Note the analogy between this expression and the inverse kinematic relation,

$$\dot{\theta} = J^{-1}(\theta)\dot{x}.$$

Again, we have a linear relationship between two variables, for a given spatial orientation of the robot, as given by the vector θ. Hence the matrix $M(\theta)J^{-1}(\theta)$ represents the context, and the linear expression can be obtained using a function network. Clearly, the discussion on context sensitive networks applies to this simplified version of the inverse dynamics problem. Using equation (33) for inverse dynamic control poses a requirement for fast computation of the inverse Jacobian matrix.

While considerable work has been done on the applications of neural networks to kinematic control, considerably less work has been reported in the general area of dynamic control. In spite of the analogy between the inverse problems, the inverse dynamics problem presents considerably greater difficulty, due to nonlinearity, coupling between variables, and lack of knowledge of the plant. Traditional proportional-integral-derivative (PID) controllers are not satisfactory for robot control, since they assume fixed plant parameters. Robot parameters are dependent on the joint configuration. This difficulty can be overcome by using some form of adaptive control based on identification of the robot parameters and consequent adjustment of controller gains. While this technique is feasible, it is complex and may be too slow for real time applications. Hence, techniques based on learning without accurate knowledge of the parameters are very appealing.

The question of control architecture arises quickly, since the most popular learning algorithms for neural networks (like backpropagation) are only useful with feed-forward networks. The question of control architecture was addressed by Psaltis, et al [33], who proposed two configurations termed "specialized learning architecture" and "generalized learning architecture" respectively, as illustrated in Fig. 14. The specialized architecture appears to be particularly well suited for robot control, since the neural controller is placed at the input of the controlled system, acting as a feedforward filter. In this structure the training of the system is based on comparing the desired response of the system d with the actual response y. The goal is to achieve zero error for all given inputs. Since the learning signal will track any variations in system response, this structure is suitable for on-line learning and real-time adaptation. However, such an application is only useful if the control network is near it correct operating point to insure stable and well-behaved operation. Hence, starting the training of the system of Fig. 14b with random weights may not be suitable. Psaltis et al suggest using the structure of Fig. 14a, which they term a "generalized learning architecture" for off-line learning. Once convergence has been obtained, the feedback network can be placed in the forward path to provide on-line adaptive control. Another difficulty which arises with specialized learning is that back-propagation cannot be used directly since the error arises at the output of the controlled system and not at the output of the neural network. It has been suggested that a solution to this problem is to have the errors propagate through the plant as if it were an additional layer with unmodifiable weights [33]. This leads back to the question of plant identification.

A complete solution to the inverse dynamics problem has been presented by Kawato et al [17, 31]. Their work is based on studies of movement control in the neuromuscular system. Basically, they assume that within the central nervous system an internal neural model of the inverse dynamics of the musculo-skeletal system is acquired while monitoring a desired trajectory and the associated motor commands. They simulated control and learning performance of a robot manipulator in order to test their hypotheses on the behavior of the biological system. A 3-link direct drive manipulator was studied, using the overall model of Fig. 15. The terms $T(t)$, $T_i(t)$ and $T_f(t)$ denote the torque input to the manipulator, the torque computed by the inverse dynamics model and the feedback torque respectively. The inverse dynamics model receives as its input the set of joint angles representing the desired trajectory and it monitors the total torque input to the manipulator. As learning proceeds, it would be desirable for the actual manipulator trajectory to approach the desired trajectory and hence for the feedback torque torque term $T_f(t)$ to approach zero. In this event the architecture of the system will approach the specialized architecture of [33].

The internal structure of inverse dynamics of the manipulator was obtained by representing dynamics using the Lagrangian formulation [7] which leads to the expression

$$\mathbf{R}(\mathbf{q})\ddot{\mathbf{q}} - (\sum_k \dot{\mathbf{q}}_k \partial \mathbf{R}/\partial \mathbf{q}_k)\dot{\mathbf{q}} - (1/2)\dot{\mathbf{q}}^T(\partial \mathbf{R}/\partial \mathbf{q})\dot{\mathbf{q}} + \mathbf{B}\dot{\mathbf{q}} + \mathbf{G}(\mathbf{q}) = \mathbf{T}(t) \quad (34)$$

where, $\mathbf{q} = (q_1, q_2, q_3)^T$ is the generalized displacement vector and $\mathbf{T} = (T_1, T_2, T_3)^T$ is the torque. $\mathbf{R}(\mathbf{q})$ is a 3x3 inertia matrix (which is not diagonal), $\mathbf{B}$ is a 3x3 diagonal matrix of damping coefficients and $\mathbf{G}(\mathbf{q})$ represents the gravitational forces. In this equation the first term on

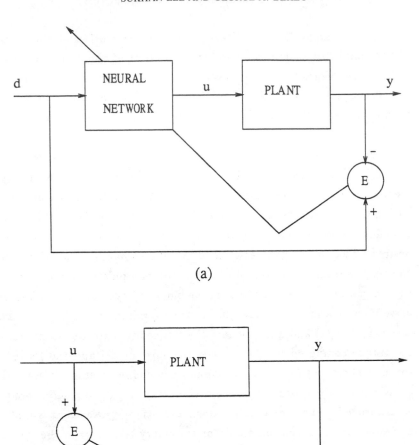

Figure 14: Alternative architectures for robot control.

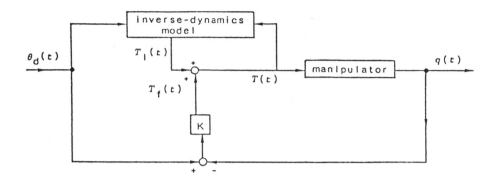

Figure 15: Block diagram of neural network for robot control, from [17].

the right hand side represents the inertial torques, the second and third terms the centripetal and Coriolis torques, the fourth term the damping torques and the fifth the torques due to gravity. The detailed structure of the inverse dynamics represented by equation (34) is shown in Fig. 16. The three joint angle inputs, the three torque outputs and the feedback torques are indicated. The blocks in this figure represent the expansion of equation (34) into 26 nonlinear filters f($\cdot$) and g($\cdot$). Detailed knowledge of the nature of the equations is required to generate these filters, which produce the inverse-dynamics model output torque in the form

$$\mathbf{T}_{i1}(t) = \sum_{l=1}^{13} w_{1l} f_l(q_{d1}(t), q_{d2}(t), q_{d3}(t))$$

$$\mathbf{T}_{ik}(t) = \sum_{l=1}^{13} w_{kl} g_l(q_{d1}(t), q_{d2}(t), q_{d3}(t)), \quad (k = 2, 3) \qquad (35)$$

In these expressions the terms w_{1l} and w_{kl} are the weights of synapses from a given subsystem to the output neuron. In other words, the neural network used by Kawato includes only one layer of modifiable weights. If a perfect inverse dynamics model is found, then equations (34) and (35) will be identical.

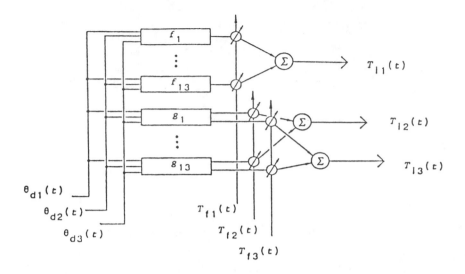

Figure 16: Detailed structure of inverse dynamics model, from [17].

The performance of Kawato's system is excellent. Both the inverse and forward dynamics models were acquired during typical movements. As learning continued, the inverse dynamics model gradually replaced any external feedback. The final system was able to generalize to control of movements quite different from those used for training. Finally, while significant knowledge is required to construct the nonlinear filters, they need not be completely accurate. The system converged even when extraneous filters were present. Nevertheless, the structure of the filters will depend on the particular robot being controlled and parameter values must be known or estimated. In view of the computational complexity of the implementation, this approach is probably best suited to parallel processing, when significant knowledge is available. Clearly, the method uses knowledge to reduce the number of unknown weights in the neural network portion of the model. The tradeoff between knowledge and neural network complexity needs further investigation.

We now present an approach to the inverse dynamics problem which are based on the concept of decomposition, as with the divide and conquer strategy of Section II, A. We begin with a decomposition of the basic equations and then consider a method which combines feedforward and feedback control.

A.   Functional Decomposition of Dynamic Equations

It is well known that when the functional relationships to be learned by a neural network are complex and highly nonlinear the learning time may become prohibitively long [35]. Part of the difficulty arises from the size of the sample training space. In the case of robot dynamics this situation is very apparent. Consider the simple case of a 2 DOF arm. The torque depends on joint position and velocity as well as on the Cartesian acceleration. If we require, for example, 10 samples in each dimension a total of 1,000,000 training samples will be required. This is clearly unrealistic. The number of required samples becomes astronomically large for 6 DOF. (It should be noted that this problem arises in the work of Kuperstein [22], who needs training samples distributed throughout the whole space in order to learn how to move a robot arm to an arbitrary desired location). It is apparent that the solution is to find ways to reduce the number of samples needed to learn the relationships. The work on context-sensitive networks discussed above represents one answer to this problem, which characterizes robotic systems. The work of Kawato et al [17] also attempted to reduce convergence time by decomposing the dynamical equations into a large number of subsystems. The method proposed by [3] is based on decomposing the dynamic equations into their basic component relations, training on these simpler relations and then recomposing to obtain the complete input- output mapping. Thus, with reference to the general dynamics equation (32), the torque can be decomposed into terms which represent the Cartesian acceleration, the Coriolis and centripetal torques, the gravitational effects and the torques due to damping. If the gravitational and damping terms are omitted for simplicity, the reduced dynamics equation becomes

$$\tau(\theta, \dot{\theta}, \ddot{\mathbf{x}}) = \mathbf{M}(\theta)\mathbf{J}^{-1}(\theta)[\ddot{\mathbf{x}} - [\mathbf{H}(\theta)\dot{\theta}]\dot{\theta}] + \tau_v(\theta, \dot{\theta}) \tag{36}$$

which can be decomposed into three terms as follows:

$$\tau_x(\theta, \ddot{\mathbf{x}}_0) = \mathbf{M}(\theta)\mathbf{J}^{-1}(\theta)\ddot{\mathbf{x}}_0$$

$$\tau_v(\theta, \dot{\theta}) = \tau_v(\theta, \dot{\theta}) \tag{37}$$

$$\ddot{\mathbf{x}}_c(\theta, \dot{\theta}) = -[\mathbf{H}(\theta)\dot{\theta}]\dot{\theta}$$

The implementation of these relationships by means of the generalized learning architecture requires some additional computation. The behavior of the system, as seen in the above equations, depends on the entire state and not simply on the joint or Cartesian position, as provided by the network. In other words, the neural network cannot control the robot with position information alone, but must be provided with velocity and acceleration information as well. Hence, we have modified the generalized architecture by adding differentiators as required by the equations. Figure 17 shows the modified architecture and its decomposition into the three terms of equation (36).

There are clear advantages to the approach outlined above. First, it can be applied to any manipulator, since the equations are used in their most generic form. Second, since the learning is performed on simpler equations with fewer dependent variables, the number of required training samples is drastically reduced. Third, the networks themselves are simpler, requiring fewer units for implementation to achieve a given accuracy. The functional decomposition takes advantage of the natural structure of of the model of a manipulator.

The technique was applied to a model of a 2-DOF arm. The three networks designed to provide the relationships of equation (36) were trained using back propagation, with data from a simulated manipulator. Two hidden layers were used in each network, with 81 training points. The results of learning the first term of equation (36), i.e., the dependence

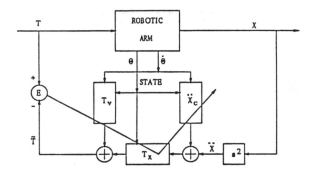

Figure 17: Generalized learning architecture using functional decomposition.

of the torque on the Cartesian acceleration, is shown in Fig. 18. The rms error was reduced to 2.8% after 2000 sweeps over the training set. The "spikes" on the learning error curve are due to a high value of the learning rate; reducing the rate eliminated the spikes but slowed down the convergence while higher values led to instability. Similar results were obtained for the other terms in the torque equation. Further research is needed to find a suitable parallel method of adjusting all three networks simultaneously, by on-line learning, thus allowing robot control while keeping track of system variations at the same time. A block diagram of a controller showing the interconnections of the three neural networks for on-line control is shown in Fig. 19.

B.   Combined Feedforward/Feedback Controllers

As indicated earlier in this chapter, the application of neural network techniques to robotics presents two major difficulties. First, the representation of strongly nonlinear mappings often results in large networks requiring large numbers of training examples and very slow convergence. Second, neural network representations of complex functional mappings are approximate by definition, and the degree of approximation is diffi-

SUKHAN LEE AND GEORGE A. BEKEY

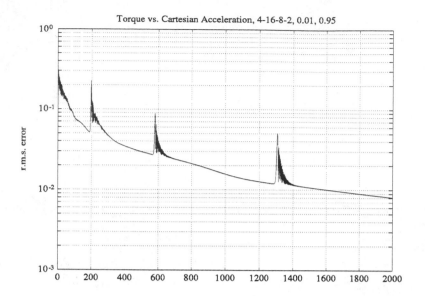

Figure 18: Learning of torque vs acceleration term.

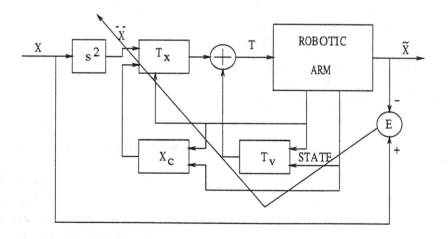

Figure 19: Robotic controller based on functional decomposition.

cult to predict. This section presents the results of a preliminary study which attempts to address both of these difficulties.

Functional decomposition methods, as outlined above, may lead to practical size networks and acceptable numbers of samples in obtaining a feedforward controller. In order to reduce the error to desirable bounds, Bassi and Bekey [4] propose to use feedback. Specifically, they propose to use, recurrently, an optimal estimate of the Cartesian trajectory in order to keep the end effector position within predetermined bounds.

The technique basically consists of the following steps. (1) An optimum trajectory function is obtained. Optimality is used to select among the infinity of possible trajectories which connect any two given points along the trajectory, and may be based on such criteria as minimum time, minimum acceleration or minimum jerk, while satisfying other constraints. (2) Functional decomposition is used to learn the inverse dynamics parameters using a generalized learning architecture. However, since feedback is used to correct for imprecisions in the feedforward control process, only the torque/Cartesian acceleration portion of the torque equation was used. (3) The resulting network is now used as a feedforward controller (see Fig. 19). (4) The loop is now closed with the optimal trajectory control, which calculates the acceleration required to drive the manipulator to the current optimal trajectory. A block diagram of this system is shown in Fig. 20.

The method was applied to a 2 DOF arm performing a variety of movements. As an example, Fig. 21 shows a start-stop movement of the arm, beginning with the end effector located (0.5,0) and ending at (0.5,0.5). This is a very large movement, equal in displacement to the total length of the arm. The goal was to achieve a straight line trajectory. It can be seen that the trajectory deviates slightly from a straight line, possibly due to the fact that only the linear terms were included in the torque equation. The final position error can be made as small as desired by a choice of control period [4].

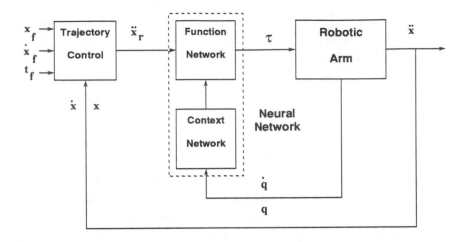

Figure 20: Connectionist feedforward and Cartesian feedback control.

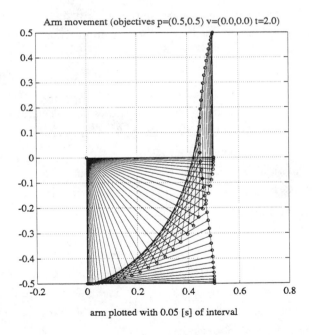

Figure 21: Combined feedforward/feedback control of large arm movement.

IV.   Neural Computation for
Path Planning

The goal of automatic collision-free path planning is to find a continuous path of a moving object from the initial position to the goal position, while avoiding collision with the obstacles. Conventional approaches to the solution of this problem consists of two main steps: 1) building a data structure to represent the geometry of workspace or the geometric constraints and 2) searching the data structure to find a collision-free path. Lozano-perez [27, 28] characterizes the position and orientation of an object as a single point, and an obstacle as a forbidden region in a configuration space. Brooks [5] represents a free space as a union of possibly overlapping generalized cones, and finds a path following the spines of the generalized cones and changing from cone to cone at spine intersection points. Kambhampati et al. [16] introduces a hierarchical path-searching method using a quadtree representation of the workspace to speed up the path planning process. In general, the above algorithms based on building and searching the data structure of work space geometry are computationally expensive, making them inefficient for real-time on-line path planning. Meanwhile, Khatib [18] incorporates the path planning into low level control based on the potential field representation of obstacles to achieve real-time performance.

This section presents a massively parallel connectionist network for real-time collision-free path planning [32]. The network is based on representing a path as a series of via points or beads connected by elastic strings which are subject to displacement due to a potential field or a collision penalty function generated by obstacles. Mathematically, this is equivalent to optimizing a cost function, defined in terms of the total path length and the collision penalty function, by moving the via points simultaneously but individually in the direction that minimizes the cost function. Massive parallelism comes mainly from (1) the connectionist model representation of obstacles and (2) the parallel computation of individual via-point motions with only local information. The network

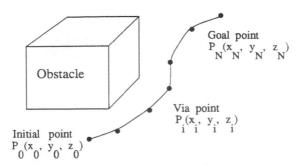

Figure 22: The via point representation of a path.

has power to effectively deal with path planning of three dimensional objects with translational and rotational motions. Finally, the network incorporates simulated annealing to solve a local minimum problem.

A.   Representation of A Collision-Free Path

We first introduce the via point representation of a path, where a path is described by a set of via points and is approximated by the line segments between the adjacent via points, as shown in Figure 22. This representation has two advantages: 1) we can achieve an arbitrary level of precision of a path by assigning a sufficient number of via points, 2) we can decompose the original problem into a set of uniform, small-sized tasks in which the relation between a point and obstacles is of a major concern, and 3) we can localize a path planning problem at individual via points, allowing massive parallel and distributed computing.

To quantify the collision between a path and obstacles, the collision penalty of a path is defined as the sum of individual collision penalties of all the via points, where the collision penalty of a via point is obtained from the connectionist network representation of individual obstacles. The connectionist network representation of the penalty function associated with an obstacle is based on the following observation: a polyhedral obstacle can be represented by a set of linear inequalities, where a point inside the obstacle should satisfy all the inequality constraints.

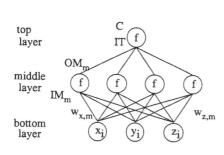

$$C = f(IT)$$
$$IT = \sum_{m=1,M} OM_m + \Theta T$$
$$OM_m = f(IM_m)$$
$$IM_m = w_{x,m} * x_i + w_{y,m} * y_i + w_{z,m} * z_i + \Theta M_m$$

C: output of the top layer unit

IT: input of the top layer unit

ΘT: threshold of the top layer unit

OM_m: output of the mth middle layer unit

IM_m: input of the mth middle layer unit

ΘM_m: threshold of the mth middle layer unit

$w_{x,m}, w_{y,m}, w_{z,m}$: coefficients in the mth constraint

Figure 23: A connectionist network to compute the penalty function associated with an obstacle.

In the connectionist network shown in Figure 23, each of the three units in the bottom layer represents respectively the x, y, z coordinate of the given point. Each unit in the middle layer corresponds to one inequality constraint of the obstacle: the connections between the bottom layer and the middle layer have their weights equal to the coefficients of x, y, z of the corresponding inequality constraint, and the threshold of each unit in the middle layer is assigned to the constant term of the corresponding inequality constraint. The connections between the top layer and the middle layer have their weights equal to 1 and the threshold of the top layer unit is assigned to a value which is 0.5 less than the number of inequality constraints or the number of units in the middle layer.

When a point is given to the bottom layer units, each of the middle layer units decides whether the given point satisfies its constraint. The top layer unit is "on" only if all of the middle layer units are "on", i.e. all the constraints are satisfied. This means that the network outputs "on" only when the point is inside the obstacle. Thus, a path collides with the obstacle if any of its via points forces the network to output

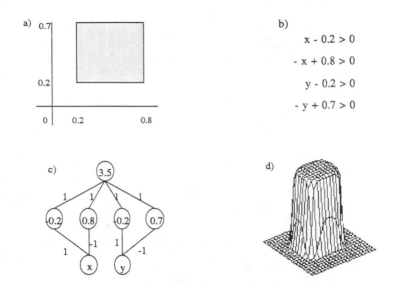

Figure 24: An example of the connectionist network for a rectangular obstacle: a) The obstacle b) Mathematical inequality constraints c) Network representation d) 3-D shape of penalty function with T=0.05.

"on". By using the sigmoid function

$$f(x) = \frac{1}{1 + e^{-x/T}} \tag{38}$$

as the activation function at each unit rather than using the step-wise threshold function, this network produces the continuous value from 0 to 1 according to the degree that a given point collides with the obstacle.

An example for an artificial potential field is shown in Figure 24. The rectangular obstacle in (a) has four inequality constraints as shown in (b). The corresponding network is illustrated in (c), and the three dimensional view of the collision penalty function is shown in (d). Note that the shape of the collision penalty function can be changed by adjusting the parameter T in the activation function. This parameter T in the activation function can play an important role in improving the performance of the algorithm as will be discussed in subsection 4.4.

B. Path Planning of A Point Object

A collision-free path planning problem is equivalent to the optimization problem with two constraints. One is to avoid the collision between the object and obstacles, and the other is to minimize the length of the path. These two constraints can be quantified based on network representation of the penalty function associated with an obstacle. The path planning algorithm developed here is based on minimizing the energy in terms of the path length and the collision penalty.

For an object represented by a point, the collision penalty of a path is defined as the sum of the collision penalties of all the via points, computed by the collision penalty connectionist networks. The energy associated with the collision penalty is thus defined by:

$$E_{collision} = \sum_{i=1}^{N} \sum_{k=1}^{K} C_i^k \tag{39}$$

where K is the number of obstacles, N is the number of via points, and C_i^k is the collision penalty of the ith via point P_i due to the kth obstacle.

The structure of the connectionist network for the computation of $E_{collision}$ is shown in Figure 25. Note that, in this connectionist network, the computation of the penalty functions for via points are carried out concurrently, so are the computations for individual obstacles.

The energy due to path length is defined as the sum of the squares of all the lengths of the line segments connecting via points, $P_i(x_i, y_i, z_i)$, for $i = 1, 2, \cdots, N$:

$$E_{length} = \sum_{i=1}^{N} L_i^2 = \sum_{i=1}^{N} [(x_i - x_{i-1})^2 + (y_i - y_{i-1})^2 + (z_i - z_{i-1})^2] \tag{40}$$

where L_i is the length of the ith line segment. The shorter the path, the smaller E_{length} is.

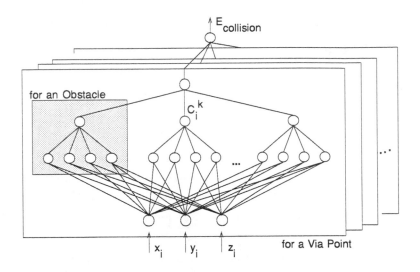

Figure 25: The structure of the collision penalty connectionist network.

The total energy of a path E is now defined as:

$$E = w_l E_{length} + w_c E_{collision} \qquad (41)$$

where w_l and w_c represent weights for each component.

By minimizing the total energy, we can have a path with less collision and shorter length. This implies that, since the total energy is a function of all the via points, individual via points should move in the directions that minimize the energy. This can be done by making the time derivative of the total energy, E, negative along the trajectory defined by the dynamic equation of each via point $P_i(x_i, y_i, z_i)$. The time derivative of E with respect to time is

$$
\begin{aligned}
\dot{E} &= \sum_i (\nabla_{P_i} E)^t \dot{P}_i \\
&= \sum_i \{ [w_l(\frac{\partial L_i}{\partial x_i} + \frac{\partial L_{i-1}}{\partial x_i}) + w_c \sum_k \frac{\partial C_i^k}{\partial x_i}] \dot{x}_i + [w_l(\frac{\partial L_i}{\partial y_i} + \frac{\partial L_{i-1}}{\partial y_i}) + \\
&\quad w_c \sum_k \frac{\partial C_i^k}{\partial y_i}] \dot{y}_i + [w_l(\frac{\partial L_i}{\partial z_i} + \frac{\partial L_{i-1}}{\partial z_i}) + w_c \sum_k \frac{\partial C_i^k}{\partial z_i}] \dot{z}_i \}
\end{aligned} \qquad (42)
$$

So, by choosing the time derivative of each coordinate as:

$$\dot{x}_i = -\eta[w_l(\frac{\partial L_i}{\partial x_i} + \frac{\partial L_{i-1}}{\partial x_i}) + w_c \sum_k \frac{\partial C_i^{\prime k}}{\partial x_i}]$$

$$\dot{y}_i = -\eta[w_l(\frac{\partial L_i}{\partial y_i} + \frac{\partial L_{i-1}}{\partial y_i}) + w_c \sum_k \frac{\partial C_i^{\prime k}}{\partial y_i}]$$

$$\dot{z}_i = -\eta[w_l(\frac{\partial L_i}{\partial z_i} + \frac{\partial L_{i-1}}{\partial z_i}) + w_c \sum_k \frac{\partial C_i^{\prime k}}{\partial z_i}] \tag{43}$$

with positive gain η, the time derivative of energy becomes:

$$\dot{E} = -\eta \sum_i (\dot{x}_i^2 + \dot{y}_i^2 + \dot{z}_i^2) < 0 \tag{44}$$

along the trajectory, and $\dot{E} = 0$ if and only if $\dot{x}_i = 0$, $\dot{y}_i = 0$, and $\dot{z}_i = 0$. This means that all via points are rearranged to decrease the energy, and finally reach equilibrium positions. From Figure 23.

$$\frac{\partial C_i^{\prime k}}{\partial x_i} = (\frac{\partial C_i^{\prime k}}{\partial IT_i^k})(\frac{\partial IT_i^k}{\partial x_i})$$

$$= (\frac{\partial C_i^{\prime k}}{\partial IT_i^k}) \sum_{m=1}^{M} (\frac{\partial OM_{i,m}^k}{\partial IM_{i,m}^k})(\frac{\partial IM_{i,m}^k}{\partial x_i})$$

$$= f'(IT_i^k) \sum_{m=1}^{M} f'(IM_{i,m}^k) w_{x,m}^k \tag{45}$$

where M is the number of constraints. Thus, from equations (43) and (45), the dynamic equations of the via point $P_i(x_i, y_i, z_i)$ is derived as:

$$\dot{x}_i = -\eta[w_l(2x_i - x_{i-1} - x_{i+1}) + w_c \sum_k f'(IT_i^k) \sum_{m=1}^{M} f'(IM_{i,m}^k) w_{x,m}^k]$$

$$\dot{y}_i = -\eta[w_l(2y_i - y_{i-1} - y_{i+1}) + w_c \sum_k f'(IT_i^k) \sum_{m=1}^{M} f'(IM_{i,m}^k) w_{y,m}^k]$$

$$\dot{z}_i = -\eta[w_l(2z_i - z_{i-1} - z_{i+1}) + w_c \sum_k f'(IT_i^k) \sum_{m=1}^{M} f'(IM_{i,m}^k) w_{z,m}^k]$$

where η is a positive gain and $f'(\cdot)$ is equal to $\frac{1}{T}f(\cdot)[1 - f(\cdot)]$. This analysis is similar to that of the back-propagation algorithm [35]. The difference is that in this algorithm, the input is varying, while in the

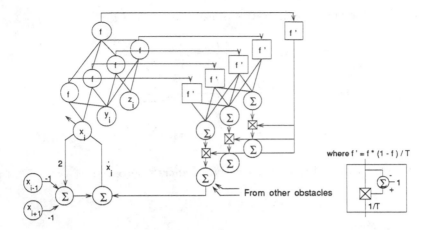

Figure 26: The computational flow of the dynamic equations.

back-propagation algorithm, the connection weights are varying. The computational flow for the dynamic equations of one via point is shown in Figure 26. Notice the parallelism involved in the computations associated with obstacles or via points.

C.   Path Planning of A Polyhedral Object

The path planning algorithm developed for a point object can be easily extended to the case of a polyhedral object. However, there are two distinctive features to be considered in the path planning of a polyhedral object: First, for an object with volume, rotation as well as translation can occur along the path. Thus, in order to represent the exact configuration of the object, the orientation as well as the position should be specified. Second, it is necessary to consider a number of object points to compute the collision penalty at of the object.

To represent the object configuration, the frame assigned to an object is represented with respect to the world frame, as shown in Figure 27. We use the roll, pitch and yaw angles to represent the orientation of the frame, while using the term "via point" to refer to the origin of each frame.

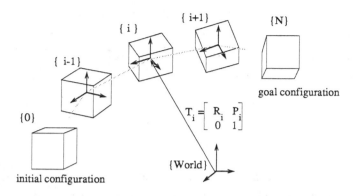

Figure 27: Representation of a via configuration of the object along the path.

For the determination of the degree of collision between the polyhedral object and the obstacles, we select a set of test points from the object. The collision penalty of the object is defined as the sum of the collision penalties at all the test points. These test points are fixed with reference to the assigned object frame, such that their actual positions with respect to the world frame can be determined by the position and orientation of the corresponding object frame as follows:

$$Q_{i,j} = R_i P_j^{test} + P_i \tag{46}$$

where $P_j^{test} \equiv [x_j^{test}, y_j^{test}, z_j^{test}]$ represents the position vector of the jth test point with respect to the object frame, $Q_{i,j} \equiv [X_{i,j}, Y_{i,j}, Z_{i,j}]$ represents the position vector of the jth test point at the ith via point with respect to the world frame, $P_i(x_i, y_i, z_i)$ is the position of the origin of the object frame at the ith via point, and R_i represents the rotational matrix corresponding to the orientation of the object frame at the ith via point. R_i can be obtained by:

$$R_i = \begin{bmatrix} \cos\alpha_i \cos\beta_i & \cos\alpha_i \sin\beta_i \sin\gamma_i - \sin\alpha_i \cos\gamma_i & \cos\alpha_i \sin\beta_i \cos\gamma_i + \sin\alpha_i \sin\gamma_i \\ \sin\alpha_i \cos\beta_i & \sin\alpha_i \sin\beta_i \sin\gamma_i + \cos\alpha_i \cos\gamma_i & \sin\alpha_i \sin\beta_i \cos\gamma_i - \cos\alpha_i \sin\gamma_i \\ -\sin\beta_i & \cos\beta_i \sin\gamma_i & \cos\beta_i \cos\gamma_i \end{bmatrix}$$

for given roll γ_i, pitch β_i, yaw α_i, angles of the object frame at the ith via point with respect to the world frame. By applying the actual position of a test point to the connectionist network as an input, the collision penalty of the test point can be directly computed.

Based on the above considerations, the total energy is now defined for the case of a polyhedral object as follows:

$$
\begin{aligned}
E &= w_l E_{length} + w_c E_{collision} \\
&= \sum_{i=1}^{N} L_i^2 + \sum_{i=1}^{N} \sum_{j=1}^{J} \sum_{k=1}^{K} C_{i,j}^k
\end{aligned}
\tag{47}
$$

where L_i is the distance between the origins of the object frame at the ith and i-1 th via points, and $C_{i,j}^k$ is the collision penalty of the jth test point at the ith via point, due to the kth obstacle. Similarly to the case of a point object, we can establish a set of dynamic equations that allows each variable to move toward minimizing the total energy:

$$
\begin{aligned}
\dot{x}_i &= -\eta_t [w_l(2x_i - x_{i-1} - x_{i+1}) + w_c \sum_j \sum_k \frac{\partial C_{i,j}^k}{\partial x_i}] \\
\dot{y}_i &= -\eta_t [w_l(2y_i - y_{i-1} - y_{i+1}) + w_c \sum_j \sum_k \frac{\partial C_{i,j}^k}{\partial y_i}] \\
\dot{z}_i &= -\eta_t [w_l(2z_i - z_{i-1} - z_{i+1}) + w_c \sum_j \sum_k \frac{\partial C_{i,j}^k}{\partial z_i}] \\
\dot{\alpha}_i &= -\eta_r [w_c \sum_j \sum_k (\nabla_{Q_{i,j}} C_{i,j}^k)^t R_\alpha P_j^{test}] \\
\dot{\beta}_i &= -\eta_r [w_c \sum_j \sum_k (\nabla_{Q_{i,j}} C_{i,j}^k)^t R_\beta P_j^{test}] \\
\dot{\gamma}_i &= -\eta_r [w_c \sum_j \sum_k (\nabla_{Q_{i,j}} C_{i,j}^k)^t R_\gamma P_j^{test}]
\end{aligned}
\tag{48}
$$

where, $\nabla_{Q_{i,j}} C_{i,j}^k = [\frac{\partial C_{i,j}^k}{\partial X_{i,j}}, \frac{\partial C_{i,j}^k}{\partial Y_{i,j}}, \frac{\partial C_{i,j}^k}{\partial Z_{i,j}}]^t$, $\frac{\partial C_{i,j}^k}{\partial X_{i,j}} = f'(IT_{i,j}^k) \sum_m f'(IM_{i,m,j}^k) w_{x,m}^k$, $\frac{\partial C_{i,j}^k}{\partial Y_{i,j}} = f'(IT_{i,j}^k) \sum_m f'(IM_{i,m,j}^k) w_{y,m}^k$, $\frac{\partial C_{i,j}^k}{\partial Z_{i,j}} = f'(IT_{i,j}^k) \sum_m f'(IM_{i,m,j}^k) w_{z,m}^k$,

$$
\begin{aligned}
R_\alpha &= \frac{\partial R_i}{\partial \alpha_i} \\
&= \begin{bmatrix}
-\sin\alpha_i \cos\beta_i & -\sin\alpha_i \sin\beta_i \sin\gamma_i - \cos\alpha_i \cos\gamma_i & -\sin\alpha_i \sin\beta_i \cos\gamma_i + \cos\alpha_i \sin\gamma_i \\
\cos\alpha_i \sin\beta_i & \cos\alpha_i \sin\beta_i \sin\gamma_i - \sin\alpha_i \cos\gamma_i & \cos\alpha_i \sin\beta_i \cos\gamma_i + \sin\alpha_i \sin\gamma_i \\
0 & 0 & 0
\end{bmatrix}
\end{aligned}
$$

$$
\begin{aligned}
R_\beta &= \frac{\partial R_i}{\partial \beta_i} \\
&= \begin{bmatrix}
-\cos\alpha_i \sin\beta_i & \cos\alpha_i \cos\beta_i \sin\gamma_i & \cos\alpha_i \cos\beta_i \cos\gamma_i \\
-\sin\alpha_i \sin\beta_i & \sin\alpha_i \cos\beta_i \sin\gamma_i & \sin\alpha_i \cos\beta_i \cos\gamma_i \\
-\cos\beta_i & -\sin\beta_i \sin\gamma_i & -\sin\beta_i \cos\gamma_i
\end{bmatrix} \quad \text{and}
\end{aligned}
$$

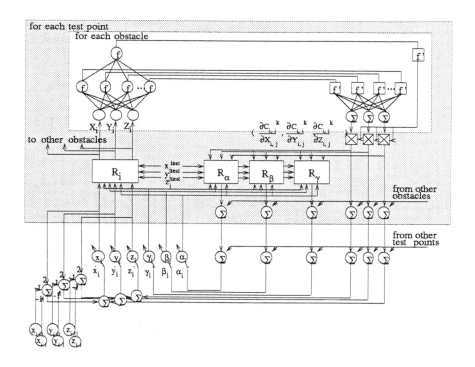

Figure 28: Computational flow of the path planning algorithm for a polyhedral object.

$$
\begin{aligned}
R_\gamma &= \frac{\partial R_i}{\partial \gamma_i} \\
&= \begin{bmatrix}
0 & \cos\alpha_i \sin\beta_i \cos\gamma_i + \sin\alpha_i \sin\gamma_i & -\cos\alpha_i \sin\beta_i \sin\gamma_i + \sin\alpha_i \cos\gamma_i \\
0 & \sin\alpha_i \sin\beta_i \cos\gamma_i - \cos\alpha_i \sin\gamma_i & -\sin\alpha_i \sin\beta_i \sin\gamma_i - \cos\alpha_i \cos\gamma_i \\
0 & \cos\alpha_i \cos\gamma_i & -\cos\beta_i \sin\gamma_i
\end{bmatrix}
\end{aligned}
$$

Figure 28 shows the connectionist architecture for computing the dynamic equations at a single via point. Note that matrix multiplication blocks are included in the architecture on account of the rotations involved in (46). However, but the parallelism in terms of the obstacles, the test points, and the via points is well maintained.

D.   Simulated Annealing for
A Local Minimum Problem

There exists a local minima problem in the algorithm developed in the previous subsections. In other words, the total energy defined in terms of the path length and the collision penalty has multiple extrema, and the dynamic equations which move via points toward minimizing the total energy cannot guarantee to find the global minimum and may be stuck into one of the local minima in the energy function. The path formed at local minima may be much longer than the optimal one, and may not be totally free from collisions with obstacles. This prompts a need to develop a techniques to escape from local minima through tunneling or hill-climbing.

The method of simulated annealing [19, 37] provides such a technique of avoiding local minima, and has effectively solved many optimization problems including the traveling salesman problem. The concept of simulated annealing is something to do with thermodynamics, especially with the process of freezing and crystallizing of a liquid, or cooling and annealing of a metal. At high temperature, the molecules of a liquid move freely with respect to one another. In the process of cooling the liquid, its thermal mobility is gradually lost, but the molecules are often able to line themselves up and form a pure crystal which corresponds to the state of minimum energy. The essence of this process is slow cooling, allowing enough time for redistribution of the molecules as they lose mobility, and thus ensuring that a low energy state will be achieved.

The concept of simulated annealing can be applied to the path planning algorithm to avoid local minima. This can be done by changing penalty function associated with obstacles slowly toward their final forms, starting from the smooth functions approximately representing the obstacles, as shown in Figure 32. Note that the penalty function, or the output of a collision penalty network, depends on the parameter T of a sigmoid function. With a large value of T, the penalty function can have a smooth surface with gradual transition across obstacle boundaries. In this case, the collision penalty network represents the obstacle not accurately but abstractly, in such a way that via points inside obstacles lie at the slope of the surface of a collision penalty function, allowing via points to move down along the surface. In the case where T is very small, the network acts like a switch: it produces "1" for a via point inside the obstacle, and "0" for a via point outside the obstacle. The shape of the surface of a collision penalty function becomes almost flat everywhere except near the obstacle boundary where a cliff appears. This makes via points hard to move along the surface. Using the relationship between the parameter T and the shape of the surface of a penalty function, we achieve the effect of simulated annealing by slowly decreasing T from the sufficiently high temperature.

Geman and Geman [10] have provided an annealing schedule which guarantees the convergence to the global minimum as:

$$\frac{T_a(t)}{T_0} = \frac{1}{\log(1+t)} \tag{49}$$

where T_a is the artificial temperature and T_0 is a sufficiently high value of the initial temperature. Since this schedule (49) results in very slow convergence, Szu [36] suggests an efficient schedule for simulated annealing,

$$\frac{T_a(t)}{T_0} = \frac{1}{1+t} \tag{50}$$

as a trade-off between the convergence time and the global optimality. The fast simulated annealing schedule, (50), allows the path planning to be done with reasonable computation time.

E.   Simulation

The proposed neural network for path planning has been tested by computer simulation. Figure 29 shows the simulation result for path planning of a point object with two obstacles. The initial path is arbitrarily chosen as a straight line from the initial position to the goal position as shown in Figure 29a). The shape of the collision penalty function associated with the obstacles is shown in Figure 29b). Notice that, the

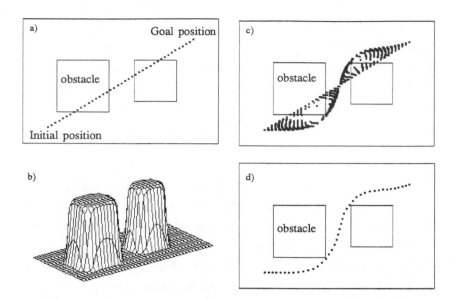

Figure 29: Path planning of a point object with two obstacles: a) Initial path b) 3D view of the collision penalty function c) Trajectories of the via points d) Final path.

via points near the boundaries of the obstacles move much faster than the others, because the slope of the collision penalty function is steep at the boundary. Figure 29c) illustrates the trajectories of via points during convergence, while Figure 29d) shows the final collision-free path.

Figure 30 illustrates a situation where the network converges to a local minimum without using simulated annealing, where path planning of a point object is carried out in a cluttered environment with five obstacles and with the fixed collision penalty function (T = 0.1) shown in Figure 30a). Figure 30b) illustrates the final path obtained by the network, which is stuck into a local minimum, because the via points initially assigned inside obstacle are placed on the flat surface of the collision penalty function, and thus do not have enough mobility to get out of the obstacle. By applying the simulated annealing to the same path planning problem, i.e., by decreasing the value of T gradually from

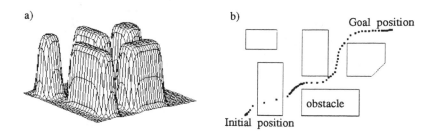

Figure 30: Path planning converges to a local minimum without simulated annealing.

the sufficiently high initial value based on a simulated annealing schedule, all the via points were able to get out of the obstacles to converge to the optimal collision-free path. Figure 31 illustrates the shape of the collision penalty function along with the fast annealing schedule, (50), with $T_0 = 0.5$, and the corresponding motion of via points with respect to time.

The path planning network has been also applied to path planning of a polyhedral object as shown in Figure 32. Figure 32a) represents the planned path of a square object when only translation is allowed; Figure 32b) is the planned path of the rectangular object when rotation as well as translation are allowed. Note that, since the collision is examined at only several test points of an object, we need to increase the number of test points to obtain a safer path.

To emphasize the fact that the network is equally effective for the path planning of a 3D object with the presence of both rotation and transition, we consider the path planning of a brick object, as shown in Figure 33. Nine test points are chosen at its center and eight vertices. The value of T is set to 1 at the beginning, and gradually changed to 0.2 at t=1. The via configurations are initially assigned so that the orientational and positional differences between adjacent configurations are uniformly distributed from the initial configuration to the goal configuration. Figure 33 illustrates the resulting path from two different view points. Notice the gradual variation of the orientation as well as the

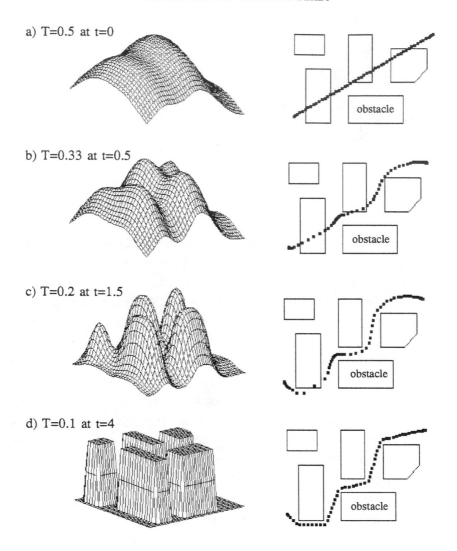

a) T=0.5 at t=0

b) T=0.33 at t=0.5

c) T=0.2 at t=1.5

d) T=0.1 at t=4

Figure 31: The variation of the shape of collision penalty function and the corresponding movement of the via points along with the fast simulated annealing schedule with $T_0 = 0.5$.

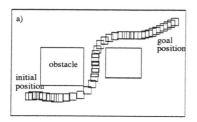 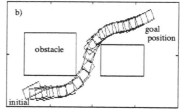

Figure 32: a) The planned path for a square object, allowing only translation b) The planned path for a rectangular object, allowing rotation as well as translation.

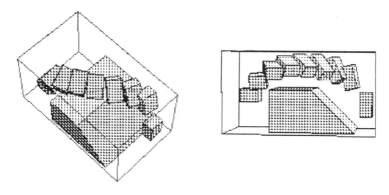

Figure 33: The planned path of a brick object in the three dimensional space: from two different view points.

position along the final path.

F.   Discussion

The proposed neural network approach to path planning provides massive parallelism in computation. This comes from the fact that the computation for the motion of each via point, as well as the penalty function of each obstacle, is totally localized, and thus can be executed in parallel. As a result, the total computation time is constant irrespective of the number of obstacles and the number of via points, although the size of network increases linearly with them.

An arbitrary degree of precision of a path may be achieved by applying a sufficient number of via points. In other words, the number of via points should be chosen in such a way that the final path approximated by its via positions should not collide with obstacles while passing from one via position to its adjacent via position. Similarly, the required precision of a path for a polyhedral object can be obtained by using sufficient number of test points which approximate the object. To ensure the safer path, avoiding clipping effect at the corner of the obstacles, the virtual boundary can be set for individual objects.

The application of the network to three dimension path planning requires to handle one more variable for the position and two more variables for the orientation than the two dimensional path planning. Since the computations for all the positional and orientational variables can be done in parallel, this does not incur extra computation time. So, by treating the two dimensional case as the special case in which one positional and two orientational variables are fixed, the algorithm can be applied to both of two and three dimensions without any modification. This is quite different from the conventional path planning algorithms which fail to work or need excessive computational cost, for dealing with the three dimensional case.

Note that, since the fast simulated annealing guarantees only the semiglobal minimum, the algorithm might still get stuck into local minima. The final path at a local minimum may not be free from collisions, especially, in a heavily cluttered environment. By detecting this situation based on the length of adjacent via positions (a collision-free path tends to have these lengths uniformly distributed along the path, and the path may be decomposed into collision-free subpaths and non-collision-free subpaths), the algorithm can be recursively applied to the non-collision-free subpaths to obtain the final collision-free path. It is also possible to apply the algorithm with the different initial path assignment.

Finally, the collision penalty network developed can be used for other geometrical modeling problems. For example, Khatib [18] used the n-ellipsoids to model the obstacle, and sets up the artificial potential field around the obstacle based on the distance to it. The collision penalty network can replace this obstacle modeling and construct the potential field directly from its output.

In summary, the neural network approach to path planning has the following advantages: (1) The algorithm is inherently massively parallel and can make good use of parallel hardware. Thus, the path planning can be done in real-time regardless of the number of obstacles, the number of via points, and number of test points of the object. (2) This massive parallelism, in turn, enables the algorithm to achieve the precision of the path to an arbitrary level without increasing the computation time. (3) The algorithm can be applied to three dimensions as well as to two dimensions without any modification or increasing the computation time. (4) The method of simulated annealing can be used in combination with the algorithm to solve the local minima problem.

The neural network approach presented here is based on the assumption that all the information about the environment is given apriori, and the obstacles are stationary. In the real world, it may be necessary to plan a path with only a partial information, or the obstacle can be moving as well as the object. So, it is worth extending this approach to more general situation utilizing above useful characteristics.

V.   Conclusion

This chapter has presented the applications of artificial neural networks to robot control, with an emphasis on inverse kinematics and dynamics and path planning. This chapter is focused on methods developed to overcome some of the major sources of difficulty in these applications, namely, slow convergence, large sample sizes and lack of accuracy. More details on works presented in this chapter can be found in [3, 4] [24, 25, 26] [32] [40, 41]. While the problems of kinematics, dynamics and path planning represent a major share of this field, there are a number of other applications. For example, neural networks have been used in sensor-based robot control [30]. In this application there are multiple sensors and multiple command variables. The network (implemented using the CMAC architecture [1]) learns the nonlinear relationship between the sensor output and the command variables over particular regions of the system state space. A recent survey by Kung and Hwang [21] lists applications in path planning, use of stereo vision in task planning and sensor/motor control. The authors major objective is to present systolic architectures suitable for problems in robotics.

Yet, in spite of a substantial number of publications, it is not yet clear whether neural network methods will find substantial applications in industrial robotic systems. The idea that either kinematic or dynamic control could be achieved without a detailed knowledge of either the governing equations of robots or their parameters is extremely appealing. Yet, this appeal is tempered by the realization that training times may be prohibitively long, and that there is no current theoretical basis for selecting the number of hidden layers or the number of units in each layer to achieve a given accuracy. Furthermore, while robot control does indeed require complex computations, current algorithms allow for such computing to be done in real time. Furthermore, the increasing availability of parallel algorithms (e.g., [9]) and processors promises to further reduce the time required to perform the computations needed for on-line control. As a consequence, at least for those situations where the robot model and its parameters are known or can be estimated, there is less need for the power of neural networks to approximate arbitrary input-output relationships.

There is a further problem with the acceptance of neural networks, which is partially related to their power. It is possible, as Kuperstein [22] has done, to allow a network to learn eye-hand mappings for robots beginning with total ignorance. Yet, the price paid for this ability is long convergence, with large numbers of training examples obtained from the whole space. On the other hand, traditional robot control methods require complete knowledge, and suffer when accurate models or precise parameter values are not available. Only a few investigators, such as

Handelman et al. [13], have attempted to combine knowledge-based methods with neural network techniques in robotics. It may be that further development of such hybrid methods, which combine knowledge, heuristics and connectionist methods will provide the tools for the development of new generations of intelligent, highly adaptive robotic systems.

References

[1] J. Albus. A new approach to manipulator control: the cerebellar model articulation controller. *Journal of Dynamic Systems, Measurement and Control*, 97:270–277, 1975.

[2] J. Barhen, S. Gulati, and M. Zak. Neural learning of constrained nonlinear transformations. *IEEE Computer Magazine*, pages 67–76, June 1989.

[3] D. F. Bassi and G. A. Bekey. Decomposition of neural network mod els of robot dynamics: A feasibility study. In W. Webster, editor, *Simulation and AI*, pages 8–13. Society for Computer Simulation, 1989.

[4] D. F. Bassi and G. A. Bekey. High precision position control by cartesian trajectory feedback and connectionist inverse dynamics feedforward. *IEEE International Joint Conference on Neural Networks*, 2:325–332, 1989.

[5] R. A. Brooks. Solving the find-path problem by good representation of free space. *IEEE Transactions on System, Man and Cybernetics*, 13:190–197, 1983.

[6] P. R. Chang and C.S.G. Lee. Residue arithmetic vlsi array architecture for manipulator pseudo-inverse jacobian computation. *IEEE Transactions on Robotics and Automation*, 5:569–582, 1989.

[7] J. J. Craig. *Introduction to Robotics*. Addison-Wesley, 1986.

[8] R. K. Elsley. A learning architecture for control based on back-propagation neural networks. *IEEE International Conference on Neural Networks*, pages 587–594, 1988.

[9] A. Fijany and A. K. Bejczy. A class of parallel algorithms for computation of the manipulator inertia matrix. *IEEE Transactions on Robotics and Automation*, 5:600–615, 1989.

[10] S. Geman and D. Geman. Stochastic relaxation, gibbs distribution, and the bayesian restoration of images. *IEEE Transactions on Pattern Recognition and Machine Intelligence*, 6:721–741, 1984.

[11] S. Grossberg and M. Kuperstein. *Neural Dynamics of Adaptive Sensory Motor Control*. Elsevier North-Holland Press, 1986.

[12] A. Guez and Z. Ahmad. Solution to the inverse problem in robotics by neural networks. *IEEE International Conference on Neural Networks*, pages 617–624, 1988.

[13] D. A. Handelman, S. H. Lane, and J. J. Gelfand. Integration of knowledge-based system and neural network techniques for robotic control. *Proceedings of IEEE International Conference on Robotics and Automation*, pages 1454–1460, 1989.

[14] G. E. Hinton. A parallel computation that assigns canonical object-based frames of reference. *Proceedings of International Joint Conference on Artificial Intelligence*, pages 683–685, 1981.

[15] G. Josin, D. Charney, and D. White. Robot control using neural networks. *IEEE International Joint Conference on Neural Networks*, 2:625–631, 1987.

[16] S. Kambhampati and L. S. Davis. Multiresolution path planning for mobile robots. *IEEE Journal of Robotics and Automation*, 2:135–145, 1986.

[17] M. Kawato, Y. Uno, M. Isobe, and R. Suzuki. A hierarchical model for voluntary movement and its application to robotics. *IEEE International Conference on Neural Networks*, 4:573–582, 1987.

[18] O. Khatib. Real-time obstacle avoidance for manipulators and mobile robots. *International Journal of Robotics Research*, 5:90–98, 1986.

[19] S. Kirkpatrick, Jr. C. D. Gelatt, and M. P. Vecchi. Optimization by simulated annealing. *Science*, 220:671–680, 1983.

[20] T. Kohonen. *Self-Organization and Associative Memory*. Springer-Verlag, 1984.

[21] S. Y. Kung and J. N. Hwang. Neural network architectures for robotic applications. *IEEE Transactions on Robotics and Automation*, 5:641–657, 1989.

[22] M. Kuperstein. Adaptive visual-motor coordination in multijoint robots using parallel architecture. *IEEE International Conference on Robotics and Automation*, pages 1595–1602, 1987.

[23] M. B. Leahy. Efficient puma manipulator jacobian calculation and inversion. *Journal of Robotic Systems*, 4:63–75, 1987.

[24] S. Lee and R. M. Kil. Multilayer feedforward potential function network. *IEEE International Conference on Neural Networks*, 1:161–171, 1988.

[25] S. Lee and R. M. Kil. Bidirectional continuous associator based on gaussian potential function network. *IEEE International Joint Conference on Neural Networks*, 1:45–53, 1989.

[26] S. Lee and R. M. Kil. Robot kinematic control based on bidirectional mapping neural network. *IEEE International Joint Conference on Neural Networks*, 3:327–335, 1990.

[27] T. Lozano-Perez. Spatial planning: A configuration space approach. *IEEE Transactions on Computers*, 32:108–120, 1983.

[28] T. Lozano-Perez and M. A. Wesley. An algorithm for planning collision-free paths among polyhedral obstacles. *Communications of ACM*, 22:560–570, 1979.

[29] T. M. Martinez, H. J. Ritter, and K. J. Schulten. Three-dimensional neural net for learning visuomotor coordination of a robot arm. *IEEE Transactions on Neural Networks*, 1:131–136, 1990.

[30] W. T. Miller. Real-time application of neural networks for sensor-based control of robots. *IEEE Transactions on System, Man and Cybernetics*, 19:825–831, 1989.

[31] H. Miyamoto, M. Kawato, T. Setoyama, and R. Suzuki. Feedback error learning neural networks for trajectory control of a robotic manipulator. *Neural Networks*, 1:251–265, 1988.

[32] J. Park and S. Lee. Neural computation for collision-free path planning. *IEEE International Joint Conference on Neural Networks*, 2:229–232, 1990.

[33] D. Psaltis, A. Sideris, and A. Yamamura. Neural controllers. *IEEE International Conference on Neural Networks*, pages 551–558, 1987.

[34] H. J. Ritter, T. M. Martinetz, and K. J. Schulten. Topology-conserving maps for learning visuo-motor-coordination. *Neural Networks*, 2(3):159–168, 1989.

[35] D. E. Rumelhart, G. E. Hinton, and R. J. Williams. *Parallel Distributed Processing*, volume 1, pages 318–362. MIT Press/Bradford Books, 1986.

[36] H. Szu. Fast simulated annealing. *American Institue of Physics*, pages 420–425, 1986.

[37] P. J. M. van Laarhoven and E. H. L. Aarts. *Simulated Annealing: Theory and Applications.* D. Reidel Publishing Company, 1987.

[38] B. Widrow and M. E. Hoff. Adaptive switching circuits. *WESCON Conv. Record*, 4:96–104, 1960.

[39] B. Widrow and S. D. Stearns. *Adaptive Signal Processing*, pages 15–116. Prentice-Hall, 1985.

[40] D. T. Yeung and G. A. Bekey. Using a context-sensitive learning network for robot arm control. *Proceedings of IEEE International Conference on Robotics and Automation*, pages 1441–1447, 1989.

[41] D.T. Yeung. *Handling dimensionality and nonlinearity in connectionist learning.* PhD thesis, University of Southern California, 1989.

[42] M. Zak. Terminal attractors for addressable memory in neural networks. *Physics Letters A*, 133:218–222, 1988.

Appendix: Proofs of the Lemma and the Theorems

A. Lemma 1

Let us consider the norm of $\dot{\mathbf{x}}$ as $\mathbf{x}$ approaches $\mathbf{x}^*$:

$$\lim_{\mathbf{x}\to\mathbf{x}^*} \|\dot{\mathbf{x}}\| = \lim_{\mathbf{x}\to\mathbf{x}^*} \frac{1}{2} \frac{\|\tilde{\mathbf{y}}\|^2}{\|\mathbf{J}^t\tilde{\mathbf{y}} + \tilde{\mathbf{x}}\|}$$

If $\tilde{\mathbf{x}} \neq \mathbf{0}$, the above equation becomes 0 but if $\tilde{\mathbf{x}} = \mathbf{0}$, both denominator and numerator of (V.) approach 0. Let us define V as

$$V = \frac{1}{2} \|\tilde{\mathbf{y}}\|^2$$

and consider the squared norm of $\dot{\mathbf{x}}$ as $\mathbf{x}$ approaches to $\mathbf{x}^*$:

$$\lim_{\mathbf{x}\to\mathbf{x}^*} \|\dot{\mathbf{x}}\|^2 = \lim_{\mathbf{x}\to\mathbf{x}^*} \frac{V^2}{\|\frac{\partial V}{\partial \mathbf{x}}\|^2}$$

By the Taylor series approximations of V and $\frac{\partial V}{\partial \mathbf{x}}$ around $\mathbf{x}^*$,

$$\lim_{\mathbf{x}\to\mathbf{x}^*} \|\dot{\mathbf{x}}\|^2 \approx \frac{1}{4} \frac{[(\mathbf{x}-\mathbf{x}^*)^t \frac{\partial^2 V}{\partial \mathbf{x}^2}(\mathbf{x}^*)(\mathbf{x}-\mathbf{x}^*)]^2}{\|\frac{\partial^2 V}{\partial \mathbf{x}^2}(\mathbf{x}^*)(\mathbf{x}-\mathbf{x}^*)\|^2}$$
$$\approx \frac{1}{4} \frac{[(\mathbf{x}-\mathbf{x}^*)^t \boldsymbol{\Sigma}\boldsymbol{\Lambda}\boldsymbol{\Sigma}^t(\mathbf{x}-\mathbf{x}^*)]^2}{(\mathbf{x}-\mathbf{x}^*)^t \boldsymbol{\Sigma}\boldsymbol{\Lambda}^2\boldsymbol{\Sigma}^t(\mathbf{x}-\mathbf{x}^*)}$$

where $\boldsymbol{\Sigma}$ is a matrix in which each column represents an eigenvector and $\boldsymbol{\Lambda}$ is a diagonal matrix in which each diagonal element represents an eigenvalue of $\frac{\partial^2 V}{\partial \mathbf{x}^2}(\mathbf{x}^*)$.

Note that the matrix $\frac{\partial^2 V}{\partial \mathbf{x}^2}(\mathbf{x}^*)$ is symmetric and assumed to be positive definite for the convexity of V around $\mathbf{x}^*$. By the assumption of convexity of V,

$$\lim_{\mathbf{x}\to\mathbf{x}^*} \|\dot{\mathbf{x}}\|^2 \leq \lim_{\mathbf{x}\to\mathbf{x}^*} \frac{1}{4}\left(\frac{\lambda_{max}}{\lambda_{min}}\right)^2 \|\boldsymbol{\Sigma}^t(\mathbf{x}-\mathbf{x}^*)\|^2 = 0$$

where λ_{min} and λ_{max} respectively represent the minimum and the maximum eigenvalues of $\frac{\partial^2 V}{\partial \mathbf{x}^2}(\mathbf{x}^*)$.

Therefore, $\mathbf{x}^*$ is an equilibrium point of (14). If $\mathbf{x} \neq \mathbf{x}^*$, $\|\tilde{\mathbf{y}}\| > 0$ and two cases can occur:

1. if $\mathbf{J}^t\tilde{\mathbf{y}} + \tilde{\mathbf{x}} \neq \mathbf{0}$, $\| \dot{\mathbf{x}} \| > 0$ and

2. if $\mathbf{J}^t\tilde{\mathbf{y}} + \tilde{\mathbf{x}} = \mathbf{0}$, $\| \dot{\mathbf{x}} \| \to \infty$.

This implies that

$$\dot{\mathbf{x}} = \mathbf{F}(\mathbf{x}^*) = \mathbf{0} \;\; iff \;\; \tilde{\mathbf{y}}(\mathbf{x}^*) = \mathbf{0}. \;\; \text{Q.E.D.}$$

B. Theorem 1

By lemma 1, (17) reaches the equilibrium state if and only if $\tilde{\mathbf{y}} = \mathbf{0}$. By applying (17) to (13), we have

$$\dot{V} = -\frac{1}{2} \| \tilde{\mathbf{y}} \|^2 \leq 0,$$

where $\dot{V} < 0 \;\; \forall \tilde{\mathbf{y}} \neq \mathbf{0}$ and $\dot{V} = 0 \;\; iff \; \tilde{\mathbf{y}} = \mathbf{0}$. Q.E.D.

C. Theorem 2

When $\mathbf{x}$ approaches the solution of $\tilde{\mathbf{y}}$, λ approaches ∞ and $\dot{\mathbf{x}}$ in (21) is approximated by

$$\dot{\mathbf{x}} \approx \frac{1}{2} \frac{\| \tilde{\mathbf{y}} \|^2}{\| \mathbf{J}^t\tilde{\mathbf{y}} \|^2} (\mathbf{J}^t\tilde{\mathbf{y}}).$$

By lemma 1, the above equation reaches the equilibrium state if and only if $\tilde{\mathbf{y}} - \mathbf{0}$. Since the time derivative of the Liapunov function is given by

$$\dot{V} = \frac{\partial V}{\partial x} \dot{\mathbf{x}} + \frac{\partial V}{\partial \lambda} \dot{\lambda},$$

we have

$$\dot{V} = -(\lambda \tilde{\mathbf{y}}^t \mathbf{J} + \tilde{\mathbf{x}}^t)\dot{\mathbf{x}} + \frac{1}{2} \| \tilde{\mathbf{y}} \|^2 \dot{\lambda}.$$

By applying (21) to $\dot{\mathbf{x}}$ and $\dot{\lambda}$, we have

$$\dot{V} = -\frac{\lambda}{2} \| \tilde{\mathbf{y}} \|^2 \leq 0$$

where $\dot{V} < 0 \;\; \forall \tilde{\mathbf{y}} \neq \mathbf{0}$ and $\dot{V} = 0 \;\; iff \; \tilde{\mathbf{y}} = \mathbf{0}$. Q.E.D.

A UNIFIED APPROACH TO KINEMATIC MODELING, IDENTIFICATION AND COMPENSATION FOR ROBOT CALIBRATION

HANQI ZHUANG AND ZVI S. ROTH

Robotics Center, College of Engineering
Florida Atlantic University
Boca Raton, FL 33431

I. INTRODUCTION

Robot Calibration is the process of enhancing the accuracy of a robot manipulator through modification of the robot control software. Humble as the name "calibration" may sound, it encompasses three distinct actions, none of which is trivial:

Step 1: Measurement of the position and orientation of the robot end-effector in world coordinates (*Pose Measurement*).
Step 2: Identification of the relationship between joint angles and end-point positions (*Kinematic Identification*).
Step 3: Modification of control commands to allow a successful completion of a programmed task (*Kinematic Compensation*).

The need for robot calibration arises in many applications which require off-line programming and situations that require multiple robots to share the same application software. Example to the first are assembly operations, in which costly hard-automation to compensate for robot inaccuracies may be avoided through the use of calibration, as shown in Fig. 1. An example to the latter is robot replacement, where

CONTROL AND DYNAMIC SYSTEMS, VOL. 39

calibration is an alternative to robot reprogramming, as shown in Fig. 2.

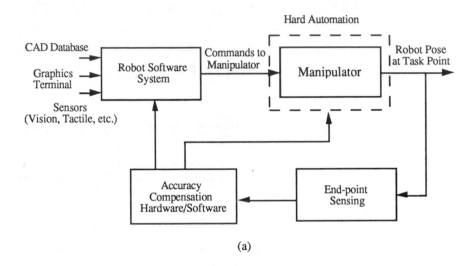

(a)

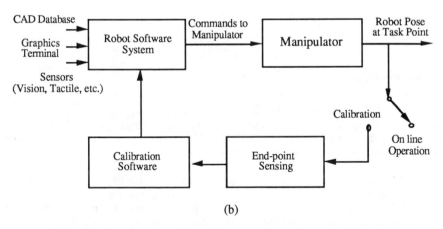

(b)

Fig. 1. Assembly Operations: (a) Without robot calibration
(b) With robot calibration.

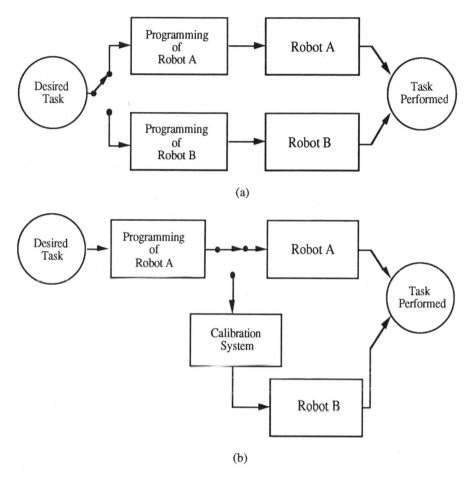

Fig 2. Robot replacement: (a) Individual programming of each robot
(b) With calibration.

Without calibration, robots which share application programs may
experience significant accuracy degradation.   The need to reprogram
the machine upon replacement (or upon other maintenance actions
that may cause permanent changes in the machine geometry) may
result in a significant process down-time.   Robots should be calibrated
in a time period which is a fraction of the reprogramming time for
calibration to be economically justifiable.

The growing importance of robot calibration as a research area
has been evidenced by a large number of publications in recent years,
including books and survey papers.   Readers interested in surveys of
robot calibration and detailed reference list are referred to [1-5].
Readers are also referred to excellent survey papers on robot metrology
such as [6, 7] which cover robot coordinate measuring techniques,

many of which are highly applicable to the calibration problem.

Modeling for robot calibration has been an active research area ever since the shortcomings of the Denavit-Hartenberg (D-H) kinematic modeling convention became noticeable [8, 9]. One of the earliest alternative models to overcome the D-H model singularities in cases of parallel consecutive joints axes was proposed by Hayati [9], and thereafter there has been a flood of publications offering a variety of models with varying number of link parameters. None of these methods however have yet become commonly accepted.

There have been two main approaches to the Measurement/Identification phases in robot calibration: *Error-Model Based Kinematic Identification* and *Joint Axes Identification*. In the first, a linearized error model, relating pose errors to kinematic parameter errors is to be constructed, yielding what has been denoted as the *Identification Jacobian* [10]. Measurements for the identification of the unknown kinematic parameter errors are done through sensing of the end-effector's world coordinates at a sufficiently large number of random robot poses "output" accompanied by joint position measurements at such configurations "input". The measured data together with the known nominal kinematic model enables standard least squares kinematic parameter identification [1, 2, 11-16]. In the latter approach, the robot joints are moved one at a time and the robot joint axes are then estimated directly from tracking the coordinates of a moving target attached to the robot. The least squares fitting of the joint axis lines is then followed by a construction of the robot actual kinematic model [4, 17-18]. As yet, there is no clear verdict which approach is superior. Comparative criteria should include

1. The number of data points required for kinematic identification.
2. The calibration measurement work space.
3. Relative accuracy of both approaches when using the same sensory equipment.
4. Ease of implementation.

An important aspect of the robot kinematic identification is the so-called "robot registration" problem, namely locating the robot base with respect to the environment or vice versa, and determining the tool transformation.

Another topic of growing interest is that of observation strategy planning for robot calibration, where one of the main criteria is the observability (or identifiability) of different kinematic parameters, including base and tool parameters [14, 19-22]. As a research area, it is far from being mature. While optimal robot measurement configurations may be found through exhaustive off-line simulation studies, using quantitative performance measures such as the condition number of the Identification Jacobian, very little theoretical work has been done to date to gain more insight about the nature of preferred observation configurations.

Whitney and his co-workers [23, 24] should be credited for being the first to produce actual _experimental_ robot calibration results, setting a standard of excellence for other researchers to follow. They were also the first to distinct between geometric errors, resulting from variations in link length, joint axis orientation and base location, and nongeometric effects, such as gear eccentricity, backlash and joint compliance. There is much debate in the robot calibration literature over the relative importance of these two error types. As yet there are no clear conclusions, which reemphasizes the importance of experimentation to verify the theoretical work.

Among all aspects of robot calibration, the compensation phase is probably the youngest research area. Surprisingly though, model-based methods have in quite a short time attained a fairly high level of maturity. Detailed algorithms for kinematic compensation have been developed [13, 25-27]. For a thorough review of these techniques readers are particularly referred to [28].

This book chapter is not intended as another survey. Instead, a unified approach to all phases of model-based calibration of rigid manipulators is attempted. The work is primarily based on [16]. The foundation to the unified approach is a new kinematic modeling convention, termed *the CPC model*. The CPC model is <u>parametrically continuous</u>. It is also <u>complete</u> in the sense that it allows the base and tool to be modeled using the same modeling convention, as that used for the "internal" manipulator links. Furthermore, error models can be constructed employing the minimum number of kinematic error parameters. These error parameters are independent and, when used in a forward kinematic model, span the entire error space. These error models, in addition to being useful for kinematic identification, may be used for kinematic compensation as well. One advantage of the unified approach is that this work will hopefully unveil some of the "mystery" presently shrouding aspects of robot registration, identifiability issues and the treatment of robot joint offsets.

The CPC modeling convention is summarized in Section II. all of this material did already appear in print. Readers who are interested in more details such as proof of the parametric continuity of the model and algorithms to map kinematic parameters from the D-H model to the CPC model and vice versa are referred to [29, 16]. The construction of the CPC error model is based on a generic linearized error model which relates robot pose errors to individual link pose errors. Section III starts by presenting a description of this generic model and the insight it brings about the identifiability problem of base and tool parameters. This material has already appeared in a more detailed form in [22]. The remainder of Section 3, however, is new. It consists of a systematic construction of the CPC error model and the linearized base and tool error models as particular cases.

The CPC model and CPC error model are then applied to the kinematic identification and compensation phases, as described in Sections IV and V. The identification algorithms presented in Section 4 are quite standard. The compensation method described in Section V is a brief summary of a more general treatment which has been described in [25, 28].

Experimental results which demonstrate the unified calibration approach using the CPC model on the Puma 560 robot are presented and discussed in Section VI.

II. KINEMATIC MODELING FOR ROBOT CALIBRATION

A. COMPLETENESS AND PARAMETRICALLY CONTINUITY OF KINEMATIC MODELS

Throughout this chapter, it is assumed that a manipulator is composed of rigid mechanical links and ideal revolute or prismatic joints. Let T_n be a 4x4 homogeneous transformation mapping the robot tool coordinate frame to the world coordinate frame,

$$T_n = A_0 A_1 A_2 \cdots\cdots A_{n-1} A_n. \tag{1}$$

where A_i, i = 1, 2, ..., n, is the homogeneous transformation from the ith link frame to the (i-1)th link frame, and it is of the form given by

$$A_i = \begin{bmatrix} n_{i,x}^a & o_{i,x}^a & a_{i,x}^a & p_{i,x}^a \\ n_{i,y}^a & o_{i,y}^a & a_{i,y}^a & p_{i,y}^a \\ n_{i,z}^a & o_{i,z}^a & a_{i,z}^a & p_{i,z}^a \\ 0 & 0 & 0 & 1 \end{bmatrix}$$

where the supperscript "a" indicates that the respective entry is an element of the A_i matrix.

1. Completeness

The concept of completeness, originally defined in [19], characterizes the capability of a kinematic model to describe all possible geometric changes in a robot.

Definition 1: A kinematic model is *complete* if it has a sufficient number of independent parameters to completely specify the geometry and motion of a robot in a reference coordinate frame.

It has been shown in [19, 30] that the number of independent parameters N in a complete model of a non-flexible robot arm must satisfy the following inequality

$$N \leq 4n - 2p + 6 \tag{2}$$

where n is the number of degrees of freedom and p is the number of prismatic joints of the robot. This inequality indicates that two link parameters are needed to be used for a prismatic joint whereas four parameters are needed for a revolute joint. The six additional parameters account for an arbitrary assignment of the world and tool frames, to be independent of any individual robot geometry. If the tool and world frames are assigned arbitrarily, equality in Eq. (2) can be attained.

2. Parametric continuity

If every small change of the robot geometry results in a small change of its model parameters, the model is said to be "proportional" [19]. Mathemati- cally propotionality means parametric continuity.

Consider two consecutive joint axes in which the ith joint axis is coincident with the z axis of the i-1th link frame. The i+1th joint axis line can be represented by a point p_i on the axis and a direction cosines vector b_i of the axis in the i-1th link frame. The pair of vectors $\{p_i, b_i\}$ is referred to as a *representation* (or *pose*) of the i+1th joint axis in the i-1th link coordinate frame. Let ρ_i be a vector containing all the geometric parameters of A_i.

Definition 2: The i+1th joint axis is at a *singular location* of A_i if one of the following occurs at this location:
1. At least one of the elements of ρ_i in A_i cannot be written as a continuous function of the elements of A_i. This singular location is termed a *model singularity of the first type*.
2. At least one of the elements of A_i cannot be written as a continuous function of the pose $\{p_i, b_i\}$. This singular location is termed a *model singularity of the second type*.

Definition 3: A_i is said to be *parametrically continuous* if there is no singular location of A_i in the space of all possible joint axis poses. A robot kinematic model is said to be *parametrically continuous* if all of its link transformations are parametrically continuous.

B. THE CPC KINEMATIC MODEL

1. A singularity-free line representation

A line $\mathcal{B}$ in 3-D space may be expressed in terms of four parameters by means of the following method [31]. Let the orientation of the line be specified by two direction cosines in a suitable reference coordinate frame $\{x, y, z\}$. Let the position of the line be specified as follows: Cut a plane which is perpendicular to the line and passes through the origin of $\{x, y, z\}$. This plane is named the $\mathcal{B}$-plane (Fig. 3). A 2-D cartesian coordinate frame is defined on the $\mathcal{B}$-plane, the origin of which is taken to be coincident with that of the reference frame. The remaining two parameters are taken to be the coordinate values in that local frame of

Point P, the intersection of Line $\mathbb{B}$ with the $\mathbb{B}$-plane. Line $\mathbb{B}$ can then be represented by a four tuple $\{b_x, b_y, l_x, l_y\}$, where b_x and b_y are the x and y components of the direction cosine vector $\boldsymbol{b}$; l_x and l_y are the coordinates of Point P in the 2-D coordinate frame defined on the $\mathbb{B}$-plane (Fig. 3). Note that b_z, the z component of the direction cosine vector, depends on b_x and b_y as follows

$$b_z = (1 - b_x^2 - b_y^2)^{1/2} \tag{3}$$

By definition (through adopting the plus sign in Eq. (3)) the direction cosine vector $\boldsymbol{b}$ is forced to lie in the upper half-space of the reference frame.

A convention for choosing the coordinate axes of the local 2-D frame remains to be defined.   As suggested by Roberts [31], the projection of the x axis of the reference frame on the $\mathbb{B}$-plane is taken to be the x axis of the local frame.   Let z be a unit vector along the z axis of the reference frame.   The reference frame is thus rotated by an angle of φ about an axis $\boldsymbol{j}$, where

$$\varphi \equiv \arccos(z \cdot \boldsymbol{b}) = \arccos(b_z) \tag{4}$$

and

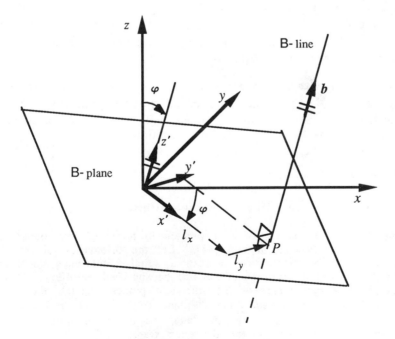

Fig. 3 A line representation.

$$j = \frac{z \times b}{\|z \times b\|} = \begin{bmatrix} \dfrac{-b_y}{\sqrt{b_x^2 + b_y^2}} \\[3mm] \dfrac{b_x}{\sqrt{b_x^2 + b_y^2}} \\[3mm] 0 \end{bmatrix} \tag{5}$$

$\|\cdot\|$ denotes the Euclidean norm.   The rotation matrix R is

$$R = Rot(j, \varphi) \tag{6}$$

R is undefined from Eq. (6) when $b = z$. R however can be defined without explicit reference to $z \times b$.   Substitution of Eqs. (4)-(5) into Eq. (5) (refer to [32]) and use of Eq. (3) to simplify the result yield

$$R = \begin{bmatrix} 1 - \dfrac{b_x^2}{1 + b_z} & \dfrac{-b_x b_y}{1 + b_z} & b_x \\[3mm] \dfrac{-b_x b_y}{1 + b_z} & 1 - \dfrac{b_y^2}{1 + b_z} & b_y \\[3mm] -b_x & -b_y & b_z \end{bmatrix} \tag{7}$$

In the case of $b = z$, that is, $[b_x, b_y, b_z]^T = [0, 0, 1]^T$, R becomes the identity matrix.   Since b_z is by definition nonnegative, R is well defined for any direction of the line in 3-D space.   Thus the unit basis vectors, x', y', z', for the new coordinate frame are given by the first, second and third columns of R, respectively.

2.   The CPC modeling convention

Let the 4x4 homogeneous transformation T_n relating the position and orientation of an end-effector to the world coordinates be

$$T_n = B_0 B_1 B_2 \ldots B_{n-1} B_n \tag{8a}$$

Refer to Fig. 4, in which joints are shown as revolute.

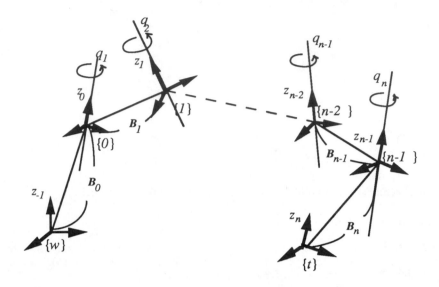

Fig. 4 Coordinate frame assignment.

Each joint is assumed to be either ideally revolute or ideally prismatic. A cartesian coordinate frame $\{x_i, y_i, z_i\}$, i = -1, 0,···, n, is established for each link. The world, base and tool frames are denoted as the -1th, 0th and nth link frames, respectively.

BASE, the transformation from the base to the world frames, is fixed and defined as

$$BASE \equiv B_0 \qquad\qquad (8b)$$

Robot calibration often necessitates the definition of a so-called "flange frame" located on the mounting surface of the robot end-effector. **FLANGE**, the transformation from that frame to the n-1th frame, contains the joint variable q_n while **TOOL**, the transformation from the tool to the flange frames, is fixed. Thus

$$FLANGE \cdot TOOL \equiv B_n \qquad\qquad (8c)$$

In the CPC model, the position and orientation of the world and tool frames can be assigned arbitrarily with the only exception that the z axis of the world frame must not lie opposite to the z axis of the base frame and the z axis of the tool frame must not be opposite to the z axis of the n-1th link frame. All other link frames are established based on the following convention:

1. The z_i axis must be on the i+1th joint axis for a revolute joint and parallel to it for a prismatic joint.

2. The coordinate frame $\{x_i, y_i, z_i\}$ forms an orthonormal right-hand system.

The CPC modeling convention virtually allows arbitrary assignment of link coordinate frames. To achieve this, two parameters, β_i and $l_{i,z}$ are added to Roberts' line parameters. $Rot(z, \beta_i)$ is introduced to allow an arbitrary orientation of the ith link frame, and $Trans(0, 0, l_{i,z})$ is introduced to allow an arbitrary positioning of the ith link frame. β_i and $l_{i,z}$ are redundant for i = 1, 2, ⋯, n-1, and thus can always be set to zero for these internal link transformations.

In the case of a revolute joint, the link transformation matrix B_i is a function of seven link parameters $\{b_{i,x}, b_{i,y}, b_{i,z}, l_{i,x}, l_{i,y}, l_{i,z}, \beta_i\}$ and one joint variable denoted as θ_i, while in the case of a prismatic joint, the joint variable is denoted as d_i. Let $l_i = [l_{i,x}, l_{i,y}, l_{i,z}]^T$ and $b_i = [b_{i,x}, b_{i,y}, b_{i,z}]^T$. The assignment rules for link parameters and joint variables are as follows. If the ith joint is revolute, then (refer to Fig. 5)

1. b_i is the direction cosines vector of the i+1th joint axis represented in the i-1th link frame.
2. β_i depends on the arbitrary direction of the x_i axis.
3. $-l_i$ is the origin of the i-1th link frame represented in the ith frame.
4. The zero position of joint variable θ_i (for i = 1, ... , n) corresponds to the zero reading of the ith joint position transducer.

Remark:
1. The joint axis index is always greater by one than that of link frame. That is why the i+1th joint axis is represented in the i-1th link frame by b_i.
2. It is important to define the translational parameter vector l_i in terms of the ith link frame rather than the i-1th link frame. This is equivalent to performing a rotation first, followed by a translation. Only in this way, are the translational parameters $l_{i,x}$ and $l_{i,y}$ sufficient to specify the position of the i+1th joint axis, leaving $l_{i,z}$ redundant. This greatly facilitates model reduction and error model reduction tasks for kinematic identification.

If the ith joint is prismatic, $l_{i,x}$ and $l_{i,y}$ are set to zero (refer to Fig. 6). The zero position of the joint variable d_i corresponds to the zero reading of the ith joint position encoder.

Define a 4x4 rotation matrix R_i as in Eq. (6). R_i is a function of $b_{i,x}$, $b_{i,y}$ and $b_{i,z}$. Specifically,

$$R_i = \left\{ \begin{array}{ll} Rot(j_i, \varphi_i) & \text{if } z_i \text{ is not parallel to } z_{i-1}, \\ & \qquad\qquad\qquad\qquad\qquad i = 0, 1, \cdots, n \qquad (9a) \\ I_{4 \times 4} & \text{if } z_i \text{ is parallel to } z_{i-1}, \end{array} \right.$$

where $j_i = e_3 \times b_i / \| e_3 \times b_i \|$ and $\varphi_i = \arccos(e_3 \cdot b_i)$; $e_3 = [0, 0, 1]^T$. By Eq. (7),

$$R_i = \begin{bmatrix} 1 - \dfrac{b_{i,x}^2}{1 + b_{i,z}} & \dfrac{-b_{i,x} b_{i,y}}{1 + b_{i,z}} & b_{i,x} & 0 \\[3mm] \dfrac{-b_{i,x} b_{i,y}}{1 + b_{i,z}} & 1 - \dfrac{b_{i,y}^2}{1 + b_{i,z}} & b_{i,y} & 0 \\[3mm] -b_{i,x} & -b_{i,y} & b_{i,z} & 0 \\[3mm] 0 & 0 & 0 & 1 \end{bmatrix} \qquad (9b)$$

where $b_{i,z} \geq 0$.

Remark: In a kinematic parameter identification process, $b_{i,z}$ is allowed to take negative values as long as the condition $b_{i,z} > -1$ is met.

Assuming that the robot is at its "home position"; i.e., the configuration at which all joint position transducers yield zero reading, the link transformation B_i is defined as follows: Cut a plane $\mathbb{B}_i$ perpendicular to the i+1th joint axis which passes through the origin of the i-1th link frame.   Then

1. Perform R_i, as defined in Eq. (9).   Denote the resulting intermediate frame as $\{x_{i-1}', y_{i-1}', z_{i-1}'\}$. z_{i-1}' becomes parallel to z_i and the other two coordinate axes lie on the $\mathbb{B}_i$ plane.
2. Perform $Rot(z, \beta_i)$.   Denote the resulting intermediate frame as $\{x_{i-1}'', y_{i-1}'', z_{i-1}''\}$, where $z_{i-1}'' = z_{i-1}'$.   The resulting frame has the same orientation as the ith link frame.
3. Perform $Trans(l_{i,x}, l_{i,y}, l_{i,z})$.   The resulting frame becomes coincident with the ith link frame.

$$R_i = \begin{bmatrix} 1 - \dfrac{b_{i,x}^2}{1 + b_{i,z}} & \dfrac{-b_{i,x}b_{i,y}}{1 + b_{i,z}} & b_{i,x} & 0 \\[3mm] \dfrac{-b_{i,x}b_{i,y}}{1 + b_{i,z}} & 1 - \dfrac{b_{i,y}^2}{1 + b_{i,z}} & b_{i,y} & 0 \\[3mm] - b_{i,x} & - b_{i,y} & b_{i,z} & 0 \\[2mm] 0 & 0 & 0 & 1 \end{bmatrix} \qquad (9b)$$

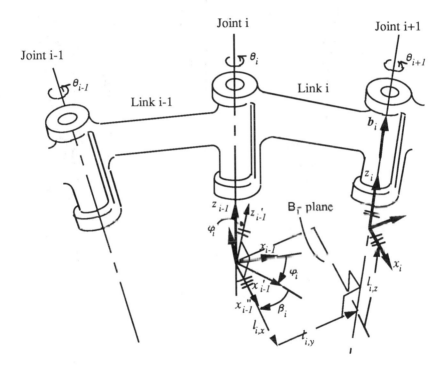

Fig. 5 The CPC modeling convention for a revolute joint.

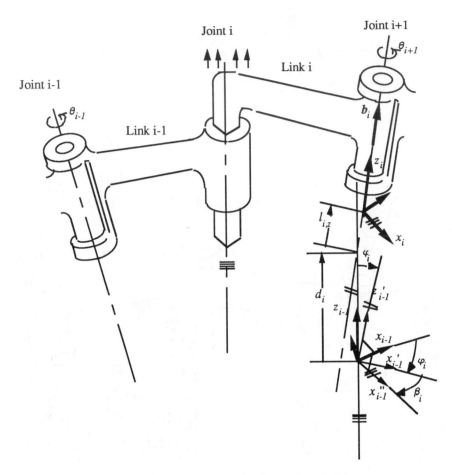

Fig. 6 The CPC modeling convention for a prismatic joint.

The link transformation B_i is therefore given by

$$B_i = Q_i V_i \qquad\qquad\qquad i = 0, 1, \cdots, n \qquad (10)$$

where Q_i represents the transformation determined by the ith joint variable and is given by

$$Q_i = \begin{cases} Rot\ (z, \theta_i) & \text{for a revolute joint} \\ \\ Trans(0,0,d_i) & \text{for a prismatic joint} \end{cases} \qquad (11)$$

and

$$V_i = R_i Rot(z, \beta_i) Trans(l_{i,x}, l_{i,y}, l_{i,z}) \tag{12}$$

By definition $\theta_0 \equiv 0$ and $d_0 \equiv 0$. V_i is specified by all the fixed link parameters in the ith link transformation B_i, and is independent of the ith joint variable.

The assignment of link frames and link parameters for a "simple" robot (i.e, two consecutive links are either perpendicular or parallel each other) can be greatly simplified. For the sake of convenience, let $\beta_i = 0$, which is always so for internal links. Two additional and convenient assignment rules are as follows:

1. If B_i-plane is the xy-plane or the xz-plane of the (i-1)th link coordinate frame, let x_i have the same direction as x_{i-1}.
2. If B_i-plane is the yz-plane of the (i-1)th link coordinate frame, let y_i have the same direction as y_{i-1}.

3. Examples of the CPC model

Example 1: Let the z_i axis be parallel to the x_{i-1} axis. In this case, $b_i = [b_{i,x}, b_{i,y}, b_{i,z}]^T = [1, 0, 0]^T$. By Eq. (9) R_i is

$$R_i = \begin{bmatrix} 0 & 0 & 1 & 0 \\ 0 & 1 & 0 & 0 \\ -1 & 0 & 0 & 0 \\ 0 & 0 & 0 & 1 \end{bmatrix}$$

Again β_i is chosen to be zero. The assignment of link frame and parameters is shown in Fig. 7.

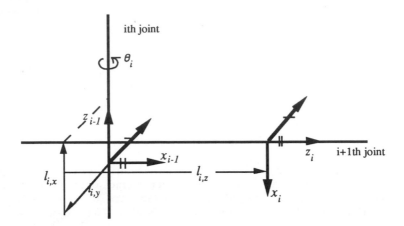

Fig. 7   CPC modeling convention for $b_i = [1, 0, 0]^T$

Example 2:   The PUMA 560 robot.

The assigned link frames are shown in Fig. 8. The link parameters are given in Table 1. The parameters of the 0th and 6th link transformations are not listed as these depend on the particular set-up of the world and tool frames.

Table 1   CPC parameters of the PUMA 560

i	$b_{i,x}$	$b_{i,y}$	$b_{i,z}$	$l_{i,x}(mm)$	$l_{i,y}(mm)$	$l_{i,z}(mm)$	β_i
0	$b_{0,x}$	$b_{0,y}$	$b_{0,z}$	$l_{0,x}$	$l_{0,y}$	$l_{0,z}$	β_0
1	0	1	0	0	0	149.09	0
2	0	0	1	431.82	0	0	0
3	0	-1	0	-20.31	0	433.05	0
4	0	1	0	0	0	0	0
5	0	-1	0	0	0	0	0
6	$b_{6,x}$	$b_{6,y}$	$b_{6,z}$	$l_{6,x}$	$l_{6,y}$	$l_{6,z}$	β_6

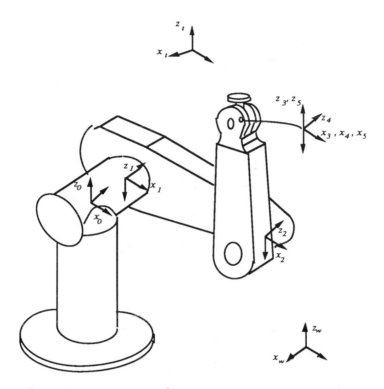

Fig. 8 Link frame assignment of PUMA arm in the CPC convention.

Example 3: Modeling of joint offsets.

Consider a case in which the reading of a joint variable is set to zero, while the actual (physical) joint angle is not at the zero position. The deviation from the zero joint position is called a "joint offset". Assume that the only geometrical error of the ith link is its joint offset. In the D-H model, it is modeled by $d\theta_i$ or dd_i. In the CPC model, nominal joint variables are always used to describe the motion of the robot. Joint offsets can be accounted for through dl_i. Fig. 9 shows such an example (in the figure the superscript "0" denotes a nominal entity). The joint offset here may be expressed in terms of $dl_{i,x}$ and $dl_{i,y}$.

If geometrical errors involve both joint offsets as well as axis misalignment, it is important to stress that $d\theta_i$ or dd_i in the D-H model no longer represents the ith joint offset as the joint offset depends also on other parameter errors. Similar observation can be made about the CPC model.

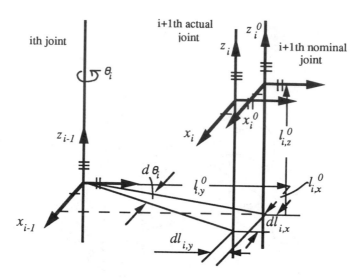

Fig. 9 Modeling a joint offset.

III. LINEARIZED KINEMATIC ERROR MODELS

A. A GENERIC LINEARIZED ERROR MODEL

1. Differential   transformations

Let dT be the *additive differential transformation* of T

$$dT = T - T^0 \tag{13}$$

where T^0 is the transformation that corresponds to the nominal kinematic parameters and T corresponds to the actual kinematic parameters. The *right multiplicative differential transformation* (RMDT) of T, ΔT^u, is defined as

$$T^0 \Delta T^u = T \tag{14}$$

The superscript "u" indicates that the respective entity is associated with the U_i matrix to be defined shortly. The additive and right multiplicative differential transformations are related through

$$\Delta T^u = I + (T^0)^{-1} dT \equiv I + \delta T^u \tag{15}$$

where δT^u has the following structure [32],

$$\delta T = \begin{bmatrix} 0 & -\delta z & \delta y & dx \\ \delta z & 0 & -\delta x & dy \\ -\delta y & \delta x & 0 & dz \\ 0 & 0 & 0 & 0 \end{bmatrix} \qquad (16)$$

where $d = [dx, dy, dz]^T$ are the translational errors and $\delta = [\delta x, \delta y, \delta z]^T$ are the rotational errors.

Define an arbitrary 4x4 homogeneous transformation T by

$$T = \begin{bmatrix} R & p \\ 0_{1 \times 3} & 1 \end{bmatrix}$$

where R is a 3x3 orthonormal matrix and p is a 3x1 vector. A vector $v = [v_x, v_y, v_z]^T$ may be represented by a skew-sysmetric matrix Ω_v,

$$\Omega_v = \begin{bmatrix} 0 & -v_z & v_y \\ v_z & 0 & -v_x \\ -v_y & v_x & 0 \end{bmatrix}$$

Thus Eq. (16) can be rewritten as

$$\delta T = \begin{bmatrix} \Omega_\delta & d \\ 0_{1 \times 3} & 1 \end{bmatrix} \qquad (16')$$

2. A generic linearized error model and error model reduction

The linearized relationship between the pose errors of individual links and those of the end-effector using RMDT have been derived in [10, 22]. Let

$$U_i \equiv A_i A_{i+1} \cdots A_{n-1} A_n, \qquad\qquad i = 0, 1, \cdots, n \quad (17)$$

Thus, $U_0 = T_n$ and $U_{n+1} \equiv I$.

Let $y^u = [d^{uT}, \delta^{uT}]^T = [dx^u, dy^u, dz^u, \delta x^u, \delta y^u, \delta z^u]^T$ be the vector of cartesian errors of the end-effector using the RMDT and $x^u = [d^u{}_0{}^T, \delta^u{}_0{}^T, \cdots, d^u{}_n{}^T, \delta^u{}_n{}^T]^T = [dx^u{}_0, dy^u{}_0, dz^u{}_0, \delta x^u{}_0, \delta y^u{}_0, \delta z^u{}_0, \cdots, dx^u{}_n, dy^u{}_n, dz^u{}_n, \delta x^u{}_n, \delta y^u{}_n, \delta z^u{}_n]^T$ be the vector of cartesian errors of every link frame of the robot using the RMDT.

Proposition 1 [22]: Given that a kinematic model T_n is parametrically continuous and differentiable, then the linearized relationship between the cartesian errors of individual links and those of the end-effector is given by

$$\delta^{\,u} = \sum_{i=0}^{n} R^{\,u}_{\,i+1}{}^{T} \; \delta^{\,u}_{\,i} \tag{18a}$$

and

$$d^{\,u} = \sum_{i=0}^{n} -R^{\,u}_{\,i+1}{}^{T} \Omega^{\,u}_{\,p,i+1} \; \delta^{\,u}_{\,i} + R^{\,u}_{\,i+1}{}^{T} d^{\,u}_{\,i} \tag{18b}$$

where $\Omega^{u}_{p,i}$, $i = 1, ..., n$, is a skew-symmetric matrix whose elements are $p^{u}_{i,x}, p^{u}_{i,y}$ and $p^{u}_{i,z}$, the elements of p^{u}_{i} associated with U_{i}. R^{u}_{i} is the rotation matrix associated with U_{i}. Eq. (18) can be arranged in a matrix form as

$$y^{\,u} = L^{u}x^{u} \tag{19}$$

where the linear mapping $L^{u}: R^{6(n+1)} \rightarrow R^{6}$ is

$$L^{\,u} = \begin{bmatrix} R^{u}_{1}{}^{T} & -R^{u}_{1}{}^{T}\Omega^{u}_{p,1} & \cdots\cdots & R^{u}_{n}{}^{T} & -R^{u}_{n}{}^{T}\Omega^{u}_{p,n} & I_{3x3} & 0_{3x3} \\ 0_{3x3} & R^{u}_{1}{}^{T} & \cdots\cdots & 0_{3x3} & R^{u}_{n}{}^{T} & 0_{3x3} & I_{3x3} \end{bmatrix} \tag{20}$$

The CPC modeling convention fits the right differential multiplicative transformation formalism [22]. Since only RMDT is used in the rest of the text, the superscript "u" is dropped if no confusion is caused. The same type of model may be developed based on left multiplicative differential transformations [22] and is applicable to kinematic models such as Craig's version of the D-H modeling convention [33].

Recall that ρ is the vector of kinematic parameters in a given kinematic model; $\rho \in R^{m}$, m being the number of link parameters. Assuming that the kinematic model is differentiable, then $x = Kd\rho$, where $K: R^{m} \rightarrow R^{6(n+1)}$, x is the vector of cartesian errors of every robot link frame, and $d\rho$ is the parameter error vector. By Eq. (20), the pose error vector y is related to the parameter deviations through

$$y = LK d\rho \tag{21}$$

The matrix structure of K depends on the particular choice of the kinematic modeling convention. More specifically, K has the following block diagonal form,

$$K = diag(K_{0}, K_{1}, .., K_{n}) \tag{22}$$

where $K_{i}: R^{m_{i}} \rightarrow R^{6(n+1)}$; m_{i} is the number of link parameters in A_{i}. For a given kinematic model, K_{i} can be derived by using (15).

Example 4:   Hayati's modification to the Denavit-Hartenberg modeling convention is through post-multiplying the ith link transformation A_i by $\text{Rot}(y, \beta_i)$ [9, 13], where A_i is as in the D-H convention [32]. The vector of the ith link parameters is $[d_i, a_i, \theta_i, \alpha_i, \beta_i]^T$. Thus the formula of K_i is

$$K_i = \begin{bmatrix} -s\beta_i c\alpha_i & c\beta_i & a_i s\beta_i s\alpha_i & 0 & 0 \\ s\alpha_i & 0 & a_i c\alpha_i & 0 & 0 \\ c\beta_i c\alpha_i & s\beta_i & -a_i c\beta_i s\alpha_i & 0 & 0 \\ 0 & 0 & -s\beta_i c\alpha_i & c\beta_i & 0 \\ 0 & 0 & s\alpha_i & 0 & 1 \\ 0 & 0 & c\beta_i c\alpha_i & s\beta_i & 0 \end{bmatrix} \qquad i = 1, 2, ..., n-1$$

Notice that the cases of i = 0, n are excluded since the modified D-H convention cannot be used to model the 0th and nth link transformations.

A necessary condition for the matrix LK to be full-rank is that the elements of $d\rho$ are independent. The linearized error model of a parametrically continuous and differentiable kinematic model is *irreducible* if all link parameters are independent. If $d\rho$ contains dependent elements, it is possible to find a linear mapping relating $d\rho$ to z whose elements are independent. The problem can be stated as follows: Find a linear mapping M, where $M : R^{m'} \to R^m$, such that $d\rho = Mz$, where $z \in R^{m'}$; m' ($<$ m) is the number of independent parameters in the kinematic model. M has the following form,

$$M = \text{diag}(M_0, M_1, ..., M_n) \tag{23}$$

where $M_i: R^{m_i'} \to R^{m_i}$; m_i' is the number of independent parameters in A_i.

Example 5:   Considering again the Hayati's modification to the D-H model. Whenever the (i+1)th joint axis is not nominally parallel to the ith axis, β_i is redundant. Thus

$$M_i = \begin{bmatrix} 1 & 0 & 0 & 0 \\ 0 & 1 & 0 & 0 \\ 0 & 0 & 1 & 0 \\ 0 & 0 & 0 & 1 \\ 0 & 0 & 0 & 0 \end{bmatrix} \qquad i = 1, 2, ..., n-1$$

yielding a reduced-order parameter error vector $[dd_i, da_i, d\theta_i, d\alpha_i]^T$. Whenever the (i+1)th joint is parallel to the ith joint, d_i is redundant.

Thus

$$M_i = \begin{bmatrix} 0 & 0 & 0 & 0 \\ 1 & 0 & 0 & 0 \\ 0 & 1 & 0 & 0 \\ 0 & 0 & 1 & 0 \\ 0 & 0 & 0 & 1 \end{bmatrix} \qquad i = 1,2,...,\ n\text{-}1$$

yielding a reduced-order parameter error vector $[da_i, d\theta_i, d\alpha_i, d\beta_i]^T$.

Combining Eq. (23) with Eq. (21) yields

$$KM = \text{diag}(K_0 M_0, K_1 M_1, .., K_n M_n) \tag{24}$$

Consequently

$$y = LKM\ z = J\ z \tag{25}$$

where $J = LKM : \mathbb{R}^{m'} \to \mathbb{R}^6$. J is termed the *Identification Jacobian*.

B. OBSERVABILITY OF KINEMATIC ERROR PARAMETERS

1. The number of observable error parameters

Let s be the number of different measurement configurations taken to calibrate the robot. To find the inverse solution of Eq. (25), sufficiently many measurements need to be taken. Let J_i be defined as in Eq. (25); i.e., $J_i \colon \mathbb{R}^{m'} \to \mathbb{R}^6$ for i=1,2, ··, s, are the identification Jacobians at each measurement configuration. Define an aggregated Jacobian matrix

$$G = \begin{bmatrix} J_1 \\ J_2 \\ \vdots \\ J_s \end{bmatrix} \tag{26}$$

where $G \colon \mathbb{R}^{m'} \to \mathbb{R}^{6s}$.

The number of independent parameters N in a complete model of a rigid robot is determined by Eq. (2). The minimum number of measurement configurations can then be determined accordingly. For example, 30 independent kinematic parameters are required to model a 6 degree-of-freedom robot such as the PUMA. For a unique solution of the parameter error vector $d\boldsymbol{\rho}$, 5 measurement configurations must be chosen as each configuration provides 6 independent equations. If more than 5 measurement configurations are used, least squares methods are employed to solve the overdetermined system.

Definition 4: The vector of independent kinematic error parameters is said to be *observable* if the rank of $G^T G$ is N.

The observability of kinematic error parameters depends on both the kinematic modeling convention as well as the selection of measurement configurations.

Proposition 2: Assume that the right multiplicative differential transformations are used to model kinematic errors. If calibration measurements do not contain information about the orientational errors of the manipulator end-effector, then all orientational parameters of A_n are unobservable. In this case, the number N of observable kinematic error parameters must satisfy the inequality

$$N \leq 4n - 2p + 6 - o_n \qquad (27)$$

where o_n is the number of independent orientational parameters in A_n. If in addition the last joint of the manipulator is revolute and the origin of the tool frame lies on the last joint axis, then the number N of observable kinematic error parameters must satisfy the inequality

$$N \leq 4n - 2p + 6 - o_n - ot_{n-1} \qquad (28)$$

where $ot_{n-1} = \min\{o_{n-1}, t_{n-1}\}$; o_{n-1} and t_{n-1} are the number of independent orientational and translational parameters in A_{n-1}, respectively.
Proof: See [22]. The unobservability of tool parameters stems from the structure of L in Eq. (20).

Remark: Proposition 2 explains why only 25 parameters are independent when the PUMA arm was calibrated using only positioning errors of the end-effector, as was done in [11, 16]. Other researchers have also observed the same phenomenon [30, 23].

2.   Observability   measures

An observability measure has been proposed in [20] using the singular values of G:

$$O(G) = (\sigma_1 \sigma_2 \cdots \sigma_m)^{1/m} q^{-1/2} \qquad (29)$$

where q is the number of rows in G, m is the number of parameters in the kinematic model and σ_i, i=1,2,$\cdots$, m, are the eigenvalues of $G^T G$. It was pointed out in [20] that whenever O is zero, the error parameters are unobservable from the measurements performed. As O increases, the contribution of geometric errors to the overall robot positioning errors dominates the effects of nongeometric and other unmodeled errors; consequently better estimation of the geometric error parameters is expected. Optimal measurement configurations for robot calibration can be found through exhaustive simulations using such a quantitative measure.

Another observability measure suggested in [21] is the condition number of G:

$$\text{Cond}(G) = \sigma_{max}/\sigma_{min} \qquad\qquad (30)$$

One advantage of $\text{Cond}(G)$ over $O(G)$ is that $\text{Cond}(G)$ is invariant under scaling of G; i.e, $\text{Cond}(kG) = \text{Cond}(G)$ where k is a scalar. $\text{Cond}(G)$ ≥ 1 whereas $\text{Cond}(G) = 1$ is ideal. $\text{Cond}(G) = \infty$ indicates that the error parameters are unobservable.

Proposition 3 [22]: Consider a parametrically continuous and differen- tiable kinematic model. If the linearized error model is reducible, then

$$\text{Cond}(G) = \infty$$
and
$$O(G) = 0.$$

Observability measures are meaningless in cases of reducible models as in such cases, these numbers are always zero or infinity respectively no matter how many measurements are taken and how well the configurations are chosen. The matrix $G^T G$ may also become singular at any iteration step of a numerical identification process even if the kinematic model itself has no singularities. Robust optimization techniques such as the Levenberg- Marquardt algorithm have been successfully applied to cope with the problem [14, 15]. These techniques can provide an identification solution even if $\text{Cond}(G) = \infty$ and $O(G) = 0$.

C. THE CPC ERROR MODELS

The derivation of the CPC error model follows three steps. First, a linear mapping K_i relating the cartesian errors to the CPC link parameter errors of each individual link is obtained. The redundant parameter errors are then eliminated by another linear mapping M_i. Finally, the CPC error model is constructed from Eq. (25).

1. Linear mapping relating pose errors to CPC parameter errors of individual links

Let $[d_i^T, \delta_i^T]^T$ be a pose error vector of the ith link and $[dl_i, db_i, d\beta_i]^T$

be the corresponding CPC parameter error vector, where $[d_i{}^T, \delta_i{}^T]^T \in R^6$ and $[dl_i, db_i, d\beta_i]^T \in R^7$. Let R_i denote the upper-left 3x3 submatrix of R_i of Eq. (9). $R_i{}^T dR_i$ is a skew-symmetric matrix. The vector of three nontrivial elements in $R_i{}^T dR_i$, denoted by δ^r_i, can be written as

$$\delta^r_i = k^r_{i,x} db_{i,x} + k^r_{i,y} db_{i,y} + k^r_{i,z} db_{i,z} \qquad i = 0,1, \cdots, n \qquad (31a)$$

The coefficients $k^r_{i,x}, k^r_{i,y}$ and $k^r_{i,z}$ are 3x1 vectors given by

$$k^r_{i,x} = [b_{i,x}b_{i,y}w_i, \; 1-b_{i,y}{}^2 w_i, \; (1+b_{i,z}+b_{i,x}{}^2)b_{i,y}w_i{}^2]^T; \qquad (31b)$$
$$k^r_{i,y} = [-1+b_{i,y}{}^2 w_i, \; -b_{i,x}b_{i,y}w_i, \; -(1+b_{i,z}-b_{i,y}{}^2)b_{i,x}w_i{}^2]^T; \qquad (31c)$$
$$k^r_{i,z} = [b_{i,y}, \; -b_{i,x}, \; b_{i,x}b_{i,y} \; b_{i,z}w_i{}^2]^T. \qquad (31d)$$

for $i = 0,1, ..., n$, where $w_i = 1/(1+b_{i,z})$. Eq. (31) will be used later.

Let us now derive the linear mapping K_i. From Eq. (10), $B_i = Q_i V_i$, and keeping in mind that joint offsets are modeled in terms of the CPC error parameters,

$$\delta B_i = B_i^{-1} dB_i = V_i^{-1} dV_i$$

Then
$$\delta B_i = Trans(-l_{i,x},-l_{i,y},-l_{i,z})Rot(z_i,-\beta_i)R_i^{-1}d\{R_iRot(z_i,\beta_i)Trans(l_{i,x},l_{i,y},l_{i,z})\}$$
$$= Trans(-l_{i,x},-l_{i,y},-l_{i,z})Rot(z_i,-\beta_i)R_i^{-1}dR_i \; Rot(z_i,\beta_i)Trans(l_{i,x},l_{i,y},l_{i,z})$$
$$+ Trans(-l_{i,x},-l_{i,y},-l_{i,z})Rot(z_i,-\beta_i)d\{Rot(z_i,\beta_i)\}Trans(l_{i,x},l_{i,y},l_{i,z})$$
$$+ Trans(-l_{i,x},-l_{i,y},-l_{i,z})d\{Trans(l_{i,x},l_{i,y},l_{i,z})\} \qquad (32)$$

It is now convenient to break δB_i into two parts: the upper-left 3x3 submatrix which contains the elements of the orientational error vector δ_i, and the last column which is the positioning error vector d_i. Then

$$\delta_i = \delta_{i,1} + \delta_{i,2} + \delta_{i,3} \qquad (33a)$$
and
$$d_i = d_{i,1} + d_{i,2} + d_{i,3} \qquad (33b)$$

where $\delta_{i,j}$ and $d_{i,j}$, $j = 1, 2, 3$, are the orientational and positioning error vector corresponding to the jth additive term of the right-hand side of (32). Simplifying each term in the right-hand side of (32) yields

$$\delta_{i,1} = Rot(z_i,-\beta_i)_{3x3} \delta^r_i \qquad (34a)$$
$$\delta_{i,2} = k_{i,\beta} d\beta_i \qquad (34b)$$
$$\delta_{i,3} = 0 \qquad (34c)$$

where $\delta^r{}_i$ is given in Eq. (31), and

$$k_{i,\beta} = [0,0,1]^T \tag{34d}$$

By Eq. (32) and Eqs. (34a)-(34d), one obtains

$$d_{i,1} = \delta_{i,1} \times l_i \tag{34e}$$
$$d_{i,2} = k_{i,\beta} \times l_i \, d\beta_i \tag{34f}$$
$$d_{i,3} = dl_i \tag{34g}$$

where "x" denotes a vector cross product.   Substitution of Eq. (34) into Eq. (33) yields

$$\delta_i = k_{i,x} db_{i,x} + k_{i,y} db_{i,y} + k_{i,z} db_{i,z} + k_{i,\beta} d\beta_i \quad i = 0,1, \cdots, n \tag{35a}$$
$$d_i = dl_i + k_{i,x} \times l_i db_{i,x} + k_{i,y} \times l_i db_{i,y} + k_{i,z} \times l_i db_{i,z} + k_{i,\beta} \times l_i d\beta_i$$
$$i = 0,1, \cdots, n \tag{35b}$$

In Eq. (35), $k_{i,\beta}$ are given by Eq. (34d), and

$$k_{i,j} = Rot(z_i,-\beta_i)_{3x3} \, k^r{}_{i,j} \qquad\qquad j \in \{x,y,z\} \tag{36}$$

where $k^r{}_{i,j}$ are given in Eq. (31).   For the robot internal links, β_i is often set to zero, thus $k_{i,j} = k^r{}_{i,j}$. Eq. (35) expresses   the cartesian errors in terms of the CPC   parameter errors for each individual link.   The coefficients in Eq. (35) define the linear mapping K_i as follows

$$K_i = \begin{bmatrix} I_{3x3} & k_{i,x} \times l_i & k_{i,y} \times l_i & k_{i,z} \times l_i & k_{i,\beta} \times l_i \\ 0_{3x3} & k_{i,x} & k_{i,y} & k_{i,z} & k_{i,\beta} \end{bmatrix} \tag{37}$$

for i = 0, 1, ..., n.

2. Linear mapping relating pose errors to independent CPC parameter errors of individual links

The   linearized   error   model   of   the   CPC   model,   derived   by differentiating T_n with respect to the parameter vector ρ, is obviously reducible.   A linear mapping M can be constructed to transform the linearized error model to an irreducible one.   One of the orientational parameters $\{b_{i,x}, b_{i,y}, b_{i,z}\}$ is redundant.   Among the translational parameters $\{l_{i,x}, l_{i,y}, l_{i,z}\}$, $l_{i,z}$ is introduced to allow an arbitrary assignment of the ith link frame along the (i+1)th joint axis. $l_{i,z}$ is redundant for i = 0, ..., n-1.   Likewise β_i, i = 0 , ..., n-1, are redundant. Both $l_{n,z}$ and β_n are introduced to provide 6 degrees of freedom to the tool frame.   Thus four independent parameters remain for each revolute joint and two independent parameters for each prismatic joint. In the next a few paragraphs, the subscript "i" is dropped for convenience.

Care should be taken when eliminating an error parameter from the set $\{db_x, db_y, db_z\}$. If, for instance, db_x and db_y are chosen as the independent parameter errors, then

$$db_z = -\frac{b_x db_x + b_y db_y}{b_z} \qquad (38)$$

By using Eq. (38), δ_i is made related to db_x and db_y. Eq. (38) however has a singular point $b_z = 0$. Physically this happens when two consecutive axes are perpendicular to each other. The following convention for selecting the independent parameters can be used to keep the CPC model differentiable.

1. Eliminate the error parameter corresponding to $\max\{|b_x|, |b_y|, |b_z|\}$.
2. The direction cosines vector b shall be kept a unit vector at every step of an identification process.

The linear mappings M_i, relating the CPC redundant error parameters to the CPC independent error parameters, are constructed taking into account the rules stated above.

Let $j \neq k \neq h$ be elements in the set $\{x, y, z\}$ such that $b_{i,j}$ and $b_{i,k}$ in b_i are chosen as independent parameters, and $b_{i,h}$ in b_i is to be eliminated. δ_i is related only to the orientational error parameters,

$$\delta_i = km_{i,j,h} db_{i,j} + km_{i,k,h}\, db_{i,k} \quad i \in [0,1, \cdots, n\text{-}1] \qquad (39a)$$
$$\delta_n = km_{n,j,h} db_{n,j} + km_{n,k,h}\, db_{n,k} + k_{n,\beta} d\beta_n \qquad (39b)$$

where $km_{i,j,h}$ is a 3x1 vector, and the same about the other coefficients. d_i however is in general a function of both the orientational and translational error parameters,

$$d_i = dl_i + km_{i,j,h} \times l_i\, db_{i,j} + km_{i,k,h} \times l_i\, db_{i,k} \quad i \in [0,1, \cdots, n\text{-}1] \qquad (39c)$$
$$d_n = dl_n + km_{n,j,h} \times l_i\, db_{n,j} + km_{n,k,h} \times l_i\, db_{n,k} + k_{n,\beta} \times l_i\, d\beta_n \qquad (39d)$$

where $dl_n = [dl_{n,x}, dl_{n,y}, dl_{n,z}]^T$ and $dl_i = [dl_{i,x}, dl_{i,y}, 0]^T$ for $i \in [0,1, \cdots, n\text{-}1]$.

The coefficients in Eq. (39) define the linear mapping $K_i M_i$. For $i \in [0, 1, \cdots, n\text{-}1]$, the reduced parameter error vector is $[dl_{i,x}, dl_{i,y}, db_{i,j}, db_{i,k}]^T$. Then by Eq. (37)

$$K_i M_i = \begin{bmatrix} I_{2 \times 2} & km_{i,j,h} \times l_i & km_{i,k,h} \times l_i \\ 0_{4 \times 2} & km_{i,j,h} & km_{i,k,h} \end{bmatrix} \quad i = 0, 1, ..., n\text{-}1 \qquad (40a)$$

For i = n, the reduced parameter error vector is $[dl_{n,x}, dl_{n,y}, dl_{n,z}, db_{n,j},$ $db_{n,k}, d\beta_n]^T$. Then

$$K_n M_n = \begin{bmatrix} I_{3x3} & km_{n,j,h} \times l_i & km_{n,k,h} \times l_i & k_{n,\beta} \times l_i \\ 0_{3x3} & km_{n,j,h} & km_{n,k,h} & k_{n,\beta} \end{bmatrix} \tag{40b}$$

The coefficients of Eq. (39) can be obtained by straightforward derivations.

Assume for example that db_x and db_y are chosen as the independent error parameters, then

$$\delta_i = km_{i,x,z} db_{i,x} + km_{i,y,z} db_{i,y} \tag{41a}$$
$$d_i = dl_i + km_{i,x,z} \times l_i db_{i,x} + km_{i,y,z} \times l_i db_{i,y} \tag{41b}$$

Substituting Eq. (38) into Eq. (35) and comparing the result with Eq. (41), one obtains

$$km_{i,x,z} = Rot(z,\beta_i)_{3x3} [-b_{i,x}b_{i,y}w_i/b_{i,z}, \ 1 + b_{i,x}^2 w_i/b_{i,z}, \ b_{i,y}w_i]^T \tag{42a}$$
$$km_{i,y,z} = Rot(z,\beta_i)_{3x3} [-1 - b_{i,y}^2 w_i/b_{i,z}, b_{i,x}b_{i,y}w_i/b_{i,z}, -b_{i,x}w_i]^T \tag{42b}$$

for i = 0, 1, $\cdots$, n, where w_i is as defined in Eq. (31). Similarly, if $b_{i,x}$ and $b_{i,z}$ are chosen as the independent parameters, then

$$km_{i,x,y} = Rot(z,\beta_i)_{3x3} [b_{i,x}/b_{i,y}, \ 1, \ (1- b_{i,z})/b_{i,y}]^T \tag{43a}$$
$$km_{i,z,y} = Rot(z,\beta_i)_{3x3} [b_{i,z}/b_{i,y} + b_{i,y}w_i, \ -b_{i,x}w_i, b_{i,x}b_{i,z}w_i/b_{i,y}]^T \tag{43b}$$

for i = 0, 1, $\cdots$, n. Finally, if $b_{i,y}$ and $b_{i,z}$ are chosen as the independent parameters, then

$$km_{i,y,x} = Rot(z,\beta_i)_{3x3} [-1, \ -b_{i,y}/b_{i,x}, -(1- b_{i,z})/b_{i,x}]^T \tag{44a}$$
$$km_{i,z,x} = Rot(z,\beta_i)_{3x3} [b_{i,y}w_i, -b_{i,z}/b_{i,x} -b_{i,x}w_i, -b_{i,y}b_{i,z}w_i/b_{i,x}]^T \tag{44b}$$

for i = 0, 1, $\cdots$, n. Furthermore, $k_{n,\beta}$ is given as in Eq. (34d).

Remark: The *BASE* error model can be obtained from Eq. (39) by changing the subscript "n" to "0".

3.   Linearized CPC error model

Recall that $[d^T, \delta^T]^T$ is a cartesian error vector representing pose errors of the manipulator. Let $d\rho$ be the corresponding CPC parameter error vector. The linearized CPC error model can be obtained by plugging Eq. (20) and Eq. (40) into Eq. (25).   An alternative method which reduces the computational cost is described next.   Denote U_i of Eq. (17) as

$$U_i = \begin{bmatrix} n_i^u & o_i^u & a_i^u & p_i^u \\ 0 & 0 & 0 & 1 \end{bmatrix}$$

where n^u_i, o^u_i, a^u_i and p^u_i are 3x1 vectors.

Proposition 4: The *linearized CPC error model* can be written as

$$d = Tb_j\, db_j + Tb_k\, db_k + Tl_x\, dl_x + Tl_y\, dl_y + [Tb_z]_n\, dl_{n,z} + k_{n,\beta}\, xl_n\, d\beta_n \quad (45a)$$
$$\delta = Rb_j\, db_j + Rb_k\, db_k + k_{n,\beta}\, d\beta_n \quad\quad\quad\quad\quad (45b)$$

w h e r e
$$db_j = [db_{0,j}\ , db_{1,j}\ , \cdots, db_{n,j}]^T\ ,$$

$$db_k = [db_{0,k}\ , db_{1,k}\ , \cdots, db_{n,k}]^T\ ,$$

$$dl_x = [dl_{o,x}, dl_{1,x}, \cdots, dl_{n,x}]^T\ ,$$

$$dl_y = [dl_{o,y}, dl_{1,y}, \cdots, dl_{n,y}]^T\ ;$$

and Tb_j, Tb_k, Tl_x, Tl_y, Rb_j and Rb_k are all 3x(n+1) matrices whose components are functions of link parameters and joint variables. Specifically, the ith column of these matrices are:

$$[Tb_j]_i = \begin{bmatrix} ((p^u_{i+1} + l_i) \times n^u_{i+1}) \cdot km_{i,j,h} \\ ((p^u_{i+1} + l_i) \times o^u_{i+1}) \cdot km_{i,j,h} \\ ((p^u_{i+1} + l_i) \times a^u_{i+1}) \cdot km_{i,j,h} \end{bmatrix} \quad\quad (46a)$$

$$[Tb_k]_i = \begin{bmatrix} ((p^u_{i+1} + l_i) \times n^u_{i+1}) \cdot km_{i,k,h} \\ ((p^u_{i+1} + l_i) \times o^u_{i+1}) \cdot km_{i,k,h} \\ ((p^u_{i+1} + l_i) \times a^u_{i+1}) \cdot km_{i,k,h} \end{bmatrix} \quad\quad (46b)$$

$$[Tl_x]_i = [\, n^u_{i+1,x}, o^u_{i+1,x}, a^u_{i+1,x}]^T \quad\quad\quad\quad (46c)$$
$$[Tl_y]_i = [\, n^u_{i+1,y}, o^u_{i+1,y}, a^u_{i+1,y}]^T \quad\quad\quad\quad (46d)$$

$$[Rb_j]_i = \begin{bmatrix} n^u_{i+1} \cdot km_{i,j,h} \\ o^u_{i+1} \cdot km_{i,j,h} \\ a^u_{i+1} \cdot km_{i,j,h} \end{bmatrix} \quad\quad\quad\quad (46e)$$

$$[Rb_k]_i = \begin{bmatrix} n^u_{i+1} \cdot km_{i,k,h} \\ o^u_{i+1} \cdot km_{i,k,h} \\ a^u_{i+1} \cdot km_{i,k,h} \end{bmatrix} \qquad (46f)$$

and

$$[Tb_z]_n = [\, n^u_{n,x},\, o^u_{n,x},\, a^u_{n,x}\,]^T \qquad (46g)$$

The coefficients $km_{i,j,h}$ and $km_{i,k,h}$ in Eq. (46) are listed in Eqs. (42)-(44).

Proof: Substitution of Eqs. (39a) and (39b) into Eq. (18a) yields

$$\delta = \sum_{i=0}^{n} R^u_{i+1}{}^T \{ km_{i,j,h}d\,b_{i,j} + km_{i,k,h}d\,b_{i,k}\} + k_{n,\beta}d\beta_n \qquad (47a)$$

since $R^u_{n+1} = I$. Substitution of Eq. (39c) and Eq. (39d) into Eq. (18b), yields

$$d = \sum_{i=0}^{n} -R^u_{i+1}{}^T \Omega^u_{p,i+1} \{ km_{i,j,h}db_{i,j} + km_{i,k,h}d\,b_{i,k}\}$$

$$+ \sum_{i=0}^{n} R^u_{i+1}{}^T \{ km_{i,j,h} \times l_i db_{i,j} + km_{i,k,h} \times l_i db_{i,k} + dl_i\} + k_{n,\beta} \times l_n \, d\beta_n$$

$$= \sum_{i=0}^{n} R^u_{i+1}{}^T \{ km_{i,j,h} \times (p^u_{i+1} + l_i)d\,b_{i,j} + km_{i,k,h} \times (p^u_{i+1} + l_i)db_{i,k} + dl_i\} + k_{n,\beta} \times l_n d\beta_n$$

$$(47b)$$

using $\Omega^u_{p,i+n} = 0$, $\Omega^u_{p,i+1} \times km_{i,j,h} = p^u_{i+1} \times km_{i,j,h}$ and $\Omega^u_{p,i+1} \times km_{i,k,h} = p^u_{i+1} \times km_{i,k,h}$. Eq. (45) is obtained from rearrangement of Eq. (47).

$$[]$$

The coefficients in Eq. (45) define the linear mapping from the space of independent CPC error parameters to the space of pose errors. The cartesian errors dT_n represented in the world frame are related to $y = [d^T, \delta^T]^T$ by

$$\delta T_n = T_n^{-1} dT_n \qquad (48)$$

where the structure of δT_n is given in Eq. (16). dT_n is found from pose measurements of the manipulator's end-effector. T_n is evaluated with nominal parameters. Thus the right-hand side of Eq. (38) is known. The above equation serves as the measurement equation when all the parameters in the CPC kinematic model are to be identified.

4. Linearized *BASE* error model

Several methods exist for determining the nominal transformation *BASE* (denoted also as ^wT_b or B_0). Obviously, if a tool frame can be established on the end-effector such that three points along each axis of the tool frame can be measured in world coordinates, the determination of the nominal *BASE* becomes trivial.

Alternatively, by taking three non-colinear points in world coordinates, one can establish an orthonormal tool frame $\{^wn_t, \ ^wo_t, \ ^wa_t\}$. The intersection point of the three vectors can be chosen as the origin of the tool frame. It is represented in the world frame by ^wp_t. Thus the homogeneous transformation matrix ^wT_t has the 3x3 rotational submatrix $^wR_t = [^wn_t, \ ^wo_t, \ ^wa_t]$ and positioning vector ^wp_t. Similarly, the forward kinematics of the robot model provides the cooridnates of these three points in the base frame given the set of joint position readings at these particular poses. Similar procedures of constructing ^wT_t are then applied to obtain ^bT_t. From ^wT_t and ^bT_t,

$$^wT_b = \ ^wT_t \ ^bT_t \ ^{-1}$$

The 0th link parameters can be initialized from the nominal *BASE*. By using an error model which maps the 0th link parameter error vector to the 0th cartesian error vector these parameters can be iteratively updated.

Six independent error parameters should be used if *BASE* is to be identified. Recall that the same number of independent error parameters are used to model the nth link errors in the CPC error model. The linearized *BASE* error model can be obtained simply by changing the subscript "n" to "0" in Eqs. (39b) and (39d). The above error model needs to be modified if the orientational errors are not measured, as will be shown later.

The transformation from the tool frame to the world frame can be written as

$$T_n = B_0 U_1 \tag{49}$$

where U_1 is the transformation from the tool frame to the base frame. Assuming that the pose errors are caused only by errors in the 0th link parameters, then

$$\delta B_0 = B_0^{-1} dT_n U_1^{-1} \tag{50}$$

dT_n is found through measurements of the manipulator's pose errors.

The right-hand side of Eq. (50) then becomes known. Eq. (50) can serve as a measurement equation in the case where only the CPC *BASE* parameters are to be found.

If only the positioning error measurements are available, the fourth columns of the right-hand side of Eq. (50), which is equal to d_0, contain unknown quantities related to the orientational errors of the end-effector. In order for Eq. (39d) to be used as an error model, d_0 should be made known. The following method is designed to overcome this difficulty. From Eq. (50),

$$\delta B_0 U_1 = B_0^{-1} dT_n \qquad (51)$$

Denote by d_0' the first three elements of the 4th column of the right-hand side of Eq. (51)

$$d_0' = Rot(z, -\beta_0) R_0^T \{p^{dt} - p^b{}_0\} \qquad (52)$$

where p^{dt} and $p^b{}_0$ are respectively the first three elements of the fourth columns of dT_n and B_0. d_0' can be computed if p^{dt} is given. Also from the left-hand side of Eq. (51),

$$d_0' = \delta_0 \times p^u{}_1 + d_0 \qquad (53)$$

where $p^u{}_1$ is the first three elements of the fourth column of U_1. Substituting d_0 and δ_0, being in the forms of Eqs. (39d) and (39b), into (53) yields

$$d_0' = dl_0 + km_{0,j,h} \times (p^u{}_1 + l_0) db_{0,j} + km_{0,k,h} \times (p^u{}_1 + l_0) db_{0,k} + k_{0,\beta} \times l_0 \, d\beta_0 \qquad (54)$$

where $dl_0 = [dl_{0,x}, dl_{0,y}, dl_{0,z}]^T$. The coefficients of Eq. (54) are presented in Eqs. (42)-(44) and (34d). Eq. (53) gives the linearized *BASE* error model when the orientational errors are not available. In this case, Eq. (52) is the measurement equation.

5. Linearized *TOOL* error model

The dimensions of an end-effector can in general be measured very accurately before it is assembled. If the end-effector's mounting errors are very small, *TOOL* transformation can be treated as error-free [13]. The robot in such a case does not need to be calibrated after a tool change. $TOOL_{old}$, the old transformation from the tool frame to the flange frame, is replaced by $TOOL_{new}$, the new transformation from the tool frame to the flange frame. Both $TOOL_{old}$ and $TOOL_{new}$ are determined off-line.

If the mounting errors are nonnegligible, $TOOL$ must be calibrated. Given the nominal $TOOL$ in the CPC convention (this is equivalent to introducing an extra link transformation in the CPC model), $TOOL$ can be updated if a $TOOL$ error model is available. The linearized $TOOL$ error model is exactly in the form of Eqs. (39d) and (39b). Moreover, the measurement equation for estimating parameters in $TOOL$ is similar to that given in Eq. (48).

Six independent parameters in $BASE$ are observable if proper measurements are taken even if only positioning errors of the end-effector are provided. In such a case, however, three parameters in the $TOOL$ are not observable. This happens since the rank of the Jacobian matrix defined by the coefficients in Eq. (39d) is 3 as $km_{n,j,h}$, $km_{n,k,h}$ and $k_{n,\beta}$ are not functions of joint variables.

IV. MODEL-BASED KINEMATIC IDENTIFICATION ALGORITHMS

A. OPTIMIZATION ALGORITHMS FOR ERROR-MODEL BASED KINEMATIC IDENTIFICATION

1. Problem statement

Kinematic parameter estimation is basically a nonlinear optimization problem. Since the nominal kinematic parameters are in general known, the solution of the problem may be obtained by iteratively solving a linear least squares problem which is formulated below.

Let Y be the aggregated cartesian error vector of the end-effector. $Y = [y_1, y_2, ..., y_m]^T$, where m is the number of robot pose measurements and y_i is the ith pose error measurement. Let z be the independent kinematic parameter vector. Y is related to z through the aggregated Identification Jacobian G,

$$Y = Gz \tag{55}$$

where G has been given in Eq. (26).

2. Singular Value Decomposition (SVD) algorithm

The pseudo inverse solution of the problem can be written as

$$z \cong - (G^T G)^{-1} G^T Y$$
$$\equiv - G^+ Y \tag{56}$$

assuming that the inverse exists. When $G^T G$ is singular, the psuedo-inversion method will fail. The singular value decomposition technique [20] can however be applied to find G^+ without directly seeking matrix inversion.

Let the aggregated Identification Jacobian be written as

$$G = \mathcal{U} \Lambda \mathcal{V}^T \tag{57}$$

by singular value decomposition, where $\mathcal{U}^T \mathcal{U} = \mathcal{V}^T \mathcal{V} = I$, and $\Lambda = \text{diag}(\sigma_1, \sigma_2, \cdots, \sigma_m)$. Here σ_i, $i = 1, 2, \cdots$, m, are the singular values of G. The matrix $\mathcal{U}$ consists of m orthonormalized eigenvectors associated with the m largest eigenvalues of $G G^T$, and the matrix $\mathcal{V}$ consists of m orthonormalized eigenvectors associated with the m largest eigenvalues of $G^T G$. It is easy to verify that the pseudo inverse G^+ can be written as [20]

$$G^+ = \mathcal{V} \Lambda^+ \mathcal{U}^T \tag{58}$$

where $\Lambda^+ = \text{diag}(\sigma^+_1, \sigma^+_2, \cdots, \sigma^+_m)$

$$\sigma^+_i = \begin{cases} 1/\sigma_i & \text{for } \sigma_i > 0 \\ \\ 0 & \text{for } \sigma_i = 0 \end{cases} \tag{59}$$

In practice, a machine-dependent small positive number ϵ is introduced to replace zero as a threshold value in Eq. (59). Then the unique solution of Eq. (55) can be written as

$$z^* \cong - \mathcal{V} \Lambda^+ \mathcal{U}^T Y \tag{60}$$

3. Levenberg-Marquardt (L-M) algorithm

Another approach to accommodate cases in which $G^T G$ are singular is to minimize a modified cost function J'

$$J' \equiv (Y + Gz)^T Q (Y + Gz) + \lambda \| z \|^2 \tag{61}$$

where λ is a positive weighting coefficient. Minimizing J' by choosing optimal value of z yields the equation

$$G^T Y + (G^T G + \lambda I) z^* = 0 \tag{62}$$

where I is an identity matrix, and z^* is the solution of the above equation. It has been shown in [35] that z^* is a minimum solution due to the fact that $(G^T G + \lambda I)$ is positive definite. In other words, a least squares solution exists even when $G^T G$ is singular.

The SVD algorithm tends to produce solutions of higher precision than the L-M algorithm does if the condition number of $G^T G$ is not too large.   When $G^T G$ is singular, the SVD algorithm tends to give larger estimation   vectors.

B. JOINT AXIS IDENTIFICATION

1. Measurement   procedures   for   Joint   Axis   Identification

Two measurement phases need to be followed.   In the first phase, data is collected in order to estimate the poses of the joint axes and in the second phase, data for identifying *BASE* and *TOOL* is collected.

Phase 1.   Measurements for estimating the poses of the joint axes.
A target such as a retroreflecter or LED is attached to the end-effector of the robot. The measurement procedure consists of the following robot motions:
Starting at Joint i = n, the joint i is moved through a large portion of its travel range, while Joints 1 to n-1 are fixed at a suitable robot configuration (also termed the *Arm Signature*). As joint i moves, the world coordinates of the target point are measured and recorded.   The procedure is repeated for Joint i = n-1, keeping Joints 1 to n-2 at the original Arm Signature position, and so on with i decreasing all the way to i = 1.   The collected data points contain all the information necessary to estimate the joint axes line equations with respect to the world coordinate frame.   Namely, in the case of a revolute joint, the target moves along a circle, such that from estimating the plane of rotation and center of rotation the joint axis can be found in the least squares sense.   Likewise for a prismatic joint, the motion of the target is along a line parallel to the joint axis.

Phase 2.   Measurements for identifying BASE and TOOL
This is done in the same way as when using error model based techniques.

2. Pose   estimation   of   a   revolute   joint   axis

Denote by $\{{}^w p^1, {}^w p^2, \cdots, {}^w p^k\}$ a set of ordered trajectory points of an observed target when Joint i is in motion.   The right- superscripts are used to number the points on a circle shown in Fig. 10.   For notation simplicity, the subscript "i" is omitted.   Ideally, these points are on the circle which lies on Plane B having a normal vector ${}^w b_i \equiv b$ and the center of which is at ${}^w p_i$. The distance from the origin of the world frame $\{W\}$ to Plane B is a.   The problem is to find ${}^w p_i$ and ${}^w b_i$ given $\{{}^w p^1, {}^w p^2, \cdots, {}^w p^k\}$.   All of these quantities are represented in $\{W\}$.   To further simplify the notation, the left-superscript "w" is dropped next for those vectors represented in $\{W\}$ without ambiguity.

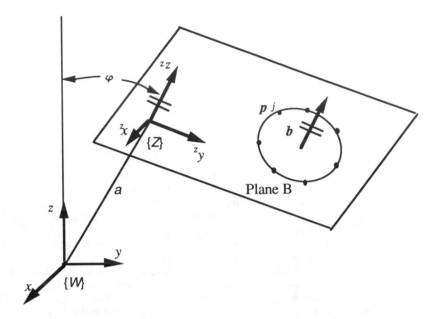

Fig. 10 The Rotation Axis and Rotation Center in $\{W\}$ and $\{Z\}$.

The solution is given in terms of the 9 steps listed below:

1. Choose three points, p^1, p^{1+q}, p^{1+2q}, from $\{{}^w p^1, {}^w p^2, \cdots, {}^w p^k\}$, where $q = \text{int}(k/3)$.
   Explanation: Ideally, the three points should form an equilateral triangle in order to obtain a good initial estimate of b. If $\{{}^w p^1, {}^w p^2, \cdots, {}^w p^k\}$ do not complete a full circle, one needs to at least make sure that p^1, p^{1+q} and p^{1+2q} are not nearly colinear.

2. Find an initial estimate b^0 to b.

$$b^0 = v_1 \times v_2 / \|v_1 \times v_2\| \tag{63}$$

   where $v_1 \equiv p^1 - p^{1+q}$, the vector from the tip of p^{1+q} to that of p^1, and $v_2 \equiv p^{1+2q} - p^{1+q}$, the vector from the tip of p^{1+q} to that of p^{1+2q}.

3. Let $b^0_h = \max\{|b^0_x|, |b^0_y|, |b^0_z|\}^T$. Then form the equation of Plane B. For example, if $b^0_h = b^0_x$, the equation can be rewritten as

$$p^j_y k_1 + p^j_z k_2 + k_3 = p^j_x \qquad\qquad j = 1, 2, \cdots, k \tag{64}$$

   where $[p^j_x, p^j_y, p^j_z]^T \equiv p^j$, and $\{k_1, k_2, k_3\}$ are to be found; refer to Fig. 4.2.1. Similar equations can be formulated when either b^0_y or

$b^0{}_z$ is the maximum element.

4. Solve for $\{k_1, k_2, k_3\}$ from Eq. (64) using a least-squares technique.
   Explanation: The least-squares solution is unique since $\{p^1, p^2, \cdots, p^k\}$ are not colinear.
5. Solve for ^wb_i and a from $\{k_1, k_2, k_3\}$ as follows

$$^wb_i \equiv b \equiv [b_x, b_y, b_z]^T = [1/g, -k_1/g, -k_2/g]^T \qquad (65)$$

and

$$a \equiv k_3/g \qquad (66)$$

where $g = \{1 + k_2{}^2 + k_3{}^2\}^{1/2}$.

6. Transform $\{p^1, p^2, \cdots, p^k\}$ to a new coordinate frame $\{Z\}$ whose x-y plane is Plane B. This is done by a rotation R and a translation t, where R is given in Eq. (7) and

$$t = [0, 0, a]^T \qquad (67)$$

$\{^zp^1, {}^zp^2, \cdots, {}^zp^k\}$, the representations of Points $p^1, p^2, \cdots, p^k$ in $\{Z\}$, are obtained from the following equations,

$$^zp^j = R^T p^j - t \qquad\qquad j = 1, 2, \cdots, k \qquad (68)$$

Explanation: Refer to Fig. 10.
7. Form the circle equation in $\{Z\}$. The circle equation in the x-y plane of $\{Z\}$ can be written as

$$(^zp^j{}_x - {}^zp^0{}_x)^2 + (^zp^j{}_y - {}^zp^0{}_y)^2 = r^2 \qquad j = 1, 2, \cdots, k \qquad (69)$$

where $[^zp^j{}_x, {}^zp^j{}_y, 0]^T \equiv {}^zp^j$, and $[^zp^0{}_x, {}^zp^0{}_y, 0]^T \equiv {}^zp$ is the center of the circle in $\{Z\}$. From Eq. (69),

$$2\,{}^zp^j{}_x\,{}^zp^0{}_x + 2\,{}^zp^0{}_y\,{}^zp^0{}_y - {}^zp^0{}_x{}^2 - {}^zp^0{}_y{}^2 + r^2 = {}^zp^j{}_x{}^2 + {}^zp^j{}_y{}^2$$
$$j = 1, 2, \cdots, k \qquad (70)$$

Let $h \equiv -{}^zp^0{}_x{}^2 - {}^zp^0{}_y{}^2 + r^2$, and $w^j \equiv {}^zp^j{}_x{}^2 + {}^zp^j{}_y{}^2$. Then from Eq. (70),

$$2\,{}^zp^j{}_x\,{}^zp^0{}_x + 2\,{}^zp^0{}_y\,{}^zp^0{}_y + h = w^j \qquad j = 1, 2, \cdots, k \qquad (71)$$

8. Solve for $\{^zp^0{}_x, {}^zp^0{}_y, h\}$ from Eq. (71) using a least-squares technique.
   Explanation: The least-squares solution is unique since $\{^zp^1, {}^zp^2, \cdots, {}^zp^k\}$ are not colinear.
9. Transfer ^zp back to $\{W\}$ using the following relationship,

$$^wp_i = R(^zp + t) \qquad (72)$$

The above procedure is repeated for i = 1, 2, $\cdots$, n. After this, the poses $\{^W p_i, ^W b_i\}$, i = 1, 2, ..., n, in the world coordinates are available. Similar analysis follows for prismatic joints.

3. Computation of D-H link parameters from identified joint axis poses

The method given next is similar to that of Chen [12] and Sklar [13]. Denote by n_i the direction vector of the common normal between the ith and i+1th joint axes. Denote by s_i and w_{i+1} the intersections of the common normal with the ith and i+1th joint axes. All of these quantities are represented in $\{W\}$. To simplify the notation, the left-superscript "w" is dropped next for n_i, s_i and w_{i+1}. The algorithm consists of the following two stages:

Stage 1. Find n_i, s_i and w_{i+1} in terms of $\{^W p_{i-1}, ^W b_{i-1}\}$ and $\{^W p_i, ^W b_i\}$

Assume that w_i is known and w_0 is at a convenient location of the 1st joint axis. s_i, w_{i+1} and n_i need to be found. Refer to Fig. 11.

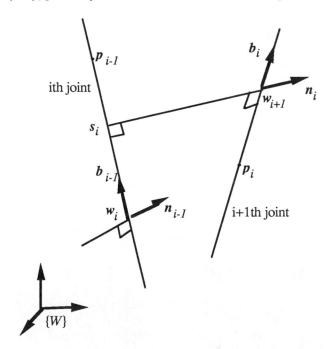

Fig. 11 Finding the intersection points of the common normal with two consecutive axes.

Case 1: The ith and the i+1th axes are non-parallel.
By the definitions of $^w b_{i-1}$, $^w b_i$ and n_i,

$$^w n_i = {}^w b_{i-1} \times {}^w b_i / \|{}^w b_{i-1} \times {}^w b_i \| \tag{73}$$

Point s_i can be represented parametrically by

$$s_i = t_{i,1}{}^w b_{i-1} + {}^w p_{i-1} \tag{74a}$$

where $t_{i,1}$ is an undetermined scalar. Similarly,

$$w_{i+1} = t_{i,2}{}^w b_i + {}^w p_i \tag{74b}$$

where $t_{i,2}$ is an undetermined scalar.

Since s_i and w_{i+1} lie on the common normal, the following equations are satisfied,

$$^w b_{i-1} \cdot (w_{i+1} - s_i) = 0 \tag{75a}$$

and

$$^w b_i \cdot (w_{i+1} - s_i) = 0 \tag{75b}$$

Substituting Eqs. (74a) and (74b) into Eq. (75) yields two linear algebraic equations for $t_{i,1}$ and $t_{i,2}$. By elimination of $t_{i,1}$ and $t_{i,2}$ from Eq. (74), one obtains

$$s_i = p_{i-1} + \frac{(p_i - p_{i-1})((b_i \cdot b_{i-1})b_i - b_{i-1})}{(b_i \cdot b_{i-1})^2 - 1} b_{i-1} \tag{76a}$$

and

$$w_{i+1} = p_i + \frac{(p_i - p_{i-1})(b_i - (b_i \cdot b_{i-1})b_{i-1})}{(b_i \cdot b_{i-1})^2 - 1} b_i \tag{76b}$$

where all vectors in the right-hand side of Eq. (76) are represented in $\{W\}$.

Case 2: The ith and the i+1th axes are parallel.
In this case, the denominators in Eq. (76) are zero. Let

$$s_i = w_i \tag{77}$$

since w_i is known. Eq. (77) implies that the offset distance between the consecutive common normal axes is set to zero. The plane perpendicular to z_{i-1} passing through w_i is given by

$$^w b_{i-1} \cdot (w_{i+1} - w_i) = 0 \tag{78}$$

where w_{i+1}, the intersection of z_i with the plane, is to be determined. Combining Eqs. (78) and (74b) and using ${}^w b_{i-1} = {}^w b_i$ yield

$$w_{i+1} = - {}^w b_i \cdot ({}^w p_i - w_i)^w b_i + {}^w p_i \qquad (79)$$

Finally,

$$n_i = (w_{i+1} - s_i)/ \|w_{i+1} - s_i\| \qquad (80)$$

Stage 2. Solve for the D-H parameters in terms of n_i, s_i and w_{i+1}
   The solution is

$$d_i = {}^w b_{i-1} \cdot (s_i - w_i) \qquad (81a)$$

If $n_{i-1} \times n_i \neq 0$,   then

$$\theta_i = \operatorname{sign}(k_1) \arccos(n_{i-1} \cdot n_i) \qquad (81b)$$

where $k_1 = (n_{i-1} \times n_i) \cdot {}^w b_{i-1}$.   Otherwise

$$\theta_i = 0 \qquad (81b')$$

If ${}^w b_{i-1} \times {}^w b_i \neq 0$,   then

$$\alpha_i = \operatorname{sign}(k_2) \arccos({}^w b_{i-1} \cdot {}^w b_i) \qquad (81c)$$

where $k_2 = ({}^w b_{i-1} \times {}^w b_i) \cdot n_{i-1}$.   Otherwise

$$\alpha_i = 0 \qquad (81c')$$

Finally

$$a_i = n_{i-1} \cdot (w_{i+1} - s_i) \qquad (81d)$$

for i = 1, 2, ..., n.

4. Construction of CPC link parameters from identified joint axis poses

If the forward and inverse kinematics of the robot is represented in the CPC modeling convention, it is desired to find the CPC link transformations directly from identified joint axis poses.

In this section, vectors without left-superscripts are represented in terms of the i-1th link frame for the ith pose. Furthermore, let $\{q_1, ..., q_n\}$ be the Arm Signature recorded in the experiment.

The procedure of determining the CPC link parameters from identified joint poses consists of the following three steps.

Step 1. Determination of the 0th link parameters

Clearly, $b_0 = {}^w b_0$. β_0 can be determined if the location of the base frame is given in terms of $\{W\}$. For example, in the case of the PUMA robot, the base frame in in general located at the intersection of the 1st and 2nd joint axes. If the x axis of the base frame is given, the **BASE** transformation can be written down. Thus,

$${}^w R_0 = R_0\, Rot(x,\, \beta_0)_{3x3} \qquad\qquad (82)$$

where ${}^w R_0$ is the rotation sub-matrix of B_0. Since R_0 is uniquely determined from b_0, β_0 can be solved from Eq. (82).

The translational parameter vector l_0 is solved from

$$l_0 = Rot(x,\, -\beta_0)_{3x3}\, R_0^T\, {}^w p_0 \qquad\qquad (83)$$

where ${}^w p_0$ is the origin of the base frame in $\{W\}$.

Step 2: Transformation of the ith joint pose from $\{W\}$ to $\{i\text{-}1\}$
This can be done by using $B_0\, B_1\, \cdots\, B_{i-1}\, Q_i$. After this stage, $\{p_i,\, b_i\}$ are obtained from $\{{}^w p_i,\, {}^w b_i\}$.

Step 3: Determination of the ith link parameters
This is similar to the method given in Step 1. The orientation parameter vectors b_i have been obtained in Step 2. The parameters β_i are redundant for internal link transformations. Without losing generality, they can be set to zero; that is $\beta_i \equiv 0$ for i = 1, 2, ..., n. The translational parameter vector l_i is solved from

$$l_i = R_i^1\, p_i \qquad\qquad i = 1, 2, \cdots, n \quad (84)$$

where p_i is the origin of the ith frame in $\{i\text{-}1\}$.

5. Identification of *BASE* and *TOOL* using error-model based technique

The methods given above identify the link transformations A_1, A_2, $\cdots, A_n$, or $B_0, B_1, \cdots, B_n$, depending upon the choice of the modeling convention. These techniques however neither involve joint measurements (except for the arm signature) nor a nominal kinematic model. The error model based technique given in Section III.C is then introduced to identify *BASE* and *TOOL*. This time a nominal model as well as joint measurements are needed. The identification of *BASE* serves two purposes. First, the joint axis identification method does not address joint offsets and such errors may sometimes be significant. Identifying *BASE* using the error model based method can provide an equivalent compensation for joint position uncertainties. Second, the world coordinate frame may be displaced from one application to

another.   In such a case only $BASE$ needs to be re-identified.
Identification of $TOOL$ helps to compensate the orientational errors of
the end-effector, since only the z axis of the tool is identified using the
joint axis identification method.   Readers are referred to Section III.C.4
for the detailed mathematical formulation and technical discussion.

V.   OPTIMAL ACCURACY COMPENSATION

A. PROBLEM FORMULATION

Denote

$$U_i = \begin{bmatrix} n_{i,x}^u & o_{i,x}^u & a_{i,x}^u & p_{i,x}^u \\ n_{i,y}^u & o_{i,y}^a & a_{i,y}^u & p_{i,y}^u \\ n_{i,z}^u & o_{i,z}^u & a_{i,z}^u & p_{i,z}^u \\ 0 & 0 & 0 & 1 \end{bmatrix}$$

where U_i has been defined in Eq. (17).   The superscript "u" indicates
that the particular entry is an element of the matrix U_i.

Let y_c be the robot pose error before compensation,   consisting of
the nontrivial elements in δT_n; that is, $y_c \equiv [d^T, \delta^T]^T \equiv [dx, dy, dz, \delta x, \delta y,$
$\delta z]^T$. y_c can be computed using the nominal and actual forward
kinematics of the robot.   Let the required joint correction vector be dq_c.
Then under the assumption of small joint corrections, a linearized
relationship between the pose errors and the joint corrections can be
written as [25, 28]

$$y_c \cong -C dq_c \qquad\qquad (85)$$

where $C \equiv [c_1, c_2, \cdots, c_n]$.   The relationship given in Eq. (85) is termed
the *Reduced- Order Accuracy Compensation Model* and C is termed the
Compensation Jacobian. $-C$ is the same as the ordinary Manipulator
Jacobian commonly used in the Robotics literature.   Eq. (85) can
conveniently be rewritten as the following discrete-time state equation

$$y_{k+1} = y_k + c_k u_k \qquad\qquad k = 1, 2, \cdots, n \qquad (86a)$$
$$y_1 = y_c \qquad\qquad\qquad\qquad\qquad (86b)$$

where $u \equiv [u_1, u_2, ..., u_n]^T \equiv dq_c$. y_{n+1} is the robot pose error vector
after   compensation.

It was shown in [32] with the D-H convention and in [25, 28] with the
CPC convention that if the ith joint is revolute, then the columns c_i of C
are

$$c_i = \begin{bmatrix} n^u_{i,x}p^u_{i,y} - n^u_{i,y}p^u_{i,x} \\ o^u_{i,x}p^u_{i,y} - o^u_{i,y}p^u_{i,x} \\ a^u_{i,x}p^u_{i,y} - a^u_{i,y}p^u_{i,x} \\ -n^u_{i,z} \\ -o^u_{i,z} \\ -a^u_{i,z} \end{bmatrix}$$ (87a)

and if the ith joint is prismatic, then

$$c_i = \begin{bmatrix} -n^u_{i,z} \\ -o^u_{i,z} \\ -a^u_{i,z} \\ 0 \\ 0 \\ 0 \end{bmatrix}$$ (87b)

Let
$$\begin{aligned} J_y &\equiv (y_c + Cdq_c)^T Q_y (y_c + Cdq_c) \\ &\equiv y_{n+1}^T Q_y y_{n+1} \end{aligned}$$ (88)

Q_y is a symmetric and non-negative matrix. The accuracy compensation problem is now reduced to that of minimizing the performance index J_y by choice of dq_c.

Remark It is suggested in [25, 28] that δT_n (or equivalently y_c) be determined using $\delta T_n \equiv T_n(\rho^0, q^0)^{-1}\{T_n(\rho^0+d\rho, q^0+dq) - T_n(\rho^0, q^0)\}$. δT_n can also be approximated by linearizing $T_n(\rho^0+d\rho, q^0+dq)$ about the nominal link parameter vector ρ^0 and the joint variable vector q^0 provided that T_n is parametrically continuous and differentiable.

B. LQR ALGORITHM

The cost function given in Eq. (88) can be augmented to include a weighted sum of squares of the control terms. The compensation problem can now be formulated as a *linear quadratic regulator* (LQR) problem. The solution of the problem is the joint variable correction vector u which minimizes the cost function $J(u, y_{n+1})$,

$$J(u, y_{n+1}) = y_{n+1}^T Q_y y_{n+1} + \sum_{k=1}^{n} \gamma_k u_k^2$$ (89)

subject to Eq. (86). Q_y is as before symmetric and nonnegative definite; and the control weights γ_k are strictly positive.

The optimal solution $u_k{}^*$ is given by the feedback control law

$$u_k{}^* = v_k\, y_k \tag{90}$$

where the gains $\{v_k\}$ are obtained from a recursive solution of a related matrix Riccati difference equation   [25, 28].

$$v_{n-k} = -\frac{c_{n-k}{}^T P_k}{c_{n-k}{}^T P_k\, c_{n-k} + \gamma_k} \qquad k = 0,\, 1,\, ...,\, n\text{-}1 \tag{91a}$$

$$P_{k+1} = P_k + P_k\, c_{n-k} v_{n-k}{}^T \tag{91b}$$

with

$$P_0 = Q_y \tag{91c}$$

The resulting minimum cost is

$$J^* \equiv y_1{}^T P_n y_1 \tag{91d}$$

Note that if $\gamma_k = 0$, then v_k may become undefined at robot singular configurations.   The choice of the modified cost function in Eq. (89) ensures the existence and uniqueness of the optimal solution at all robot configurations. Readers are referred to [25, 28] for methods of choosing Q_y and γ_k.

After $dq_c = u$ is obtained, the actual joint variable q, to be used at this particular task point, is

$$q = q^0 + dq_c \tag{91e}$$

The procedure is now repeated at every task point.

Remark:
1. The resulting joint commands may in certain cases exceed the joint travel boundaries. The solution is no longer optimal if a saturated joint command is obtained. To prevent this, one may iteratively readjust the weighting coefficients γ_k.   In addition, safety margins at the joint boundaries must be left when the task is created, leaving room for joint corrections.   This is true for any type of accuracy compensation   algorithms.
2. In general, pose errors after compensation cannot be zero, even if the identification phase of the calibration process is perfect.   This is because the state equation (86) is a linearized error model, and because the optimal cost function is not necessarily zero.

3. Since y_1 is the initial pose error vector and y_{n+1} is the final pose error vector after compensation, the compensation performance may be assessed from the ratio of $J*$ (given in Eq. (91d)) and $y_1^T Q_y y_1$.

VI. EXPERIMENTAL RESULTS USING THE CPC ERROR MODEL

A. MEASUREMENT SET-UP

The experiment system, similar to the one discussed in [19], was made up of a PUMA 560 Robot, a Mitutoyo Model CX-D2 coordinate measuring machine (CMM) and an IBM AT computer as illustrated in Fig. 12. The six degrees of freedom PUMA 560 robot has a reach of 878 *mm* to the center of the wrist, a repeatability of 0.1 *mm* and a maximum payload of 4.0 *kg*. The CMM is a parallelpiped measurement equipment with three degrees of freedom. The measurement with the CMM are made through manually bringing a touch probe, mounted on the CMM, to contact a tooling ball mounted on the robot. The CMM has an accuracy of approximately 0.1 *mm* and a work volume of 400x500x800 *mm*^{3}.

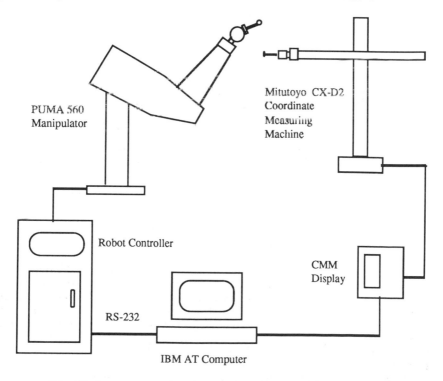

Fig. 12 Arrangement of equipment for calibration experiments.

The robot base coordinate frame was placed at the intersection of the first joint and second joint axes of the PUMA arm. The tool coordinate frame was placed at the tooling ball center of the end-effector (see Fig. 13). The CMM position indicators were set to zero after driving each axis to its travel limit. This established the coordinate frame for the CMM, which was then defined as the world coordinate frame for the measurement process.

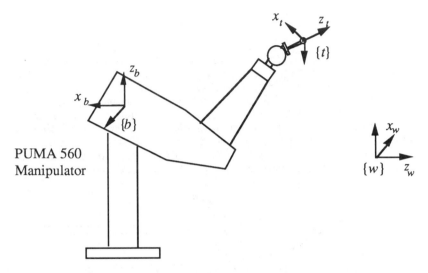

Fig. 13 Coordinate assignment for calibration experiments

To uniquely compute the center of the tooling ball, four measurements need to be performed, assuming that the radius of the ball is known. More measurements can reduce the influence of measurement noise. In this experimental process, five measurements were taken for each arm position. Evidently by using this type of end-effector, the tool orientation cannot be measured.

B. IDENTIFICATION RESULTS

Kinematic identification results were obtained using different levels of complexity of the kinematic models and different numbers of measurement points. Three types of models were of interest: nominal model, the **BASE** model identified by applying the **BASE** error model only and complete model identified by applying the complete CPC error model. The nominal model was obtained from the specifications of the PUMA arm except for the **BASE** transformation, which had to be determined by three measurements, as described in Section III.C.4. The different models were tested using different combinations of 100 positions measured throughout the work volume of the CMM. The effect of varying the number of data points on the kinematic identification results was investigated by choosing 10, 25 and 50 measurement points from left-arm configuration only as well as from both left- and right arm configurations. The identification results of the *BASE* transformation were also obtained by using 10 and 25 data points.

Since only the end-effector positions were measured, the transformation *TOOL* had three identifiable parameters. Three translational parameters, $l_{6,x}$, $l_{6,y}$ and $l_{6,z}$, were chosen as the identifiable parameters in *TOOL*. Since the origin of the tool frame lay on the 6th joint axis, two parameters in A_5 were identifiable [9,10]. Thus $b_{5,x}$ and $b_{5,y}$ were chosen as the identification parameters. A total of 25 unknown parameters were to be identified, which is the maximum number of identifiable parameters in such a set-up according to Proposition 1.

1. Identification of *BASE*

The nominal link parameters of $B_1, B_2, ..., B_5$, obtained from engineering drawings of the PUMA arm, are those given in Table 1. The nominal *BASE*, ($^{w}T_b$ or B_0), was determined using Eq. (48). Since the end-effector was not changed in both identification and verification experiments, *FLANGE* and *TOOL* can be considered as one link transformation. The nominal *BASE* and *FLANGE·TOOL* parameters are listed in Table 2. Since in this experiment the nominal z-axis of the base frame was aligned with the y-axis of the world frame (see Fig. 13), $b_{0,x}$, $b_{0,z}$, $l_{0,x}$, $l_{0,y}$, $l_{0,z}$ and β_0 were chosen as the six independent parameters in *BASE*.

Sets of 10 and 25 data points from both left- and right-arm configurations were used to identify the *BASE* transformation by applying the linearized *BASE* error model given in Section III.C.4. Since the obtained results were quite similar, only the identified parameter errors with the 10 calibration points are listed in Table 3.

Table 2   Model-1: nominal *BASE* and *FLANGE·TOOL* parameters

i	$b_{i,x}$	$b_{i,y}$	$b_{i,z}$	$l_{i,x}(mm)$	$l_{i,y}\ (mm)$	$l_{i,z}\ (mm)$	$\beta_i\ (deg)$
0	-0.00034	0.99904	-0.01047	820.81	320.30	290.80	89.97
6	0	0	1	0	0	0	94.85

Table 3   Model-2: **BASE** identified by 10 measurements
(These are parameter deviations from Model-1)

i	$db_{i,x}$	$db_{i,y}$	$db_{i,z}$	$dl_{i,x}(mm)$	$dl_{i,y}$ (mm)	$dl_{i,z}$ (mm)	$d\beta_i$ (deg)
0	-0.00081	0.00093	0.00279	-0.2811	0.2779	-0.5750	-0.0214

2. Identified   Models

Sets of 10, 25 and 50 calibration points from left-arm only and from both left- and right-arm configurations of the PUMA robot were used in the experiment.   The linearized CPC error model given in Section III.C.3 was applied for error parameter estimation. Both the SVD and the L-M algorithms, were used for identification.   Since the error model had only 25 independent parameters, then by using a large number of unknown parameters artificially created a case of a singular G^TG matrix which allowed an interesting comparison between the two algorithms. The SVD algorithm in such a case resulted in large parameter errors as well as large condition numbers and zero observability measures. The L-M algorithm however produced good results even in this case as expected.   On the other hand, if 25 parameters were used in the error model, both the methods produced similar results. Selected identification models with 25 error parameters obtained by using two iterations of the SVD algorithm are listed in Tables 4-7.   The condition numbers and observability measures associated with these models are illustrated in Fig. 14.

Names of models are given to facilitate the comparison of them.

Model-1:   the nominal model with **BASE** being identified by the 3-point set-up given in Section VI.B.1.

Model-2:   the nominal model with **BASE** being identified by applying the **BASE** error model. 10 data points were used to identify the **BASE** transformation.

Model-3:   the identified model with 10 data points taken from both of the   arm   configurations.

Model-4:   the identified model with 25 data points taken from both of the   arm   configurations.

Model-5: the identified model with 50 data points taken from both of the   arm   configurations.

Model-6:   the identified model with 25 data points taken only from the   right-arm   configuration.

Model-6 is listed here in order to emphasize the importance of configuration choices in kinematic identification.

Remark:
1.   The identified parameter errors for this particular robot are small
   (for example, the maximum identified length error is 1.8205 mm in

Model-4).

2. The identification algorithm always converged within three or less iterations.

3. Model-4 and Model-5 are very similar. Their condition numbers are also very close. This indicates that no significant improvement with the existing measurement points can be made on the identification when the number of measurement points exceeds a certain threshold value. Models obtained by using more than 50 points were virtually the same as Model-5 and therefore are not listed here.

4. Both the condition number and observability measure give good indications of identification qualities for Model-3 and Model-4. However the difference of the observability measures for Model-4 and Model-5 is very large. Observability measures may indeed give inaccurate indications from time to time.

5. There is an agreement between the condition number and observability measure that Model-6 is inferior compared to Model-4. These measures however fail to explain why Model-6 is worse even than Model-3, as will be seen in the next section.

Table 4   Model-3:   identified by 10 measurements
(These are parameter deviations from Model-1)

i	$db_{i,x}$	$db_{i,y}$	$db_{i,z}$	$dl_{i,x}(mm)$	$dl_{i,y}$ (mm)	$dl_{i,z}$ (mm)	$d\beta_i$ (deg)
0	-0.00125	0.00093	0.00298	-0.6987	1.8521	0	0
1	-0.00043	-0.00000	-0.00129	0.2483	1.1908	0	0
2	-0.00001	0.00254	-0.00000	0.4881	0.4769	0	0
3	0.00554	0.00008	-0.01166	-2.1597	5.0540	0	0
4	0.00775	-0.00065	0.03533	-0.7350	0.9815	0	0
5	-0.00618	0.00073	0.03780	0	0	0	0
6	0	0	0	-0.4391	0.0789	0.2529	0

Table 5 Model-4: identified by 25 measurements
(These are parameter deviations from Model-2)

i	$db_{i,x}$	$db_{i,y}$	$db_{i,z}$	$dl_{i,x}(mm)$	$dl_{i,y}$ (mm)	$dl_{i,z}$ (mm)	$d\beta_i$ (deg)
0	-0.00083	0.00000	0.00084	-0.3410	-1.6840	0	0
1	0.00114	-0.00000	0.00129	0.2241	0.7181	0	0
2	-0.00039	0.00012	0.00000	0.1610	-0.4311	0	0
3	0.00021	0.00002	-0.00581	-0.4431	1.8205	0	0
4	-0.01118	-0.00007	0.00288	0.0994	0.2812	0	0
5	0.01005	0.00005	0.00184	0	0	0	0
6	0	0	0	-0.0874	0.0685	-0.1557	0

Table 6 Model-5: identified by 50 measurements
(These are parameter deviations from Model-2)

i	$db_{i,x}$	$db_{i,y}$	$db_{i,z}$	$dl_{i,x}(mm)$	$dl_{i,y}$ (mm)	$dl_{i,z}$ (mm)	$d\beta_i$ (deg)
0	-0.00074	0.00000	0.00060	-0.3317	-1.8777	0	0
1	0.00105	-0.00000	-0.00005	0.2166	0.6783	0	0
2	-0.00032	0.00057	0.00000	0.1752	-0.3399	0	0
3	0.00007	0.00002	-0.00589	-0.4249	1.9211	0	0
4	-0.01020	-0.00006	0.00438	-0.0562	0.3285	0	0
5	0.00910	0.00005	0.00386	0	0	0	0
6	0	0	0	-0.0608	0.0396	0.0101	0

Table 7 Model-6: identified by 25 measurements
from only right configuration
(These are parameter deviations from Model-2)

i	$db_{i,x}$	$db_{i,y}$	$db_{i,z}$	$dl_{i,x}(mm)$	$dl_{i,y}\ (mm)$	$dl_{i,z}\ (mm)$	$d\beta_i\ (deg)$
0	-0.00034	0.00000	-0.00163	-0.4775	-2.4681	0	0
1	0.00281	-0.00000	-0.00037	-0.1639	-0.9088	0	0
2	0.00055	0.00052	0.00000	0.15732	-1.9286	0	0
3	-0.00185	0.00001	-0.00377	-1.4302	0.6858	0	0
4	-0.01147	-0.00007	0.00107	0.2191	0.2425	0	0
5	0.00821	0.00003	0.00109	0	0	0	0
6	0	0	0	-0.0572	0.0204	-0.118	0

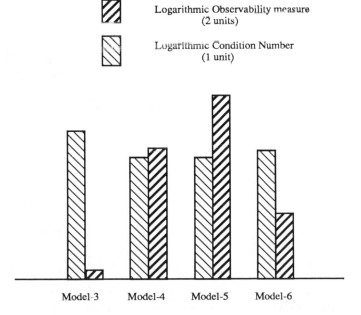

Logarithmic Observability measure
(2 units)

Logarithmic Condition Number
(1 unit)

Model-3   Model-4   Model-5   Model-6

Fig. 14 Logarithmic observability measures and condition numbers
for selected models.

C.   VERIFICATION AND COMPENSATION RESULTS

After the kinematic model of a manipulator is identified, it is necessary to verify if the model is a good mathematical representation of the manipulator.   There are basically two ways to perform a verification task.   The first method consists of measuring the poses of the manipulator at different locations followed by comparing the measured poses with the computed poses using the forward kinematics of the identified models.   Since the forward kinematics is applied in this method, the method is termed a *forward verification* method, or simply a *verification* method.   The second method consists of solving the inverse kinematics of the identified model given the poses of the manipulator followed by measuring the actual poses driven by the outputs of the inverse kinematic algorithm.   This method is termed an *inverse verification* method, or simply a *compensation* method as it essentially solves an accuracy compensation problem.   Obviously when a forward verification method is employed, sampled poses have to be different from those used for the identification of the models.   If the world coordinate frame in the verification or the compensation phase is not the same as that in the identification phase, the transformation relating the new world frame to the base frame has to be re-identified.

Standard statistical measures for assessing the efficiency of kinematic identification and compensation may be used.   The positioning error dp_i is defined as the absolute value of the difference between the ith anticipated and actual end-effector positions; i.e., $dp_i = |p_i - p_i^0|$, where p_i^0 is the anticipated position and p_i is the actual position. The mean positioning error dp_{mean} is the additive mean value of all positioning errors. The maximum positioning error dp_{max} is the maximum value among all positioning errors.   The standard deviation of the positioning errors STD_p is defined as follows:

$$STD_p = \sqrt{\frac{\sum_{i=1}^{m} dp_i^2 - \frac{(\sum_{i=1}^{m} dp_i)^2}{m}}{m - 1}} \qquad (92)$$

where m is the number of data points.

1.   Forward   Verification   Results

For the verification experiment, 40 new measurements were taken. The time span between the identification experiments and verification experiments was a few weeks.   Both the PUMA arm and the CMM were at slightly different positions in the two experiments and therefore the *BASE* transformation   had   to   be   identified   in   the   verification experiment. There are two ways to verify the efficiency of Model-3. The first is to identify the model by using 10 new measurements. The second is to identify only **BASE** by using the 10 new measurements while keeping the old transformation ^bT_t. It was found that the former method produces more accurate results.

The mean, maximum and standard deviations of the positioning errors of the selected models are illustrated in Fig. 15.

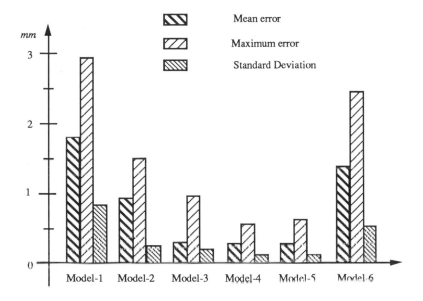

Fig. 15 Mean, Maximum and Standard Deviation of positioning errors
from the verification experiments.

Remark:
1. This particular robot had reasonably good pre-compensation accuracy compared to other PUMA arms given in the literature [15]. The 2-3 mm positioning errors produced by Model-1 as shown in Fig. 15 are however unacceptable for many applications.
2. By carefully identifying the parameters in the transformation **BASE** alone, the mean positioning errors can be reduced to about 1.0 mm (Model-2).
3. If the same number of data points are used to identify the parameters of the complete rather than those of **BASE** alone, the mean and maximum positioning errors are further reduced to about 0.4 mm and 1.0 mm respectively (Model-3).
4. The identification of **BASE** alone has quite a few merits. The identification and compensation efficiency can be improved by employing the identified **BASE** along with the nominal ^bT_t to form a

new "nominal model". This facilitates closed inverse kinematic solutions, and produces more accurate positioning results compared to the original nominal model (Model-1) whose $BASE$ is obtained from Eq. (48).

5. Model-4 and Model-5 further reduce the positioning errors to about 0.4 mm on the average. The fact that both models produce similar compensation results is consistent with the indications given by their Identification Jacobian condition numbers and their error parameters. Further improvements are difficult to obtain due to the measurement noise in the system, finite word length effects etc.

6. Large positioning errors exist with Model-6 although more data points were used to identify it compared to Model-3. This suggests in general that measurements should be taken throughout the entire reachable working volume with the robot in both arm configurations.

2. Compensation Results

The compensation experiments were performed during the same week as the forward verification experiments. Eight sample points were checked for Model-1 through Model-5. Model-6 was dropped based on the poor performance in the forward verification experiments. The Linear Quadratic Regulator algorithm was used to compute the required joint corrections. All the results were obtained without multiple iterations. The models used for checking the compensation results were listed in Tables 2-7 except for the $BASE$ transformation that was re-identified. The compensation results are shown in Fig. 16.

As had been anticipated, the compensation results matched very well with the forward verification results. Many of the comments given above apply here as well.

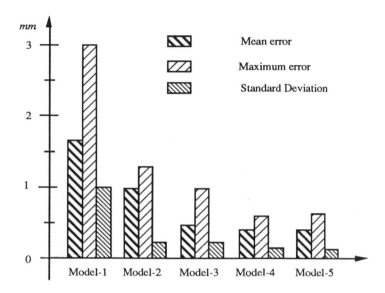

Fig. 16 Mean, Maximum and Standard Deviation of positioning errors
from the compensation experiments.

REFERENCES

1. Mooring, B. W., Z. S. Roth and M. Driels, *The Fundamentals of Robot Calibration*, Book Manuscript (in Print), to appear in Jan. 1991, John Wiley & Sons.

2. Roth, Z. S., B. W. Mooring and B. Ravani (1987), "An Overview of Robot Calibration," *IEEE J. Robotics Automation*, Vol. 3, No. 5, Oct., pp. 377-385.

3. Hollerbach, J. M., "A Survey of Kinematic Calibration," *Robotics Review*, edited by O. Khatib, J. J. Craig and T. Lozano-Perez, MIT Press, 1988.

4. Stone, H. W., *Kinematic Modeling, Identification, and Control of Robotic Manipulators*, Kluwer Academic Publishers, 1987.

5. Hayati, S. A., "Calibration,", *International Encyclopedia of Robotics, Application and Automation*, edited by R. C. Dorf and S. Y. Nof, John Wiley & Sons, Inc., 1988, Vol. 1, pp. 165-167.

6. Lau, K., N. Dagalakis and D. Myers, "Testing," *International Encyclopedia of Robotics, Application and Automation*, edited by R. C. Dorf and S. Y. Nof, John Wiley & Sons, Inc., 1988, Vol. 3, pp.1753-1769.

7. Jiang, B. C., J. T. Black and R. Duraisamy, "A Review of recent Development in Robot Metrology," J. Manufacturing Systems, Vol. 7, No. 4, 1988, pp. 339-357.

8. Mooring, B. W., "The Effect of Joint Axis Misalignment on Robot

Positioning Accuracy," *Proc. ASME Int. Computers in Eng. Conf. Exhibit*, Illinois, 1983, pp. 151-156.

9. Hayati, S. A., "Robot Arm Geometric Parameter Estimation," *Proc. 22th IEEE Int. Conf. Decision Control*, 1983, pp. 1477-1483.

10. Wu, C. H., "A Kinematic CAD Tool for the Design and Control of a Robot Manipulator," *Int. J. Robotics Research*, Vol. 3, No. 1, 1984, pp. 58-67.

11. Chen, J., and L. M. Chao, "Positioning Error Analysis for Robot Manipulators with All Rotary Joints," *Proc. IEEE Int. Conf. Robotics Automation*, 1986, pp. 1011-1016.

12. Vaishnav, R. N., and E. B. Magrab, "A General Procedure to Evaluate Robot Positioning Errors," *Int. J. Robotics Research*, Vol. 6, No. 1, 1987, pp. 59-78.

13. Veitschegger, W. K., and C. H. Wu, "Robot Calibration and Compensation," *IEEE J. Robotics Automation*, Vol. 4, No. 6, Dec. 1988, pp. 643-656.

14. Bennett, D. J. and J. M. Hollerbach, "Identifying the Kinematics of Robots and Their Tasks," *Proc. IEEE Int. Conf. Robotics Automation*, Scottsdale, Arizona, May, 1989, pp. 580-586.

15. Mooring, B. W. and S. S. Padavala, "The Effect of Kinematic Model Complexity on Manipulator Accuracy," *Proc. IEEE Int. Conf. Robotics Automation*, Scottsdale, Arizona, May, 1989, pp. 593-598.

16. Zhuang, H., *Kinematic Modeling, Identification and Compensation of Robot Manipulators*, Ph.D. Dissertation, Florida Atlantic University, December, 1989.

17. Barker, L. K., "Vector-Algebra Approach to Extract Denavit-Hartenberg Parameters of Assembled Robot Arms," *NASA Technology Paper 2191*, August, 1983.

18. Sklar, M., *Metrology and Calibration Techniques for the Performance Enhancement of Industrial Robots*, Ph.D. Dissertation, University of Texas, Austin, Texas, 1989.

19. Everett, L. J., M. Driels and B. W. Mooring, " Kinematic Modeling for Robot Calibration," *Proc. IEEE Int. Conf. Robotics Automation*, (1987), Raleigh, N. C., April, pp. 183-189.

20. Menq, C. H. and Borm, J. H.," Estimation and Observability Measure of Parameter Errors in a Robot Kinematic Model," *Proc. U.S.A.-Japan Symposium on Flexible Automation*, Minneapolis, MN, July, 1988, pp. 65-70.

21. Driels, M. R., and U. S. Pathre, "Significance of Observation Strategy on the Design of Robot Calibration Experiments," *Advances in Computing Research*, 1989.

22. Zhuang, H, Z. S. Roth and F. Hamano, "Observability Issues in Kinematic Error Parameter Identification of manipulators," *Proc. American Control Conf.*, San Diego, CA, may, 1990.

23. Whitney, D. E., C. A. Lozinski, and J. M. Rourke, "Industrial Robot Forward Calibration Method and Results," *ASME J. Dynamic Syst. Meas. Contr.*, Vol. 108, Mar., 1986, pp. 1-8.

24. Shamma, J. S. and D. E. Whitney, "A Method for Inverse Robot Calibration," *ASME J. Dynamic Systems, Measurement, and Control*, Vol. 109, March, 1987, pp. 36-43.

25. Zhuang, H., F. Hamano and Z. Roth, "Optimal Design of Robot Accuracy Compensators," *Proc. 1989 IEEE Int. Conf. Robotics Automation*, Scottsdale, Arizona, May, 1989, pp. 751-756.
26. Vuskovic, M, I, "Compensation of Kinematic Errors Using Kinematic Sensitivities," *Proc. IEEE Int. Conf. Robotics Automation*, Scottsdale, Arizona, May, 1989, pp. 745-750.
27. Huang, M. Z. and A. Gautam, "An Algorithm for On-line Compensation of Geometric Errors in Industrial Robots," *Proc. 2nd Conf. on Recent Advances in Robots*, May, FAU, Boca Raton, FL, 1989, pp. 174-177.
28. Zhuang, H., Z. S. Roth and F. Hamano, "Analysis and Design of Robotic Accuracy Compensators," to appear in *Progress in Robotics and Intelligent Systems*, edited by C. Y. Ho and G. W. Zobrist, Ablex Publishing, 1990.
29. Zhuang, H, Z. S. Roth and F. Hamano, "A Complete and Parametrically Continuous Model for robot manipulators," Proc. *IEEE Int. Conf. Robotics and Automation*, Cincinnati, Ohio, may, 1990.
30. Everett, L. J. and T. W. Hsu, "The Theory of Kinematic Parameter Identification for Industrial Robots," *ASME J. Dynamic Systems, Measurement, and Control*, Vol. 110, No. 1, 1988, pp. 96-100.
31. Roberts, K. S., "A New Representation for a Line," *Proc. Int Conf. Comput. Vision & Pattern Recog.*, 1988, pp. 635-640.
32. Paul, R. P. , Robot Manipulators: *Mathematics, Programming, and Control*, Cambridge, Mass., MIT Press, 1981.
33. Craig. J. J., *Introduction to Robotics; Mechanics and Control*, Addison Wesley, 1986.
34. Golub, G. H. and C. Reinsch, "Singular Value Decompsition and Least-Sqares Solutions," *Handbook for Automatic Computation, Vol. II,* edited by J. II. Wilkinson and C. Reinsch, Springer-Verlag, 1971, pp. 134-151.
35. Marquardt, D. W., "An Algorithm for Least-squares Estimation of Nonlinear Parameters," *J. Soc. Indust. Appl. Math.*, Vol. 11, No. 2, June, 1963, pp. 431-441.
36. Chen S., *Robot Calibration Using Stereo Vision*, M.S. Thesis, Florida Atlantic Univ., December, 1987.

NONLINEAR CONTROL ALGORITHMS
IN ROBOTIC SYSTEMS

T.J. TARN, S. GANGULY

Department of Systems Science and Mathematics

Washington University

St. Louis, MO 63130

A.K. BEJCZY

Jet Populsion Laboratory

California Institute of Technology

Pasadena, CA 91109

I   Introduction

The system aspects of robotics in general and of robot control in particular are clearly manifested through a number of technical facts. First, robot arms are multi-degree-of-freedom mechanical *systems*. Their dynamic model is a *system* of coupled nonlinear differential equations, and robot arm motion requires several actuators acting as an integrated drive *system*. Second, robot tasks are represented through

CONTROL AND DYNAMIC SYSTEMS, VOL. 39

a coupled *system* of kinematic and geometric equations, and a *system* of various sensors relates robot actions to task performance. Third, modern robot control is implemented through computer algorithms embodied in software *systems* that are housed in a *system* of microprocessors with a supplement of electronic components. Fourth, industrial robot applications typically define a *system* set-up, call for two or more robots organized in workcells cooperating in performing work within a production *system*. Moreover, in man-extension or teleoperator applications it is customary to employ two robot arms as a left-right dual-arm *system* which obviously is more than just two single arms put side by side. Fifth, in the development of robot control intelligence the main point and challenge is the definition and algorithmic implementation of hierarchical and relational *system* frames and procedures in the planning and task decomposition programs.

In this paper we focus on robot control algorithms and their implementation in real-time control to show the need and benefits of rigorous *systematic theoretic approach* to robot control problems. We use state space techniques in our treatment of robot control. The state description of a dynamic system contains all information that is necessary to determine the control action to be taken since, by definition of a dynamic system, the future evolution of the system is fully determined by its present state and by future inputs. It should be kept in mind that control theories and practices evolved from the fundamental idea that control inputs to a given system should be determined from the state of the system.

In the subsequent sections we summarize our work and results for the control of a single-arm robot and of multiple-arm robots using a new dynamic feedback control technique directly referenced to task space commands, derived through principles and techniques of differential geometric systems and control theory as described by *Tarn, Bejczy* and coworkers [1; 2; 3; 4; 5], by *Chen* [6], and by *Bejczy, Tarn* and coworkers [7; 8; 9; 10]. In this task-driven dynamic control technique the nonlinear robot arm system is feedback linearized and simultaneously output decoupled by an appropriate nonlinear feedback and a nonlinear coordinate transformation. In section VI, we briefly discuss a computational architecture for implementing dynamic feedback control of robot arms.

The use of nonlinear feedback to decouple a nonlinear system is not new to control engineers. In an earlier work, *Hemami* and *Camana* [11] applied nonlinear feedback control technique to a simple locomotion system and obtained decoupled subsystems. *Hewit* and *Burdess* [12] applied a method of active force control to the problem of obtaining dynamic decoupling of the motion trajectories of a robot arm. However, they assumed that one part of the dynamic equation is completely unknown and introduced accelerometers that provide information about the accelerations of the manipulator joints. They calculate on-line the inertia matrix of the system in order to estimate gravitational, Coriolis and friction forces, which makes the control law unnecessarily complicated. Almost at the same time, an attractive idea of how to compensate for coupling and nonlinearities through a suitable feedback was introduced by *Nicosia, Nicolo* and *Lentini* [13]. For robot systems whose number of inputs equals the number of outputs, they obtained parallel chains of double integrators by means of nonlinear feedback. However, the paper discusses formal manipulations and gives no general results.

Freund [14; 15] presented suitable nonlinear feedback methods for the control of industrial robots. The methods are based on partitioning the dynamic equations of the robot. The overall system behavior that results from the application of this nonlinear control strategy is characterized by decoupled, linear, second-order equations. But the methods have the same restrictions as the method of *Nicosia, Nicolo* and *Lentini*: [13] the number of inputs equals the number of outputs. Moreover, the linearization in general is not in terms of the state of the system but rather in terms of the output. Thus it is difficult to investigate the stability of the system for a broad class of states; and, as stated by the author, they were developed from practical experience with the design of control strategies for a few types of industrial robots.

There are other dynamic control strategies that exist in the literature. For instance, the "computed torque" technique developed by *Markiewicz* [16] and *Bejczy* [17] and called as the "inverse dynamic" technique by *Paul* [18] and by *Raibert* and *Horn* [19]; the "resolved acceleration control" technique developed by *Luh, Walker* and *Paul* [20]; the "operational space" technique developed by *Khatib* [21], and so on. It is beyond the scope of this paper to review and evaluate all of them. Each has its own advantages and disadvantages.

The purpose of this paper is to present dynamic control on a solid system theoretic foundation using techniques from differential geometric system and control theory. It will be shown that this method can thereafter be applied to the control of redundant robot arms and flexible robot arms including those with joint elasticity as well as link deformation. The concise notations enable us to analyze rather complex mathematics that generally arise with flexible robot arms in a very illustrative manner. For a recent result on the control of flexible robot arms, see [33] and [34]. In [31] and [32], the method is applied for redundant systems. We shall elaborate more on this in a subsequent section.

II   General Model of a Robot Arm

We consider a rigid robot arm mechanism with m degrees of freedom. A mathematical model of the complete mechanical system comprises the mechanical part of the system and m actuators that actuate one degree of freedom each. Using homogeneous coordinates together with the Denavit-Hartenberg four-parameter representation of robot arm kinematics, and using Lagrangian techniques to express the dynamical behavior of a rigid (nonflexing) robot arm with m joints results in an equation of the form [17; 22]

$$\tau_i = \sum_{j=1}^{m} D_{ji}(\mathbf{q})\ddot{q}_j + \sum_{j=1}^{m}\sum_{k=1}^{m} D_{jik}(\mathbf{q})\dot{q}_j\dot{q}_k + D_i(\mathbf{q}), \quad i = 1,\ldots,m, \qquad (1)$$

where τ_i is the generalized force generated by or required for the motion of the i-th mechanical joint, q_i is the coordinate of the i-th mechanical joint, $\mathbf{q} = (q_1,\ldots,q_m)$, $D_{ij}(\mathbf{q})$ is the inertia load projections to joint "i" related to the velocity of joint variables "j" and "k", and $D_i(\mathbf{q})$ is the gravity load at joint i. The function definitions of the dynamic projection functions D_i, D_{ij}, and D_{ijk} can be found in [17; 22]. Actuator models are usually given in the following form [23]

$$\ddot{q}_i = a_1\dot{q}_i + f_i\tau_i + b_iu_i, \quad i = 1,\ldots,m \qquad (2)$$

where a_i, f_i and b_i are constants, and u_i is the input for the i-th actuator, $|u_i| \leq k_i$, a positive constant.

Let $x_i = q_i$, $x_{m+i} = \dot{q}_i$, and denote $(x_1,\ldots,x_{2m})$ by x and $(x_1,\ldots,x_m)$ by $\tilde{x}$. From (1) and (2) we then obtain the state equation of the system:

$$\frac{d}{dt}\begin{bmatrix} x_1 \\ \vdots \\ x_m \\ \hline x_{m+1} \\ \vdots \\ x_{2m} \end{bmatrix} = \begin{bmatrix} 0 & I \\ \hline 0 & 0 \end{bmatrix}\begin{bmatrix} x_1 \\ \vdots \\ x_m \\ \hline x_{m+1} \\ \vdots \\ x_{2m} \end{bmatrix} + \begin{bmatrix} 0 \\ \mathbf{E}^{-1}(\tilde{x})\mathbf{H}(x) \end{bmatrix} + \begin{bmatrix} 0 \\ \mathbf{E}^{-1}(\tilde{x})\mathbf{K} \end{bmatrix}\mathbf{u} \quad (3)$$

where

$$\mathbf{E} = \begin{bmatrix} 1 - f_1 D_{11} & -f_1 D_{12} & \cdots & -f_1 D_{1m} \\ -f_2 D_{21} & 1 - f_2 D_{22} & \cdots & -f_2 D_{2m} \\ \vdots & & \vdots & \\ -f_m D_{m1} & -f_m D_{m2} & \cdots & 1 - f_m D_{mm} \end{bmatrix},$$

$$\mathbf{H} = - \begin{bmatrix} a_1 x_{m+1} + f_1\left(\sum_{j=1}^{m}\sum_{k=1}^{m} D_{1jk} x_{m+j} x_{m+k} + D_1\right) \\ \vdots \\ a_m x_{2m} + f_m\left(\sum_{j=1}^{m}\sum_{k=1}^{m} D_{mik} x_{m+j} x_{m+k} + D_m\right) \end{bmatrix},$$

$$\mathbf{K} = \begin{bmatrix} b_1 & 0 & \cdots & 0 \\ 0 & b_2 & \cdots & 0 \\ \vdots & & & \\ 0 & 0 & \cdots & b_m \end{bmatrix}.$$

For simplicity in writing, the dependence of D_i, D_{ij} and D_{ijk} on $\tilde{x}$ has been omitted.

With task description y solved in the joint space, we have the following compact representation for the robot arm control dynamics:

$$\dot{x} = f(x) + \sum_{i=1}^{m} g_i(x) u_i,$$
$$y_i = h_i(\tilde{x}), \quad i = 1,\ldots,r, \quad (4)$$

where the first equation is the state equation and the second is the output or task equation. The task or output equation y_i here is equal to the forward kinematic solution of the robot arm. The functions $f(x)$ and $g_i(x)$ in (4) can be easily identified with the corresponding terms in (3).

Note that (4) is the standard representation of a multi-input and multi-output nonlinear control dynamics problem. Note also that, in the case of a robot arm, x are joint space variables and y_i are task or external work space variables expressed in terms of joint space variables.

The robot arm control model (3) is highly nonlinear and strongly coupled. Therefore, in the case of simultaneous motion of several joints, the motion and the torque (or force) applied at one joint have a dynamic effect on the motion at other joints. Since the dynamic coefficients or projection functions are dependent on the values of the joint variables, the effect of dynamic coupling between motions at different joints will depend on the actual link configuration during motion. Furthermore, we must also notice that, as the velocity increases, we cannot neglect the centripetal and Coriolis forces.

In addition to the above a major role is played by friction and must be compensated for properly. Brush type servo motors typically have a breakaway friction that is about 4 to 6 percent of the full rated torque of the motor. That sets a lower bound on the friction values. Furthermore, hard nonlinearities and possibly destabilizing friction effects at low velocity contribute to the problem of control.

To incorporate a friction compensation, the "sticktion" is determined by exhaustive experimentation as the least amount of current required to move the arm from rest. Thereafter a friction profile was generated and then fine tuned by exhaustive experimentation. Under normal operating conditions of the PUMA 560 robot arm, it was found that the sticktion drops to about 70% of its value within 50 ms and continued at this value when the arm moved at constant velocity until the last 50 ms when the arm decelerated. The sticktion value came up to about 85% of its initial value at this point.

III   Nonlinear Feedback Applied to a Single Robot Arm

We show that for a nonlinear and coupled system with certain task, there exists a feedback such that in a suitable coordinate representation, the system can be converted to a linear and output decoupled system. We show its application to a single robot arm system (4) with functions given by (3).

For mathematical details see [1; 4; 6]. As an example, the general algorithm derived and described in the above references was applied to the second-order dynamic model of the six degree-of-freedom PUMA 560 robot arm with general three degree-of-freedom continuous positioning and three degree-of-freedom continuous orientation tasks [24]. The control dynamics of this problem satisfies the necessary and sufficient conditions for feedback linearization and simultaneous output decoupling. Therefore, there must exist:

1. A nonlinear feedback $u = \alpha(x) + \beta(x)v$ where $\beta(x)$ is invertible, and $u = (u_1, u_2, \ldots, u_6)$ is the joint force or contol vector.

2. A diffeomorphic transformation $T(x)$ such that, after the nonlinear feedback and diffeomorphic transformation, the new system is linear in the Brunovsky canonical form and the outputs are decoupled.

The application of our algorithm as described in [1; 4; 6] yields the following results:

1. The required nonlinear feedback is

$$u = \alpha(x) + \beta(x)v = \tag{5}$$

$$= -K^{-1}E(\tilde{x})J_h^{-1}\begin{bmatrix} L_f^2 h_1(\tilde{x}) \\ \vdots \\ L_f^2 h_6(\tilde{x}) \end{bmatrix} + K^{-1}E(\tilde{x})J_h^{-1}v,$$

where J_h is the Jacobian matrix of the task or output equation $h(\tilde{x})$.

2. The required diffeomorphic transformation is

$$z = T(x) = (h_1(\tilde{x}), L_f h_1(\tilde{x}), \ldots, h_6(\tilde{x}),\ L_f h_6(\tilde{x}))^T, \tag{6}$$

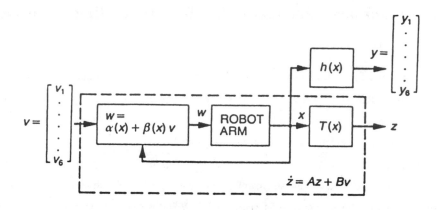

Figure 1: Externally Linearized and Decoupled Robot Arm system when Nonlinear Feedback and Diffeomorphic Transformation are introduced.

where the superscript T signified the transpose.

In the above equations L_f is the Lie derivative, $L_f h = [f, h] = \frac{\partial h}{\partial x} f - \frac{\partial f}{\partial x} h$, with f and h defined in (3) and (4). Note that $\frac{\partial h}{\partial x}$ and $\frac{\partial f}{\partial x}$ are Jacobian matrices. Note also that $L_f^2 h = L_f(L_f h) = [f, L_f h] = [f, [f, h]]$; in general, $L_f^n h = L_f(L_f^{n-1} h) = [f, L_f^{n-1} h]$.

By using the nonlinear feedback and diffeomorphic transformation given above, the six degree-of-freedom PUMA 560 nonlinear dynamic system (3) with six degree-of-freedom trajectory task or output (4) is converted into the following Brunovsky canonical form and simultaneously output decoupled (see also Figure 1):

$$\dot{z} = Az + Bv \stackrel{\triangle}{=} f(z) + \hat{g}(z)v,$$
$$y = Cz \stackrel{\triangle}{=} \hat{h}(z) \tag{7}$$

where

$$
z = \begin{bmatrix} z_1 \\ z_2 \\ \vdots \\ z_{12} \end{bmatrix}, \quad
A = \begin{bmatrix} 0 & 1 & & & & & & \\ 0 & 0 & & & & & & \\ & & \ddots & & & & & \\ & & & 0 & 1 & & & \\ & & & 0 & 0 & & & \\ & & & & & \ddots & & \\ & & & & & & 0 & 1 \\ & & & & & & 0 & 0 \end{bmatrix},
$$

$$
B = \begin{bmatrix} 0 & & & \\ 1 & & & \\ & \ddots & & \\ & & 0 & \\ & & 1 & \\ & & & \ddots \\ & & & 0 \\ & & & 1 \end{bmatrix}, \quad
C = \begin{bmatrix} 1 & 0 & & & & \\ & & \ddots & & & \\ & & 1 & 0 & & \\ & & & & \ddots & \\ & & & & 1 & 0 \end{bmatrix}.
$$

The above matrices A, B and C are of dimension 12×12, 12×6 and 6×12, respectively.

The control results stated in (5), (6) are not restricted to the control of the PUMA 560 arm only. They are much more general. They can be applied to the control of any robot arm as long as the robot arm dynamic model and task description satisfies the characteristics used in deriving (5), (6). This is apparent when one considers the details of proofs elaborated in [1; 4; 6].

We note that equation (7) consists of six $(i = 1, 2, \ldots, 6)$ independent subsystems of the following form:

$$
\dot{z}_i = \begin{bmatrix} 0 & 1 \\ 0 & 0 \end{bmatrix} z_i + \begin{bmatrix} 0 \\ 1 \end{bmatrix} v_i,
$$

$$
y_i = \begin{bmatrix} 1 \\ 0 \end{bmatrix} z_i \tag{8}
$$

where

$$z_i = \left[\begin{array}{c} z_{2i-1} \\ z_{2i} \end{array} \right].$$

We also note that each subsystem (8) has double poles at the origin; thus, the overall system is unstable. To render it stable we add a linear feedback loop F to the system and assign the poles arbitrarily by using the fact that the linear system is completely controllable. It is not difficult to see that, as long as F is a constant block diagonal matrix, the system will remain an output decoupled linear system. We call this new system L and D block in Figure 2.

To stabilize the feedback linearized and output decoupled system by pole assignment, we add a linear PD feedback controller to each subsystem. That is,

$$v_i = \bar{v}_i - F_i z_i, \ i = 1, \ldots, 6,$$

where $F_i = [f_{i1} f_{i2}]$.

Then (8) becomes:

$$\dot{z}_i = \left[\begin{array}{cc} 0 & 1 \\ 0 & 0 \end{array} \right] z_i + \left[\begin{array}{c} 0 \\ 1 \end{array} \right] (\bar{v}_i - F_i z_i) = \left[\begin{array}{cc} 0 & 1 \\ -f_{i1} & -f_{i2} \end{array} \right] z_i + \left[\begin{array}{c} 0 \\ 1 \end{array} \right] \bar{v}_i \quad (9)$$

and

$$y_i = \left[\begin{array}{cc} 1 & 0 \end{array} \right] z_i, \ i = 1, 2, \ldots, 6. \quad (10)$$

Thus the new system (L and D block in Figure 2) consists of six subsystems, each has the second order form (9) with linear output equation (10).

It can be computed easily that the poles of the above subsystem are

$$s_{1,2} = -\xi \omega_n \pm \omega_{nj} \sqrt{1 - \xi^2},$$

where ξ damping ratio, ω_n natural frequency, and $\omega_n^2 = f_{i1}$, $2\xi\omega_n = f_{i2}$.

For the purpose of controller design we consider the above mathematical model (9), (10) as the model of the real system. Thus the desired (nominal) input to each subsystem can be derived from the following equations:

$$\begin{bmatrix} \dot{z}^d_{2i-1} \\ \dot{z}^d_{2i} \end{bmatrix} = \begin{bmatrix} 0 & 1 \\ -f_{i1} & -f_{i2} \end{bmatrix} \begin{bmatrix} z^d_{2i-1} \\ z^d_{2i} \end{bmatrix} + \begin{bmatrix} 0 \\ 1 \end{bmatrix} v^d_i, \tag{11}$$

$$y^d_i = \begin{bmatrix} 1 & 0 \end{bmatrix} \begin{bmatrix} z^d_{2i-1} \\ z^d_{2i} \end{bmatrix}, \quad i = 1, 2, 3, \tag{12}$$

where y^d_i is the desired path, and the superscript 'd' denotes that these equations are the 'model' equations for a robust (or optimal) control system design.

To design a robust (or optimal) system, let us denote the output error by

$$e_i = \begin{bmatrix} e_{i1} \\ e_{i2} \end{bmatrix} = \begin{bmatrix} y_i - y^d_i \\ \dot{y}_i - \dot{y}^d_i \end{bmatrix}.$$

Thus from (9), (10) and (11), (12) we have

$$(\bar{v}_i - \bar{v}^d_i) = (\ddot{y}_i - \ddot{y}^d_i) + f_{i2}(\dot{y}_i - \dot{y}^d_i) + f_{i1}(y_i - y^d_i).$$

That is

$$\Delta \bar{v}_i = \dot{e}_{i2} + f_{i2}e_{i2} + f_{i1}e_{i1}$$

or

$$\dot{e}_i = \begin{bmatrix} 0 & 1 \\ -f_{i1} & -f_{i2} \end{bmatrix} e_i + \begin{bmatrix} 0 \\ 1 \end{bmatrix} (\bar{v}_i - \bar{v}^d_i) \stackrel{\Delta}{=} A_i e_i + b_i \Delta \bar{v}_i. \tag{13}$$

Next we introduce an optimal error-correcting loop by minimizing the following cost function for $\Delta \bar{v}_i$:

$$J(\Delta \bar{v}_i) = \int_0^T \Delta \bar{v}_i R \Delta \bar{v}_i dt + \int_0^T e_i(t)^T Q e_i(t) dt + e_i(T)^T S e_i(T) \tag{14}$$

where R positive definite, Q positive semidefinite matrix, S positive semidefinite matrix, T terminal time.

From the well-known optimal linear control theory the optimal correction is

$$\Delta \bar{v}^0_i = -R^{-1} b_i^T P(t) e_i(t), \tag{15}$$

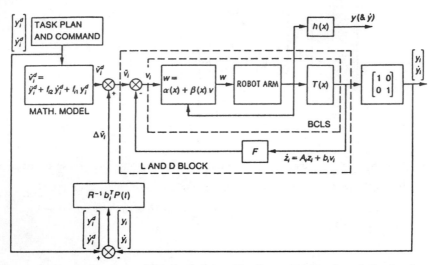

Figure 2: Dynamic Control Method Using Nonlinear Feedback and Optimal Error
Correction in the Task Space.

where

$$P(t) \;=\; \begin{bmatrix} P_{11} & P_{12} \\ P_{12} & P_{22} \end{bmatrix}$$

is a positive definite solution of the Riccati equation:

$$\dot{P}(t) \;=\; -P(t)A_i - A_i^T P(t) + P(t)b_i R^{-1} b_i^T P(t) - Q, \quad P(T) = S.$$

If we consider the steady state solution $(t \to \infty)$ of the above equation then
$\dot{P}(t) = 0$ and the Riccati equation becomes an algebraic equation

$$-PA_i - A_i^T P + Pb_i R^{-1} b_i^T P - Q = 0.$$

that is

$$2f_{i2}P_{12} + R^{-1}P_{12}^2 - Q_{11} = 0,$$

$$-P_{11} + f_{i2}P_{12} + f_{i1}P_{22} + R^{-1}P_{12}P_{22} - Q_{12} = 0,$$

$$-2(P_{12} - f_{i2}P_{22}) + R^{-1}P_{22}^2 - Q_{22} = 0.$$

Thus (15) becomes

$$\triangle \bar{v}_i^0 = -R^{-1}b_i^T P e_i(t), \tag{16}$$

that is,

$$\triangle \bar{v}_i^0 \;=\; -R^{-1}[P_{12}(y_i - y_i^d) + P_{22}(\dot{y}_i - \dot{y}_i^d)]$$

where

$$P_{12} \;=\; -Rf_{i1} \pm R\sqrt{f_{i1}^2 + Q_{11}R^{-1}},$$

$$P_{22} \;=\; -2Rf_{i2} \pm \sqrt{4f_{i1}^2 - (2f_{i1} \mp \sqrt{f_{i2}^2 + Q_{11}} - Q_{11})R^{-1}}.$$

The new strategy for dynamic control of robot arms is summarized in Figure 2 and has been evaluated by a digital simulation technique described in [1; 4; 6; 7; 8; 9].

The design of a new dynamic control strategy for robot arms discussed in this article has three major steps.

1. Convert the original nonlinear robot arm control dynamics into a feedback linearized and simultaneously output decoupled system by using the required nonlinear feedback and diffeomorphic transformation. (See (5), (6), the Brunovsky canonical form, (7), and Figure 1).

2. Stabilize the feedback linearized and simultaneously output decoupled system by designing a linear error correction PD controller for each individual (decoupled) subsystem. (See (9), (10)).

3. Render the control robust versus uncertainties in machine and task model parameter values by adding an optimal error-correcting loop to each individual (decoupled and linear) subsystem. (See (13)-(16) and Figure 2).

An extensive digital simulation program was developed to answer the following main question: How stable and robust is the new overall dynamic control scheme versus model imperfections and other causal uncertainties? Note that the nonlinear feedback is essentially based on robot arm model (geometric and inertia) parameters. In particular, we were interested in evaluating the following cases or in finding quantitative answers to the following questions:

1. If the initial condition is away from the desired path, how does the real trajectory track (or converge to) the desired path?

Table I: Simulation Results for the Control of One PUMA 560 Robot Arm.

Dynamic model parameter error		0%	+5%	−5%	+10%	−10%	+20%	−20%	+30%	−30%
max. position error (mm)	with Δv	0.008	0.23	0.23	0.46	0.46	0.91	0,91	1.37	1.37
	without Δv	0.027	0.77	0.77	1.55	1.55	3.10	3.10	4.64	4.65
max. velocity error (mm/sec)	with Δv	0.024	0,099	0.084	0.19	0.17	0.36	0.35	0.54	0.53
	without Δv	0.053	0.32	0.30	0.61	0.61	1.21	1.21	1.81	1.82
total trajectory time (sec)		6	6	6	6	6	6	6	6	6

2. In general, how far away are the real trajectory and velocity from the desired path and velocity; that is, what is the overall position accuracy of the control scheme?

3. What is the torque (or force) that is needed for each joint?

4. Considering the perturbation because of the actuators, gears and amplifiers, if we change by a certain percentage the torques (or forces) that are applied to the robot arm relative to the required torque (or force), how does this difference affect the behavior of the overall system? This question is equivalent to the situation when the actuator model is missing from the nonlinear feedback.

Table I illustrates some of the simulation results. As seen in this table, even with 20% dynamic model parameters error, the nonlinear feedback combined with optimal error correction produces less than 1 mm maximum position and less than 0.5 mm/sec maximum velocity errors. The average errors are typically less than one quarter of the maximum errors. The control system performance is indeed robust. More simulation results can be found in [8].

The control algorithm for the nonlinear feedback control of robot arms has also been experimentally implemented at the Center for Robotics and Automation at Washington University. The details of that can be found in [25]. Several tasks were performed by the dual arm system at the center and the experimental results are very promising and supplement the theory very well.

IV   Extension of Single Robot Arm Control

The feedback linearized and simultaneously output decoupled dynamic control technique for a single robot arm discussed in the previous section was derived under two assumptions: (i) the output equation 'y' is a trajectory in the task space and (ii) the system dynamics is of second order. These two assumptions, however, do not represent general restrictions for our new dynamic control technique.

A   Force Measurement and Force Control

Suppose that the robot arm not only moves but also interacts dynamically with the environment; for instance, it pushes an object while moving. Suppose also that we have a six d.o.f. force-torque sensor at the base of the robot end effector. Then we can rewrite the robot arm dynamic equation (1) in the following compact form:

$$D(q)\ddot{q} + G(q,\dot{q}) + J^T(q)F = \tau \qquad (17)$$

where F is the interactive force or torque of end effector with environment and $J^T(q)$ is the transpose of the Jacobian matrix of the direct kinematic solution $h(q)$. F vanishes in the absence of force-torque constraints. The force measurement in task space is related to joint variables and joint driving torques by the following general expression:

$$F \;=\; (J(q)J^T(q))^{-1}J(q)\tau_s$$

where τ_s is the portion of joint driving torques that corresponds to the interactive force F in the task space. To compensate for both position and force, we use the following mixed task representation

$$y = W_p h(q) + W_F(J(q)J^T(q))^{-1}J(q)\tau_s \qquad (18)$$

where W_p and W_F are weighting matrices. Equation (18) is a very general task equation. By selecting W_P and W_F properly we can obtain the position task, force task, or hybrid task respectively.

Table II: Simulation Results for Error Reduction by Force Feedback.

Dynamic model parameter error		0%	10%	20%	30%
maximum position error (mm)	without force feedback	1.08	1.21	1.35	1.72
	with force feedback	0.009	0.038	0.077	0.115
maximum orientation error (degree)	without force feedback	0.001	0.071	0.144	0.216
	with force feedback	0.0001	0.0048	0.0097	0.0147

Let $x_i = q_i$, $x_{m+1} = \dot{q}_i$ and denote $(x_1, \ldots, x_{2m})$ by x. From (17), (18) we obtain the following compact representation for the robot arm control dynamics

$$\dot{x} = f(x) + \sum_{i=1}^{m} g_i(x)u_i,$$
$$y = \tilde{h}(\tilde{x}) \tag{19}$$

where the first equation is the state equation and the second is the output or task equation, including the interactive force-torque output in the task space.

Equation (19) satisfies the sufficient conditions for feedback linearization and simultaneous output decoupling as stated in Theorem 3.7 [4]. Applying our algorithm, we can find a nonlinear feedback and a nonlinear state space transformation converting the original nonlinear system (19) into a linear and decoupled system. This linear system consists of m independent subsystems, each of which is a double integrator. Thus we can use the design method described in the previous section to problems involving force feedback. First we stabilize the feedback linearized and output decoupled system by designing a linear error correction PD controller for each individual (decoupled) subsystem. Then we render the control robust versus uncertainties in machine and task model parameter values by adding an optimal error-correcting loop to each individual (decoupled and linear) subsystem.

Computer simulation studies are being conducted to evaluate performance of our new dynamic control technique for force and hybrid position and force output tasks. Table II illustrates the effect of using force and moment measurements in a hybrid position and force control task to reduce position errors. As seen in this table, the effect is dramatic: the maximum position and orientation errors are reduced by more than an order of magnitude when force-moment measurements are applied in the control scheme.

B   Third Order Model of the Arm

Suppose now that, as input to the control system, we take armature voltage applied to actuator motors rather than torques acting on joints. The robot arm system is now described by a set of third order coupled nonlinear differential equations rather than of second order ones. It has been shown [26] that the new third order system equations satisfy the sufficient conditions for feedback linearization and simultaneous output decoupling. Thus, we can find a nonlinear feedback and a nonlinear state space transformation that convert the original third order nonlinear system into a linear and output decoupled system. The high level structure of $\alpha(x), \beta(x)$ and $T(x)$ in the nonlinear feedback (5) and nonlinear state transformation (6) is preserved although the system order increases by one. The increased system order introduces the third order Lie derivatives $L_f^3 h_i$ of h_i along vector field f in the expression of $\alpha(x)$. The diffeomorphism $T(x)$ is constructed by adding the second order Lie derivatives $L_f^2 h_1$, $L_f^2 h_2$ and $L_f^2 h_3$ after the first order Lie derivatives.

It has been shown in the reference quoted above that, for the first three links of the PUMA 560 robot arm, the nonlinear feedback $w = \alpha(x) + \beta(x)v$ and nonlinear state transformation $z = T(x)$ are constructed as follows:

$$\alpha(x) = -DJ_h^{-1} \begin{bmatrix} L_f^3 h_1 \\ Lf^3 h_2 \\ L_f^3 h_3 \end{bmatrix}, \tag{20}$$

$$\beta(x) = DJ_h^{-1} \tag{21}$$

where D and f include now the electric motor constants, and

$$
T(x) = \begin{bmatrix} z_1 \\ z_2 \\ z_3 \\ \hline z_4 \\ z_5 \\ z_6 \\ \hline z_7 \\ z_8 \\ z_9 \end{bmatrix} = \begin{bmatrix} h_1 \\ L_f h_1 \\ L_f^2 h_1 \\ \hline h_2 \\ L_f h_2 \\ L_f^2 h_2 \\ \hline h_3 \\ L_f h_3 \\ L_f^2 h_3 \end{bmatrix} . \tag{22}
$$

The above $\alpha(x)$, $\beta(x)$ and $T(x)$ convert the original third order nonlinear dynamic equation with task space trajectory output equations into the following Brunovsky canonical form

$$
\begin{aligned}
\dot{z} &= Az + Bv \stackrel{\triangle}{=} f + \hat{g}v, \\
y &= Cz \stackrel{\triangle}{=} \hat{h}
\end{aligned} \tag{23}
$$

where

$$
z = \begin{bmatrix} z_1 \\ z_2 \\ z_3 \\ -- \\ z_4 \\ z_5 \\ z_6 \\ -- \\ z_7 \\ z_8 \\ z_9 \end{bmatrix}, \ A = \left[\begin{array}{ccc|ccc|ccc} 0 & 1 & 0 & & & & & & \\ 0 & 0 & 1 & & 0 & & & 0 & \\ 0 & 0 & 0 & & & & & & \\ \hline & & & 0 & 1 & 0 & & & \\ & 0 & & 0 & 0 & 1 & & 0 & \\ & & & 0 & 0 & 0 & & & \\ \hline & & & & & & 0 & 1 & 0 \\ & 0 & & & 0 & & 0 & 0 & 1 \\ & & & & & & 0 & 0 & 0 \end{array} \right],
$$

$$B = \begin{bmatrix} 0 & 0 & 0 \\ 0 & 0 & 0 \\ 1 & 0 & 0 \\ \hline 0 & 0 & 0 \\ 0 & 0 & 0 \\ 0 & 1 & 0 \\ \hline 0 & 0 & 0 \\ 0 & 0 & 0 \\ 0 & 0 & 1 \end{bmatrix}, \ C = \begin{bmatrix} 1 & 0 & 0 & 0 & 0 & 0 & 0 & 0 & 0 \\ \hline 0 & 0 & 0 & 1 & 0 & 0 & 0 & 0 & 0 \\ \hline 0 & 0 & 0 & 0 & 0 & 0 & 1 & 0 & 0 \end{bmatrix}.$$

Note that (23) consists now of three independent linear subsystems of third order, with poles located at the origin. Thus, the overall system is unstable. Again, we render it stable by adding a linear feedback loop F to the system. We assign the poles arbitrarily by using the fact that the linear system is completely controllable. If we select F as a constant block diagonal matrix, the system will remain an output decoupled linear system. Computer simulation studies show interesting performance variations with respect to pole assignments [27]. Table III shows the simulation results when the system poles are assigned at -221.72, -20.73 $\pm$ j 3.13. Simulation results were also obtained by assigning closed-loop system poles such that the resultant system approximates a first order or a second order system. The best results are obtained with poles that lead to a closed-loop system approximating a first order system behavior. Table III also shows simulation results for this last case.

The third order model has also been currently implemented experimentally with the two arm robot system at Washington University's Center for Robotics and Automation. The experimental evaluation of the performance is still in progress and has been very promising to corroborate the theory that has been developed.

Table III: Simulation Results for Third Order System Model. (A: F and Δv Equivalence; B: First Order System Approximation).

parameter coefficient	1.00	0.95	1.05	0.90	1.10	0.80	1.20	0.70	1.30
parameter error	0	−5%	5%	−10%	10%	−20%	20%	−30%	30%
max. pos. error (mm)	0.25	1.33	1.33	2.66	2.66	5.32	5.32	7.77	7.99
max. vel. error (mm/s)	0.36	0.91	0.49	1.52	0.98	2.73	2.11	3.95	3.32
max. accel. error (mm/s)	6.84	6.02	7.91	5.63	9.08	6.20	11.55	8.95	14.12
max. req. voltage (v)	14.73	14.67	14.79	14.62	14.84	14.52	14.96	14.42	15.09
period(s)	6	6	6	6	6	6	6	6	6

B: Simulation of first order system approximation
Parameter coefficient COE = 1.20, i.e., parameter error = 20%

dominating poles	−3	−6	−10	−15	−20	−30	−40	−80	−150
max. pos. error (mm)	1.39	0.70	0.42	0.28	0.21	0.14	0.11	0.01	0.01
max. vel. error (mm/s)	0.81	0.27	0.18	0.12	0.10	0.07	0.06	0.11	0.11
max. accel. error (mm/s)	8.90	8.70	8.66	8.61	8.66	8.76	8.87	12.19	25.83
max. req. voltage (v)	15.03	15.04	15.04	15.05	15.05	15.05	15.05	15.07	15.08
period(s)	6	6	6	6	6	6	6	6	6

Remark: the integration step for dominating poles −80 and −150 is 1 ms, for the other poles is 2 ms.

C   Redundant Robot Arms

Under certain conditions, connected robot arms represent a redundant system. For the control of such systems, an appropriate nonlinear feedback technique can be used which exactly linearizes and output decouples the nonlinear, coupled and redundant robot arm system. See [31] and [32] for details of the work. Such a situation of redundancy may arise when two or more robot arms are working simultaneously on the same workpiece. This is very natural for space applications where the assembly, maintenance and servicing of the space station will require manual work of Extra Vehicular Activity (EVA) astronauts in the Initial Operational Configuration of the space station. This implies two handed manual work of EVA astronauts and, therefore represents a redundant system.

It is shown in [32] that with the application of a suitable pseudoinverse for the system, the redundant arm system becomes input-output decoupled and suitably linearized in a new coordinate frame with the application of the nonlinear feedback control methodology. Thereafter the analysis can be similar to the one applied

earlier to non redundant rigid robot arm. This extension of the proposed methodology illustrates the strength of the control laws for systems that require complex
mathematical analysis.

To illustrate the technique by which our methodology can be extended to the
control of redundant robot arms, we summarize here the major technical steps
which translate our methodology to the control of a single open-chain redundant
robot arm. The basic system and output equations are the following:

$$D(q)\ddot{q} + H(q,\dot{q})\dot{q} + G(q) = \tau, \ \dim q = \dim \tau = m > 6 \tag{24}$$

$$y = h(q), \dim y = p \le 6. \tag{25}$$

Comparing these equations to equations (1) and (4), we see that we have more
control capability ($\tau > 6$) than we strictly need ($y < 6$).

With reference to equation (5) let us define

$$M(x) = J_h E(x)^{-1} K. \tag{26}$$

We can see that the nonlinear feedback is determined from the following algebraic
equations:

$$M(x)u(x) \quad - \quad -L_f^2 h \tag{27}$$

$$M(x)\beta(x) \quad = \quad \gamma. \tag{28}$$

For a nonredundant arm, $M(x)$ is an invertible matrix. For a redundant arm, it is
not invertible because J_h is not invertible. There are two basically different ways
of treating this non invertibility problem.

(i)   Expand the output space (or task space) dimensionality to match the dimen-
      sionality of the system. That will render $M(x)$ invertible. As an example,
      we can consider a seven degree-of-freedom redundant robot arm which can
      be made nonredundant by constraining one of the joints to move along a
      prespecified trajectory.

(ii) Use the pseudo inverse of $M(x)$ to distribute the control in the redundant
      system.

The second methodology is now explained with some mathematical details. In (27), and (28), $L_f{}^2h$ is the second order Lie derivative of h along f,

$$\gamma = \begin{bmatrix} \gamma_1 & & & & \\ & \gamma_2 & & & \\ & & \ddots & & \\ & & & \ddots & \\ & & & & \gamma_p \end{bmatrix} \tag{29}$$

$\gamma_i = [\, 1 \quad 1 \quad \ldots \quad 1 \,]$ is a $1 \times m_i$ row vector with all entries equal to 1 and m_i, $i = 1,\ldots,p$ are chosen such that $m_i > 0$ and $m_1 + m_2 + \cdots + m_p = m$. The index m_i is associated with the fact that a total number of n independent actuators (inputs) are to be divided into p groups to control p outputs. The required nonlinear transformation is given by [1]

$$T(x) = \left[\, h_1 \quad L_f h_1 \quad \ldots \quad h_p \quad L_f h_p \,\right]^T. \tag{30}$$

Since both (27) and (28) are underdetermined, there are infinitely many solutions for them. Any solution serves the purpose of linearization and decoupling provided that $\beta(x)$ is invertible. A solution to (27) is given by

$$\alpha(x) = -M^+(x)L_f^2 h(x) \tag{31}$$

where $M^+ = M^T \left(M M^T \right)^{-1}$ is the generalized inverse of $M(x)$. The general solution to (28) is

$$\beta(x) = M^+(x)\gamma + (I - M^+ M)\, H \tag{32}$$

where H is an arbitrary matrix which is chosen to make $\beta(x)$ invertible.

After applying the nonlinear feedback and the nonlinear coordinate transformation given above, the original system (24) with output (25) is converted into the following linear and decoupled system

$$\begin{aligned} \dot{z} &= Az + Bv, \\ y &= Cz \end{aligned} \tag{33}$$

where

$$A = \begin{bmatrix} A_1 & & \\ & \ddots & \\ & & A_p \end{bmatrix} ; \quad B = \begin{bmatrix} B_1 & & \\ & \ddots & \\ & & B_p \end{bmatrix} ; \quad C = \begin{bmatrix} C_1 & & \\ & \ddots & \\ & & C_p \end{bmatrix} ;$$

$$A_i = \begin{bmatrix} 0 & 1 \\ 0 & 0 \end{bmatrix} ; \quad B_i = \begin{bmatrix} 0 \\ \gamma_i \end{bmatrix} ; \quad C_i = \begin{bmatrix} 1 & 0 \end{bmatrix} ; \quad i = 1, \ldots, p.$$

Note that the linear system obtained in (33) consists of p independent subsystems. The control problem of the whole mechanical system is then simplified to a design problem of individual subsystems. The ith subsystem is defined by

$$\begin{bmatrix} \dot{z}_{2i-1} \\ \dot{z}_{2i} \end{bmatrix} = \begin{bmatrix} 0 & 1 \\ 0 & 0 \end{bmatrix} \begin{bmatrix} z_{2i-1} \\ z_{2i} \end{bmatrix} + \begin{bmatrix} 0 \\ \gamma_i \end{bmatrix} v_i, \tag{34}$$

$$y_i = \begin{bmatrix} 1 & 0 \end{bmatrix} \begin{bmatrix} z_{2i-1} \\ z_{2i} \end{bmatrix}, \quad i = 1, \ldots, p, \tag{35}$$

where v_i is the ith group input with m_i components. To stabilize the subsystem (34, 35), we introduce a constant feedback $v_i = -K_i z_i + \bar{v}_i$ with

$$K_i = \begin{bmatrix} 0 & 0 \\ k_{i1} & k_{i2} \end{bmatrix} ; \quad z_i = \begin{bmatrix} z_{2i-1} \\ z_{2i} \end{bmatrix},$$

and $\bar{v}_i$ is the new reference input. With such a constant feedback, subsystem (34, 35) becomes

$$\begin{bmatrix} \dot{z}_{2i-1} \\ \dot{z}_{2i} \end{bmatrix} = \begin{bmatrix} 0 & 1 \\ -k_{i1} & -k_{i2} \end{bmatrix} \begin{bmatrix} z_{2i-1} \\ z_{2i} \end{bmatrix} + \begin{bmatrix} 0 \\ \gamma_i \end{bmatrix} \bar{v}_i, \tag{36}$$

$$y_i = \begin{bmatrix} 1 & 0 \end{bmatrix} \begin{bmatrix} z_{2i-1} \\ z_{2i} \end{bmatrix}, \quad i = 1, \ldots, p \tag{37}$$

or in compact form

$$\dot{z}_i = \bar{A}_i z_i + B_i \bar{v}_i$$
$$y_i = C_i z_i$$

where $\bar{A}_i$ can be easily identified from (36). For the above system (36, 37), the damping ratio ξ and the natural frequency ω_n are related with the feedback gains by

$$\omega_n{}^2 = k_{i1}; \quad 2\xi\omega_n = k_{i2}.$$

We now consider (36, 37) as the new mathematical model of the real system which is exactly linearized, output decoupled and stabilized. Then the desired (nominal) input to each subsystem can be derived from the following system

$$\begin{bmatrix} \dot{z}^d_{2i-1} \\ \dot{z}^d_{2i} \end{bmatrix} = \begin{bmatrix} 0 & 1 \\ -k_{i1} & -k_{i2} \end{bmatrix} \begin{bmatrix} z^d_{2i-1} \\ z^d_{2i} \end{bmatrix} + \begin{bmatrix} 0 \\ \gamma_i \end{bmatrix} \bar{v}^d_i, \tag{38}$$

$$y^d_i = \begin{bmatrix} 1 & 0 \end{bmatrix} \begin{bmatrix} z^d_{2i-1} \\ z^d_{2i} \end{bmatrix}, \quad i = 1,\ldots,p \tag{39}$$

where the superscript 'd' indicates desired quantities. From (38, 39), the desired input can be obtained in terms of the desired task space trajectory.

$$\gamma_i \bar{v}^d_i = \ddot{y}^d_i + k_{i2}\dot{y}^d_i + k_{i1}y^d_i, \quad i = 1,\ldots,p. \tag{40}$$

It is observed that the left hand side of (40) is the sum of m_i inputs in the task space computed from the planned trajectory. For a given planned trajectory at any instant of time, the right hand side of (40) is a given value. Applying the generalised inverse, we obtain,

$$\bar{v}^d_i = \gamma^T_i \left(\gamma_i \gamma^T_i\right)^{-1} \left(\ddot{y}^d_i + k_{i2}\dot{y}^d_i + k_{i1}y^d_i\right). \tag{41}$$

It can therefore be seen from (41) that the problem with redundant robot arms has been reduced to the same formulation as that of nonredundant rigid robot arms which has been discussed in section III. Further analysis of the error can be done in the same way as in equation (13) onwards and a robust optimal controller can be designed in the same manner.

D   Flexible Robot Arms

Modeling and control of robot arms with non-negligible elastical deformations have invited researches in recent years due to the demands of robot arms which are of lighter weight, move faster and consume less energy. It can be shown that with the use of a nonlinear feedback law, input-output linearization and decoupling can be achieved for a flexible robot arm including those with joint elasticities and link deformation for both rotary and prismatic joints and can even incorporate the effect of some flexible part such as flexible end effector with nonlinear elasticity.

This is an extension of the input-output linearization and decoupling of finite dimensional rigid robot arms. Stabilizability of the system under feedback can then be studied in the light of the concept of zero dynamics and perturbation theory of infinite dimensional systems. The novelty in this extension is the fact that the complex mathematics that arise from the analysis of flexible robot arms as an infinite dimensional system can be dealt with very effectively and in an illustrative manner, and can be turned into a finite dimensional system problem. This provides more flexibility for the purpose of control as well as for the analysis of the system. Mathematical details are not quoted here since the mathematical model even for the simpler cases is rather involved. Therefore the interested reader is referred to [33] and [34].

V Coordinated Control of Multiple Robot Arms

The feedback linearized and simultaneously output decoupled dynamic control technique is also applicable to the coordinated control of two or more dynamically cooperating robot arms. Dynamic cooperation implies that two (or more) robot arms are working on the same object simultaneously. For the use of our dynamic control technique for dynamic control coordination of two (or more) robot arms, two different approaches are possible. In the first approach, the cooperating arms are regarded as open loop kinematic chains with kinematic and dynamic constraints of points on the object on which they perform work. In the second approach, the two (or more) arms are regarded as one kinematic unit forming a closed loop kinematic chain. In the first approach, two different methods are possible dependent on sensor instrumentation of the robot arm. The first method uses a dynamic coordinator only acting on relative position and velocity task space errors between two (or more) robot arms. The second method also acts on relative force-torque errors between two (or more) arms as sensed at the end effectors, provided that such sensing is available.

A   Coordinated Control of Distinguished Arms

In the first control strategy approach, the robot arms are distinguished from each other, and each can be feedback linearized and output decoupled individually, thus each forms a linear and output decoupled system. Suppose we have two robot arms denoted as robot "a" and "b". Following the linearization and output decoupling by nonlinear feedback and state transformation, robot "a" is described as a linear and decoupled system [2] as

$$
\begin{aligned}
\dot{z}^a &= A^a z^a + B^a v^a, \\
y^a &= C^a z^a.
\end{aligned}
\tag{42}
$$

Similarly, robot "b" is described by

$$
\begin{aligned}
\dot{z}^b &= A^b z^b + B^b v^b, \\
y^b &= C^b z^b.
\end{aligned}
\tag{43}
$$

The first control method in the "distinguished arms" coordination strategy considers the two arms in a "leader/follower" mode. Robot "a" is assigned to be the "leader" and robot "b" the "follower". We design a feedback for robot "a" such that it tracks the desired trajectory, and an optimal coordinator for robot "b" such that robot "b" will follow robot "a" with a constant offset distance. The coordinator is operating on the difference between outputs of the two robots, minus the constant offset distance. Each system described by equations (42) and (43) consists of six independent subsystems of the following form

$$
\begin{aligned}
\dot{z}_i^j &= \begin{bmatrix} 0 & 1 \\ 0 & 0 \end{bmatrix} z_i^j + \begin{bmatrix} 0 \\ 1 \end{bmatrix} v_i^j, \\
y_i^j &= \begin{bmatrix} 1 & 0 \end{bmatrix} z_i^j, \quad i = 1, \ldots, 6, \ j = a, b,
\end{aligned}
\tag{44}
$$

where

$$
z_i^j = \begin{bmatrix} z_{2i-1}^j \\ z_{2i}^j \end{bmatrix}, \quad j = a, b.
$$

We stabilize both robots with the same constant feedback. Then the new subsystems have the following second order form:

$$\dot{z}_i^j = \begin{bmatrix} 0 & 1 \\ -f_{i1} & -f_{i2} \end{bmatrix} z_i^j + \begin{bmatrix} 0 \\ 1 \end{bmatrix} v_i^j,$$

$$y_i^j = \begin{bmatrix} 1 & 0 \end{bmatrix} z_i^j, \quad i = 1,\ldots,6, \quad j = a,b. \tag{45}$$

It is noted that the damping ratio ξ and natural frequency ω_n of this second order system are so that $\omega_n^2 = f_{i1}, 2\xi\omega_n = f_{i2}$. We can consider (45) as the new mathematical model of the real system which is now exactly linearized, output decoupled and stabilized.

The design of a feedback for master robot "a" to track a desired trajectory is the same as for a single robot arm. To design the optimal coordinator, we denote the constant offset distance between robot "a" and "b" by a vector $d = (d_1,\ldots,d_6)^T$. The error between the outputs of two robots is defined by

$$e_{2i-1}(t) = y_i^a(t) - y_i^b(t) - d_i. \tag{46}$$

The objective of coordinator design is to eliminate the error.

By differentiating the output equations of (45) combining it with (16) and letting $e_{2i} = \dot{e}_{2i-1}$, we obtain the dynamic equation for the error:

$$\begin{bmatrix} \dot{e}_{2i-1} \\ \dot{e}_{2i} \end{bmatrix} = \begin{bmatrix} 0 & 1 \\ -f_{i1} & -f_{i2} \end{bmatrix} \begin{bmatrix} e_{2i-1} \\ e_{2i} \end{bmatrix} + \begin{bmatrix} 0 \\ 1 \end{bmatrix} (v_i^a - v_i^b - f_{i1}d_i) \tag{47}$$

$$\stackrel{\triangle}{=} A_i \begin{bmatrix} e_{2i-1} \\ e_{2i} \end{bmatrix} + b_i\triangle\bar{v}_i.$$

In order to eliminate the error $e_i(t) = (e_{2i-1}(t)e_{2i}(t))^T$, we choose the following cost function

$$J = \int_0^T (\triangle\bar{v}_i R\triangle\bar{v}_i + e_i(t)^T Q e_i(t))dt + e_i(T)^T S e_i(T).$$

The optimal coordinator is then given by

$$\triangle\bar{v}_I = -R^{-1}b_i^T P(t)e_i(t)$$

Table IV: Simulation results for Coordinated Control of Two PUMA 560 Robot
Arms.

dynamic model parameter errors	robot "a" 0%		+10%	−10%	−20%	+30%
	robot "b" 0%		+10%	+10%	+20%	+30%
max. position error (mm)	with Δv	0.008	0.42	0.76	1.52	1.25
	without Δv	0.023	1.20	2.20	4.50	3.60
max. relative velocity error (mm/sec)	with Δv	0.045	0.23	0.20	0.45	0.65
	without Δv	0.137	0.63	0.61	1.20	1.80
total trajectory time (sec)		6	6	6	6	6

where $P(t)$ is a positive definite solution of the Riccati equation

$$\dot{P}(t) = - P(t)A_i - A_i^T P(t) + P(t)b_i R^{-1} b_i^T P(t) - Q,$$

with

$$P(T) = S.$$

The robustness of our new dynamic control scheme is illustrated again in Table IV which shows the effect of assigned dynamic parameter variations relative to the nominal (or "ideal") values on the tracking performance. The results are related to the simulation of the "optimal dynamic coordinator".

The second control method in the distinguished arm's coordination strategy uses the force-torque output and feedback technique briefly described by (17)-(19) in the first part of Section III of this paper. Using the force feedback and output framework we can obtain a control coordination strategy for two (or more) robot arms working on the same object in a "worker/coworker" mode.

In the case of two cooperating robot arms, we first incorporate the mass of the object into the dynamic projection parameters of one of the two robot arms, say robot "a". Let $p(q_a)$ be the position and/or orientation of the object, and let F

be the force and/or torque sensed at the end effector of robot "b". The dynamic equation of motion for robot "b" is

$$D_b(q_b)\ddot{q}_b + E_b(q_b, \dot{q}_b) + J_b^T(q_b)F = \tau_b \qquad (48)$$

and the dynamic equation of motion for robot "a" is

$$D_a(q_a)\ddot{q}_a + E_a(q_a, \dot{q}_a) - J_a^T(q_a)F = \tau_a. \qquad (49)$$

Considering the enlarged output equation of the form

$$y = \begin{bmatrix} p(q_a) \\ F \end{bmatrix}. \qquad (50)$$

we can apply the feedback linearization and decoupling method to the above system such that the inputs τ_a only will regulate the outputs $p(q_a)$ and the inputs τ_b only will regulate the output F. Note that the solution of this problem has application in cases where the second robot arm has to support *dynamically* the actions of the first robot arm which are defined in *geometric* terms.

In the simulations, the interactive force is generated by multiplying the relative position error between the two end effectors with a stiffness matrix. In the force feedback control formulation, we take an output equation in the following form as in (50):

$$y = \begin{bmatrix} P_a \\ W_p P_b + W_F F_b \end{bmatrix}.$$

The first robot is purely position servoed in the task space while the second robot is both position and force servoed. It is found that better performance can be achieved by appropriately choosing the force output weighting matrix W_F for the second robot arm. Using 10 ms integration step size, the simulations show that an improvement of five times over the relative position error is made by adopting the force feedback in different dynamic model variations. It is worth noting that the maximum position error of robot "a" is of the order of 1.0 to 4.0 mm. However, the maximum relative position error between he two robots is less than a half-millimeter in most cases. This means that robot "b" is following robot "a" very closely to reduce the interactive force error. See Table V.

Table V: Simulation Results of Force Feedback Control

coe		force feed-back ?	max. rel. position error (mm)	max. force error (Newton)	max. posi error of 'a' (mm)	max. velo. error of 'a' (mm/s)	max. torque of 'a' (Nm)	max. torque of 'b' (Nm)
'a'	'b'							
1.00	1.00	yes	0.00	0.00193	0.00	0.00	46.95	47.28
		no	0.00	0.0056	0.00	0.00	43.58	43.95
1.10	0.90	yes	0.46	2.1886	1.17	0.35	47.92	39.55
		no	2.35	11.08	1.17	0.35	47.91	39.58
1.20	0.80	yes	0.95	4.5000	2.34	0.71	52.23	33.73
		no	4.70	22.174	2.34	0.71	52.23	35.21
1.30	0.90	yes	1.04	4.925	3.91	1.06	60.34	42.18
		no	5.22	24.614	3.91	1.06	60.34	40.20
0.80	0.70	yes	0.27	1.231	2.61	2.87	37.62	33.12
		no	1.31	6.168	2.61	2.87	37.62	33.17
1.30	1.20	yes	0.27	1.277	3.51	1.06	56.55	52.73
		no	1.33	6.291	3.51	1.06	56.55	52.67
1.10	1.30	yes	0.46	2.190	1.17	0.36	47.91	56.89
		no	2.34	11.05	1.17	0.36	47.91	57.02
0.90	1.20	yes	0.69	3.278	1.17	0.35	39.25	52.71
		no	3.52	16.62	1.17	0.35	39.25	52.67

Note: integration step = 10 ms. natural frequency ω_n of overall closed-loop system = 25.0 damping ration ξ of overall closed-loop system = -.9, corresponding position feedback gain k_1 = 625.0, corresponding velocity feedback gain k_2 = 45.0. poles of each subsystem of overall closed-loop system: $s_1 = -11.60$, $s_2 = -33.40$.

B Coordinated Control of Indistinguished Arms

In the second control strategy approach, the two (or more) robot arms are indistinguished from each other; they form a closed loop kinematic chain. For tasks of lifting a heavy workpiece using robot arms, two or more robots are required if the workpiece is out of loading limit of any available robot arm. Suppose that m robot arms are used in such a task and that they all grasp on the same object (workpiece) in order to lift it, turn it, etc. Our primary concern is to obtain a dynamic model of these robots for the control purpose. Since they grasp on the same object, the dynamic behavior of one robot is not independent of the dynamic behavior of the other robots any more. Rather, a unity of mechanical system is formed by the robot arms involved and by the grasped object.

We will derive the Lagrange's equations of motion for this mechanical system.

Those equations will serve as a model of the system to design control algorithms. For the m robots under consideration, we name them robot 1, robot 2, ..., and robot m, respectively. We assume that robot i has n_i joints (a joint connects the two adjacent links). Then robot i has (n_i+1) links with the first link being the base fixed in the work space. We also assume that each robot firmly grasps the object so that there is no movement between its end effector and the object. Closed chains are formed in such a configuration by the m robot arms, the object, and the ground. Notice that the object and the last links of the robot arms become a single link. From the Kutzbach-Grubler criterion [18], the degrees of freedom of a spatial linkage structure connected by joints with each joint possessing one degree of freedom are given as follows

$$p = 6(i-1) - 5j$$

where i is the number of links and j is the number of joints. This formula reflects the fact that each moving link has six degrees of freedom and the fixed link (the ground) has none, and that each joint of one degree of freedom causes a loss of five degrees of freedom for a link. For our case of m robots, the degrees of freedom of this entire mechanical system is then

$$
\begin{aligned}
p &= 6\left[\sum_{k=1}^{m}(n_k-1)+1\right] - 5\sum_{k=1}^{m} n_k \\
&= \sum_{k=1}^{m} n_k - 6m + 6
\end{aligned}
\tag{51}
$$

where n_k is the number of joints of robot k.

The generalized coordinates q can be chosen arbitrarily as long as they are linearly independent of each other. In practice, we choose q in such a way that the inverse relation mentioned below is simple. For example, a closed chain formed by two robots, one having six joints and the other one having five joints, will form a system with five degrees of freedom. For the generalized coordinates of the closed chain system, we can choose the joint variables of the robot which has five joints. Then the inverse relation can be obtained by using the inverse kinematics of the robot with six joints. The generalized coordinates are functionally related to the joint variables θ. We denote the relation by

$$q = Q(\theta). \tag{52}$$

Knowing the generalized coordinates q, the configuration of the mechanical system, thus the joint variable θ, is uniquely determined. We denote such inverse relation by

$$\theta = \Theta(q). \tag{53}$$

It is worth mentioning that the principle of choosing the generalized coordinate q is to make the inverse relation Θ as simple as possible. With the above notations, the Lagrange's equations of motion for the closed chain mechanical system is described by

$$\left(\frac{\partial \Theta}{\partial q_i}\right)' \left(\left[\frac{\partial^2 L}{\partial \theta^2}\frac{\partial \Theta}{\partial q}q + \frac{\partial^2 L}{\partial \theta^2}\right] \begin{bmatrix} \dot{q}'\frac{\partial^2 \Theta_1}{\partial q^2}\dot{q}' \\ \vdots \\ \dot{q}'\frac{\partial^2 \Theta_n}{\partial q^2}\dot{q} \end{bmatrix} + \frac{\partial^2 L}{\partial \theta \partial \dot{\theta}}\frac{\partial \Theta}{\partial q}\dot{q} - \left(\frac{\partial L}{\partial \theta}\right)'\right) = \left(\frac{\partial \Theta}{\partial q_i}\right)' F,$$
$$i = 1, 2, \ldots p \tag{54}$$

where L is the Lagrangian of the whole mechanical system. Equation (54) is a generalization of the equations of motion of the two robot arms.

We assign a coordinate frame to each link of every robot arm. We locate a world coordinate frame in the common work space of the m robots. In the process of expressing the kinetic and potential energies of the mechanical system, we divide the mass of the object into m parts. The division can be made in accordance with the load capability of each individual robot. The geometry of the object may as well be considered in the process of dividing the mass of the object. Each robot is responsible for one part of the object mass by adding it to the mass of the last link. After carrying out the derivations of the Lagrangian function, we obtain the dynamic equations of the mechanical system

$$D(q)\ddot{q} + E(q, \dot{q}) + G(q) = J'_\theta F \tag{55}$$

where

$$D(q) = J'_\theta \tilde{D}(\Theta(q)) J_\Theta$$

$$J_\Theta = \frac{\partial \Theta}{\partial q}$$

$$\tilde{D}(\theta) = \begin{bmatrix} D^1 & & & & \\ & D^2 & & \text{\Large 0} & \\ & & \cdot & & \\ & \text{\Large 0} & & \cdot & \\ & & & & D^m \end{bmatrix}$$

where $D^r = \{D^r_{ij}\}$ is the inertia matrix of robot r

$$D^r_{ij} = \sum_{k=\max\{i,j\}}^{n_r} Tr\left(\frac{\partial T^r_k}{\partial \theta^r_i} I^r_k \frac{\partial (T^r_k)'}{\partial \theta^r_j}\right)$$

$$E(q,\dot{q}) = J'_\Theta \tilde{D}(\Theta(q)) \begin{bmatrix} \dot{q}' \frac{\partial^2 \Theta_1}{\partial q^2} \dot{q} \\ \vdots \\ \dot{q}' \frac{\partial^2 \Theta_n}{\partial q^2} \dot{q} \end{bmatrix}$$

$$+ J'_\theta \begin{bmatrix} \dot{q}' J'\Theta E_1 \\ \vdots \\ \dot{q}' J'_\Theta E_n \end{bmatrix} J_\Theta \dot{q}$$

$$E_i = \begin{bmatrix} E^1_i & & & & \\ & E^2_i & & \text{\Large 0} & \\ & & \cdot & & \\ & \text{\Large 0} & & \cdot & \\ & & & & E^m_i \end{bmatrix}, \quad i = 1,\ldots,n$$

with $E^r_i = \{E^r_{ijk}\}$ is the coefficient of centripetal ($j = k$) or Coriolis ($j \neq k$) force of robot r

$$E^r_{ijk} = \sum_{s=\max\{i,j,k\}}^{n_r} Tr\left(\frac{\partial T^r_s}{\partial \theta^r_k \partial \theta^r_j} I^r_s \frac{\partial (T^r_s)'}{\partial \theta^r_i}\right)$$

$$G(q) \;=\; -J'_{\Theta} \begin{bmatrix} G^1 \\ G^2 \\ \vdots \\ G^m \end{bmatrix}.$$

$$G^r \;=\; \begin{bmatrix} G_1^r \\ G_2^r \\ \vdots \\ G_{n_r}^r \end{bmatrix} \quad \text{is the gravity force of robot } r$$

$$G_i^r \;=\; \sum_{k=i}^{n_r} m_k^r g' \frac{\partial T_k^r}{\partial \theta_i^r} \bar{r}_k^r.$$

In the above definitions, $T_i^r = A_{01}^r A_{12}^r \ldots A_{(i-1)i}^r$, where A_{ij}^r is the Denavit - Hartenberg homogeneous transformation matrix from coordinate frame i to coordinate frame j of robot r; m_i^r is the mass of link i of robot r; $\bar{r}_i^r$ is the mass center of link i of robot r; I_i^r is the pseudoinertia matrix of link i of robot r; g is the acceleration due to gravity, defined to be a 4×1 column vector with the last component being equal to zero.

Equation (55) characterizes the dynamic behavior of the whole mechanical system. However, this equation is nonlinear, coupled, and complicated. It poses great difficulty in controller designs.

We take the position (orientation) of the object handled as the system output

$$y = h(x^1) = \Big[h_1(x^1) h_2(x^1) \ldots h_p(x^1) \Big]'. \tag{56}$$

The first half of the vector y represents the position and the second half represents the orientation of the object. To perform linearization and output block decoupling for the system (55) with output equation (56), we can now use the algorithm developed by us [1; 4; 6] to find the required nonlinear feedback and the required nonlinear coordinate transformation. The control problem of the two arm closed chain is then simplified to a design problem of linear systems.

Note that for $p = 6$ the obtained linear system consists of six independent subsystems. Since each subsystem is controllable, we may locate the poles of each subsystem by adding a constant feedback. As we have done for one arm control system, an optimal correction loop may also be designed to reduce the tracking error and to improve the robustness against model uncertainties.

This formulation has the advantage of automatically handling the coordination and load distribution between two robot arms through the dynamic equations. By choosing independent generalized coordinates, kinematic and dynamic constraints have been taken into account in the process of deriving the equations of motion.

The considerations presented above for a closed chain two-arm robot system can be extended to closed chain multiple-arm robot systems. We now consider the multiple-arm robot as a single mechanical system consisting of closed kinematic chains. The closed kinematic chains are formed through the common ground or base and through a single object to which each robot arm is rigidly connected with its end effector. Since the "m" robots grasp on the same object rigidly, the dynamic behavior of one robot is not independent of the dynamic behavior of the other robots any more. The "m" robot arms and the grasped object form now a unity of mechanical system. Our primary concern is, therefore, to obtain dynamic model of this new closed chain system for the purpose of dynamic control. (See Figure 3 and Figure 4.)

VI Implementation

As seen in Figure 2, the implementation of the nonlinear feedback requires the representation of the desired task space motion in the form of linear and decoupled differential equations. If the robot is modeled as a second order system, then the desired motion input equations are of second order. If the robot is modeled as a third order system, then the desired motion input equations are of third order.

For the purpose of illustrating the implementation of the nonlinear feedback, let us take a second order system represented for the robot arm model and let the robot end effector be required to travel along a straight line segment $\gamma(t)$ in the task space from point A at time $t_0 = 0$ to point B at time t_f such that the end effector is at rest at points A and B. This implies $\dot{\gamma}(t_0) = \dot{\gamma}(t_f) = 0$. We would also like a smooth robot motion that implies that the acceleration and its time derivative should be continuous along c and such that $\ddot{\gamma}(t_0) = \ddot{\gamma}(t_f) = 0$. We select

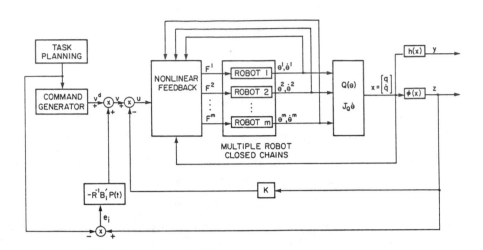

Figure 3: Schematics of Dynamic Control Method in the Closed Chain Model
Approach.

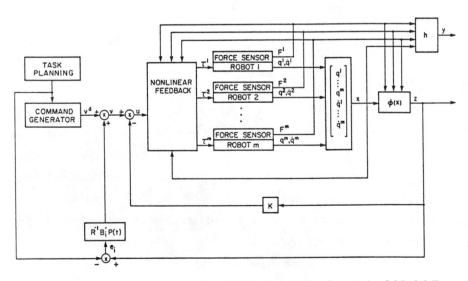

Figure 4: Schematics of Dynamic Control Method in the Constrained Model Force
Control Approach.

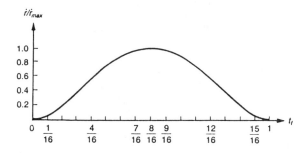

Figure 5: Velocity Profile of a Fifth Order Polynomial Used for Motion Planning.

a fifth order polynomial representation of the line segment AB. Thus, we have:

$$\tilde{\gamma}(t) = a_5 t^5 + a_4 t^4 + a_3 t^3 + a_2 t^2 + a_1 t + a_0 \tag{57}$$

with conditions

$$\tilde{\gamma}(t_0) = A, \quad \tilde{\gamma}(t_f) = B,$$
$$\dot{\tilde{\gamma}}(t_0) = \dot{\tilde{\gamma}}(t_f) = \ddot{\tilde{\gamma}}(t_0) = \ddot{\tilde{\gamma}}(t_f) = 0.$$

For simplicity in computation, we select a spherical reference frame for $\tilde{\gamma}(t)$ with origin at A. The line segment is now a ray frame with origin $S = S(A)$ running to B with representation $(r(t), \Phi_B, \theta_B)$ where Φ_B and θ_B are constants. Thus, we only have to find a fifth order polynomial representation for $r(t)$. Denoting the length of AB by L, we find from six algebraic equations in t under the specified initial and terminal conditions that

$$r(t) = \frac{6L}{t_f^5} t^5 - \frac{15L}{t_f^4} t^4 + \frac{10L}{t_f^2} t^3. \tag{58}$$

We note that this $r(t)$ has a very desirable velocity profile. (See also Figure 5). It reaches maximum value at $t_f/2$ where the acceleration becomes zero. We also have the $\ddot{r}(t) > 0$ for $t < t_f/2$ and $\ddot{r}(t) < 0$ for $t > t_f/2$. These facts also facilitate the handling of cases when, along with the value of L, the maximum velocity is given instead of t_f. In that case we have to find t_f by solving a system of four algebraic equations which always is solvable since the coefficients of the four equations form linearly independent vectors.

Having obtained $r(t)$, we can easily transform the representation of AB from the spherical frame to a Cartesian world frame by

$$
\bar{\gamma}(t) = \begin{bmatrix} x(t) \\ y(t) \\ z(t) \end{bmatrix} = \begin{bmatrix} r(t)\sin\theta_B\cos\Phi_B + x_A \\ r(t)\sin\theta_B\sin\Phi_B + y_A \\ r(t)\cos\theta_B \end{bmatrix} \tag{59}
$$

which is then the position task control input to our dynamic control scheme.

Suppose now that we have an orientation requirement for the robot end effector (e/e) while it is moving from A to B as follows:

At t_0 : e/e roll axis is parallel to x axis of task space.

For $t \leq t_1$: orientation is unchanged.

For $[t_1, t_2]$: e/e roll axis is changed to become parallel to Z axis of task space,

where $t_0 \leq t_1 \leq t_2 \leq t_f$ for given t_1 and t_2.

For $t \geq t_2$: orientation is unchanged.

This orientation requirement as stated is not representable as a closed function of time. Therefore, as proposed in [10], we assign parallel coordinate frames along AB so that the orientation can be written as

$$
v(t) = [s(t), w(t)] \text{ for } t_0 \leq t \leq t_f
$$

This description effectively projects the orientation vectors into one single $\mathbb{R}^3$ space W.

Without loss of generality, we assume that orientation vectors are unit vectors, and we choose the path of them in W to be the arc of radius 1 in the first quadrant in the $X - Z$ plane. Then we only have to find a representation for $w(t)$. We have:

$$
w = \begin{bmatrix} 1 \\ 0 \\ 0 \end{bmatrix} \text{ for } t_0 \leq t \leq t_1,
$$

$$
w = \begin{bmatrix} 0 \\ 0 \\ 1 \end{bmatrix} \text{ for } t_2 \leq t \leq t_f.
$$

To provide smooth transition in orientation, we require that

$$
\dot{w}(t_1) = \dot{w}(t_2) = \ddot{w}(t_1) = \ddot{w}(t_2) = 0.
$$

After some simple algebraic computations [30], and having introduced again a fifth order polynomial to represent $z(t)$ similar to the case of $r(t)$, we obtain for the orientation task:

$$\tilde{w} = \begin{cases} x(t) \\ y(t) \\ z(t) \end{cases} , \quad where \tag{60}$$

$$x(t) = \sqrt{1 - z^2(t)} \text{ for } 0 \le t \le t_f,$$

$$y(t) = 0 \text{ for } 0 \le t \le t_f,$$

$$z(t) = \begin{cases} 0 & \text{for } 0 \le t \le t_1 \\ (t - t_1)^3[p_5(t - t_1)^2 + p_4(t - t_1) + p_3] & \text{for } t_1 < t \le t_2 \\ 1 & \text{for } t_2 < t \le t_f \end{cases}$$

with

$$p_5 = \frac{6}{(t_2 - t_1)^5}, \quad p_4 = \frac{15}{(t_2 - t_1)^4}, \quad p_3 = \frac{10}{(t_2 - t_1)^3}.$$

Equation (60) is then the orientation task control input to our dynamic control scheme. This orientation representation has the same desirable properties as did the position representation discussed earlier.

In summary, (59) and (60) together with their time derivatives *directly drive* the robot arm through a linear system model in our feedback liearized and simultaneously output decoupled dynamic control scheme which is based on nonlinear feedback and nonlinear state transformation.

VII Computer Control Architecture

The new dynamic control technique and the corresponding new task planning procedure described in this paper are computational schemes and algorithms. Their implementation requires the use of high-performance computers in the run-time and real-time control.

Based on the type of computations involved in the nonlinear feedback, optimal error correction, dynamic coordination and task planning, a distributed but closely integrated computing architecture suggest itself, similar to the one described by

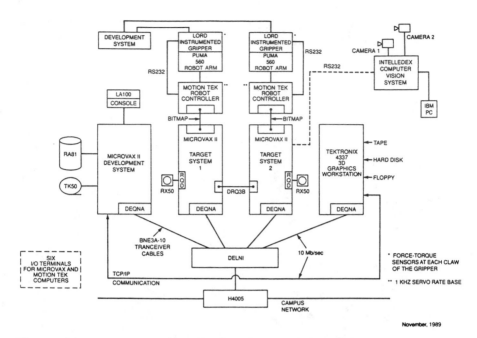

Figure 6: Computer Control Architecture for a Dual-Arm Robot System at Washington University

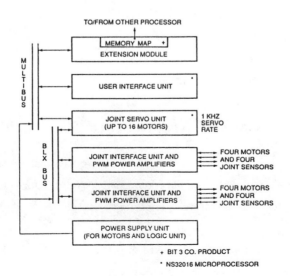

Figure 7: Motion Tek Robot Controller Architecture Schematics

Bejczy and Szakaly [28]. This architecture is presently operational at the Center for Robotics and Automation at Washington University. It uses several CPUs in a multibus frame and uses a robot motor control subsystem with direct BLX bus connection to the motor control CPU housed in the multibus as indicated in Figure 6. This figure also shows the computer control architecture for a dual arm (PUMA 560) system control augmented with LORD instrumented (force-torque sensors) grippers, Intelledex computer vision system and a Tektronix graphic workstation. In Figure 7, the Motion Tek robot controller architecture schematics is presented which clearly shows its modular architecture.

Multiple bus masters operating in a closely coupled computing environment enable sharing of information from various sources within a processing node. The ability to share memory space makes it easy to synchronize multiple processors in order to coordinate robot control and sensor data handling for control. A point in this case is also that motor control is integrated into the multibus.

The robot motor control subsystem provides much richer motor control related sensing capabilities than today's industrial robot motor controllers do. The rich sensing capabilities make it possible to calculate the following quantities in real time: joint position, joint speed, joint acceleration, motor torque, load on motor, motor back-EMF, motor electrical resistance, motor winding temperature, motor power supply status. All these quantities are either needed or desired in our dynamic control scheme.

Our dynamic control technique has just been implemented [25] and its performance is currently being evaluated. The results obtained so far corroborate our earlier simulation results very well and are very promising. However a considerable amount of work is still left before the implementation reaches a presentable level of maturity.

VIII   Conclusion

The feedback linearized and simultaneously output decoupled dynamic control technique based on nonlinear feedback and state transformation actually transforms robot arm control problems from the joint space to the task space and performs robot servoing in terms of task space variables within a linear system frame, allowing also the use of powerful techniques from optimal control of linear systems. On the joint space level, our dynamic control technique commands drive forces or torques or their equivalent quantities addressed to the joint drive system. The above control can be extended to the control of flexible robot arms including those with joint elasticity and link deformation for both rotary and prismatic joints and also to the control of redundant robot arms. That covers a very broad range and almost all categories of robot manipulators demonstrating once again the effectiveness of the method.

The "task-level" nature of the new dynamic control technique becomes apparent by examining the upper left part of Figure 2. As seen there, the expected output from a task planning program is a commanded path (or trajectory) "y" in the task space as a function of time. From this function and its time derivatives a linear mathematical model "v" is constructed which becomes a linear control input to the feedback linearized robot arm. Thus, the planned and commanded trajectory together with its time derivatives *directly drives* the robot arm through a linear system model.

We also note that the optimal error correction loop shown in the lower left part of Figure 2 directly operates on the task level and not on the joint control level. The task level errors are then decomposed by the nonlinear gain matrix $\beta(x)$ into joint force or torque drive commands as shown in Figure 2. It is also important to note that the nonlinear compensating vector $\alpha(x)$ and the nonlinear gain matrix $\beta(x)$ in the nonlinear feedback equation "w" do not change from task to task, unless the load carried by the robot hand is changing drastically relative to a nominal value assumed in the nonlinear feedback design. In that case changes in the $\alpha(x)$ vector and $\beta(x)$ matrix can be carried out by using straightforward algebraic formulas [17]. Consequently, the basic control system parameters in the new dynamic control method do not need readjustment from task. In that sense the control methods is "intelligent" since it is capable of directly responding to changing task level commands "y" embodied in the linear task control input "v".

Motivated by the new dynamic control method summarized briefly in the previous sections, our description of the robot arm motion planning problem is as follows: given certain goal data on the six d.o.f. movement of the robot hand in the task space, find a mathematical representation of a six d.o.f. robot hand trajectory in the task space in the form of a closed function of time which can be used as input to the robot arm control system to achieve a desired robot hand motion. The given goal data can include a geometric description of a curve in the 3-D space and the robot hand's orientation along it, or points in the 3-D space the robot hand has to pass through, specification of velocities at some points in the task space including initial and final velocities, etc. Thus, we are only interested in motion planning in the task space; in our motion planning no transformation is involved between task space and joint space. We point out again that this approach to robot arm motion planning is possible because of the development of the new dynamic control method described in the previous section. This new control technique only requires the formulation of functions which define the end effector trajectory and its time derivatives in the six d.o.f. task space as inputs to the robot arm controller for position and velocity control of the robot arm.

A general theoretical frame is outlined in [10] for task space motion planning, matching the requirements of our new dynamic control technique. In task space motion planning we use tools from differential geometry. We found it desirable since this sets the task space motion planning problem on the same mathematical foundation as our task driven feedback control design briefly described in the previous sections. More details on motion planning can also be found in this volume (See Chapter entitled "Kinematic and Dynamic Task Space Motion Planning for Robot Control").

References

[1] T.J. Tarn, A.K. Bejczy, A. Isidori, and Y.L. Chen: " Nonlinear Feedback in Robot Arm Control." Proc. of 23rd IEEE Conf. on Decision and Control. Las Vegas 1984.

[2] A.K. Bejczy and T.J. Tarn, "Dynamic Control of Robot Arms in Task Space Using Nonlinear Feedback," Automatisierungstechnik, Vol. 36, No. 10, pp. 374-388, 1988.

[3] T.J. Tarn, A.K. Bejczy, and X. Yun: " Dynamic Coordination of Two Robot Arms." Proc. of 25th IEEE Conf. on Decision and Control, Athens 1986.

[4] T.J. Tarn and A.K. Bejczy: " Software Elements: Intelligent Control of Robot Arms." In Dorf, R.C. (Ed.): Encyclopedia of Robotics, J. Wiley and Sons, New York 1987.

[5] T.J. Tarn, A.K. Bejczy, and X. Yun: "Modeling and Control of Two Coordinated Robot Arms." In C.I. Byrnes and A. Kurzhanski (Eds.): "Modeling and Adaptive Control," Springer-Verlag, New York 1987.

[6] Y.L. Chen: " Nonlinear Feedback and Computer Control of Robot Arms." D.Sc. Dissertation, Washington University, St. Louis 1984.

[7] A.K. Bejczy, T.J. Tarn, and Y.L. Chen:" Robot Arm Dynamic Control by Computer." Proc. of 1985 IEEE International Conf. on Robotics and Automation, St. Louis 1985.

[8] A.K. Bejczy, T.J. Tarn, Y.L. Chen:"Robust Robot Arm Control with Nonlinear Feedback." Proc. of IFAC Symposium on Robot Control, Barcelona 1985.

[9] A.K. Bejczy, T.J. Tarn, X. Yun, and S. Han: " Nonlinear Feedback Control of PUMA 560 Robot Arm by Computer." Proc. of 24th IEEE Conf. on Decision and Control, Fort Lauderdale 1985.

[10] A.K. Bejczy, T.J. Tarn, and Z.F. Li: " Task Driven Feedback Control of Robot Arms; A Step Towards Intelligent Control." Proc. of 25th IEEE Conf. on Decision and Control, Athens 1986.

[11] H. Hemami and P.C. Camana: " Nonlinear Feedback in Simple Locomotion System." IEEE Transactions on Automatic Control, AC-19, pp. 855-860.

[12] J.R. Hewit and J.R. Burdess: "Fast Dynamic Decoupled Control for Robotics, Using active Force control." Journal of Mechanism and Machine Theory 1980.

[13] S. Nicosia, F. Nicolo, and D. Lentin: " Dynamical Control of Industrial Robots with Elastic and Dissipative Joints." International Federation of Automatic Control 8th Triennial World Congres, Kyoto 1981.

[14] E. Freund and M. Syrbe: " Control of Industrial Robots by means of Microprocessors." Lecture Notes in Control and Information Sciences, Springer-Verlag,
Berlin, Heidelberg, New York 1976, pp. 167-185.

[15] E. Freund: " Fast Nonlinear Control with Arbitrary Pole-placement for Industrial Robot and Manipulators." The International Journal of Robotics Research, 1 (1982), MIT Press.

[16] B.R. Markiewicz: " Analysis of the Computed Torque Drive Method and Comparison with Conventional Position Servo for a Computer-Controlled Manipulator." Jet Propulsion Lab., Tech. Memo 33-601, 1973.

[17] A.K. Bejczy: (1974). "Robot Arm Dynamics and Control." JPL, TM 33-669.

[18] R.P. Paul: " Modeling trajectory Calculation and Servoing of a Computer
Controlled Arm." Stanford Artificial Intelligence Lab., Stanford University,
Stanford, California, A.I. Memo 177, 1972.

[19] M.H. Raibert and B.K.P. Horn: " Manipulator Control Using the Configuration Space Method." Industrial Robot 5 (1978), 69-73.

[20] J.Y.S. Luh, M.W. Walker, and R P.C. Paul: "Resolved-Acceleration Control
of Mechanical Manipulators," IEEE Trans. Automatic Control AC-25 (1980),
No. 3.

[21] O. Khatib: "The Operational space Formulation in Robot Control." Proceedings of 15th ISIR, Tokyo 1985.

[22] R.P. Paul: " Robot Manipulators: Mathematics, Programming, and Control."
The MIT Press, Cambridge, MA, 1981.

[23] M. Vukobratovic and D. Stokic: " Control of Manipulation Robots." Springer-
Verlag, Berlin 1982.

[24] T.J. Tarn, A.K. Bejczy, and X. Yun: " Nonlinear Feedback for PUMA 560
Robot Arm." Robotics Laboratory Report, SSM-RL-85-03, Washington University, St. Louis 1985.

[25] T.J. Tarn, A.K. Bejczy, S. Ganguly, and Z.F. Li: " Nonlinear Feedback Method of Robot Arm Control: A Preliminary Experimental Study." Proceedings of the 1990 IEEE Int. Conf. on Robotics and Automation, Cincinnati, OH, May 13-18, 1990, pp. 2052-2057.

[26] T.J. Tarn, A.K. Bejczy, and X. Yun: " Third Order Dynamic Equations and Nonlinear Feedback for PUMA 560 Robot Arm." Robotics Laboratory Report, SSM-RL-85-04, Washington University, St. Louis 1985.

[27] T.J. Tarn, A.K. Bejczy, and Z.F. Li: " Computer Simulation of Nonlinear Feedback for Third Order Dynamic Model of PUMA 560 Robot Arm with Alternative Pole Assignments." Robotics Laboratory Report, SSM-RL-86-07, Washington University, St. Louis 1986.

[28] A.K. Bejczy and Z. Szakaly: "Universal Computer Control System for Space Telerobots.", Proceedings of the IEEE International Conference on Robotics and Automation, Raleigh, NC, March 31 - April 2, 1987.

[29] W. Boothby: " An Introduction to Differential Manifolds and Riemanian Geometry." Academic Press, New York 1975, pp. 294-297.

[30] T.J. Tarn, A.K. Bejczy, and Z.F. Li: " Robot Arm Motion Planning: A Review and a New Approach." Robotics Laboratory Report, SSM-RL-86-06, Washington University, St. Louis 1986.

[31] T.J. Tarn, A.K. Bejczy, and Z.F. Li: " Connected Robot Arms as Redundant Systems." To appear in the Proceedings of the NATO Advanced Research Workshop on "Robots with Redundancy: Design, Sensing and Control," Salo, Italy, June 1988.

[32] T.J. Tarn, A.K. Bejczy, and X. Yun: " New Nonlinear Control Algoriyhms for Multiple Robot Arms", IEEE Transactions on Aerospace and Electronic Systems, Vol 24, No.5, September 1988.

[33] X. Ding, T.J. Tarn, A.K. Bejczy, "On the Modeling of Flexible Robot Arms." Progress in Robotics and Intelligent Systems, Y.C. Ho and G. Zobrist, Editors, to be published.

[34] X. Ding, T.J. Tarn, A.K. Bejczy, "Nonlinear Feedback Control of Flexible Robot Arms." Progress in Robotics and Intelligent Systems, Y.C. Ho and G. Zobrist, Editors, to be published.

KINEMATIC AND DYNAMIC
TASK SPACE MOTION PLANNING
FOR ROBOT CONTROL

Z. LI

Mallinckrodt Institute of Radiology

Washington University School of Medicine

St. Louis, MO 63110

T.J. TARN

Dept. of Systems Science and Mathematics

Washington University

St. Louis, MO 63130

A.K. BEJCZY

Jet Populsion Laboratory

California Institute of Technology

Pasadena, CA 91109

CONTROL AND DYNAMIC SYSTEMS, VOL. 39

I. INTRODUCTION

Traditionally, robots are designed in such a way that their motion planning and control are treated as separate issues. The compatibility between the motion planning scheme used and the controller design is not emphasized. Designer of robot manipulators concentrated on the controller design, while the motion planning is largely left as a task for the robot operator. Therefore, the availability of well-trained operators becomes a prerequisite to the use of robots. Furthermore, this leads to controller designs that are based on fixed motion trajectories only, without full incorporation of sensory detection of changes of the environment they are working in and adaptivity to them. All these tasks are again essentially left to the operating programmer.

In order to reduce the amount of work to be performed by the operators, the design of robot manipulators should incorporate as much motion planning capabilities as possible based on known capabilities of the robot manipulator itself as well as the ability to deal with a changing environment that it is working in, such that executable motion trajectories can be automatically formulated with minimal requirement on operator input. On the other hand, the prerequisite of this feature is the existence of a robot servo controller design that can interact with motion planning conveniently.

Robot motion planning can be defined as follows: given the desired task, generate a mathematical description of the robot motion trajectory for the task which can be used to formulate robot control input commands. Various constraints and criteria can be added that the robot motion planner has to take into account when generating a robot motion trajectory. Among them are: task space obstacles, robot kinematic constraints, robot dynamic constraints, motion time optimality, energy consumption optimality, and others. Based on the type of constraints considered, robot motion planning can be subdivided into two subproblems: kinematic motion planning and dynamic motion planning. Kinematic motion planning appears in the form of obstacle avoidance path planning, robot workspace analysis, multi-arm collision avoiding motion planning. Dynamic motion planning, on the other hand, deals with minimum time and minimum energy trajectory planning, and dynamic robot workspace analysis. Based on the approaches adopted, robot motion planning can also be divided into algorithmic motion planning and heuristic motion planning. The former utilizes the fact that the robot system and its environment can be modeled mathematically with tools from classical geometry, algebraic geometry, topology, etc. Results obtained with this approach have been abundant, especially in solving the obstacle avoidance path planning problem. However, these results typically are very computation intensive and may be beyond the capabilities of current computer technology. Furthermore, since the dynamic constraints on the robot motion are ignored, the robot trajectory thus produced still can not be used directly to formulate robot motion control input commands. In fact, many results only produce a task space geometric curve that avoids the obstacles, with no concern over the kinematic and dynamic properties of the motion trajectory such as the velocities and accelerations along the curve and the available drive force. Intermidiate steps have to be performed before the motion planning task is finished.

Our objective with robot motion planning development is to allow the robot operator to specify the task to be performed by the robot at a high level. The operator's input command to the robot should be in general terms such as assembling a given part with given locations of the parts, description of assembly constraints, and maximal time allowed for the assembly. The robot motion planner should be able to "fill in" automatically the missing details such as exactly how the robot should move to pick the part and assemble it. Furthermore, the robot should automatically obtain a time parametrized robot motion trajectory that satisfies both the kinematic and dynamic constraints of the robot. The end product of the robot motion planning should be a robot motion trajectory that meets the goals given by the robot operator and at the same time satisfies all the constraints and, therefore, can be used by the robot controller directly to formulate robot control input commands. Robot systems having such capabilities will be more intelligent and autonomous, and would require less operator interactions.

In this chapter, we specifically treat the kinematic and dynamic constraints of robot systems. Since any executable robot motion trajectories must satisfy these constraints, a good understanding of them is a prerequisite of robot motion plannning and control. Our aim is that our results be used directly by the robot motion planner to verify that the kinematic and dynamic constraints are satisfied and the planned robot motion trajectory is indeed executable. This constitutes a necessary first step towards the realization of fully autonomous robot systems.

Works on robot motion planning have been numerous. Here we quote them according to subproblems studied. Earlier work in robot motion planning concentrated on the generation of robot joint trajectories from given robot task space trajectories. Paul (1975,1979) proposed end effector paths made up of straight line segments. In order to travel from point A to point B in the task space along a straight line in time period T, a drive transform $D(t/T)$ is chosen that relates the homogeneous matrices P_A and P_B identifying A and B. Thus $P_B = D(1)P_A$. The entries of D are so chosen that the motion represented by $D(t/T)$ has constant velocity. The velocity and acceleration of the robot end effector are controlled by converting them into the joint space expressions and smoothed by a quadratic interpolation routine. Taylor (1979) extended Paul's work by using the dual number quarternion representation to describe the location of the end effector. Because of the properties of quarternions, transitions between the end effector locations due to the rotational operations require less computations, while no advantage is produced for the translational operations.

Chand and Dotty (1985) examined the problem of how many points on a given end effector trajectory must be available to generate smooth joint trajectories by using cubic splines. They derived a set of formulas that can be used to determine the effect of approximation errors in the velocity at the last point of the trajectory on the velocities of proceeding points, thus help choosing the total number of points on the trajectory.

Many papers have been published on the planning of obstacle avoidance paths for robot arms. Schwartz and Sharir (1983a,1983b) adopted a typical algorithmic approach to solve the problem. By identifying the topological structure of the set formulated by the kinematic constraints of the robot and the environment as a Collins set, they were able to derive conditions on the existence of an obstacle avoiding path amid obstacles. An algorithm is given to check the conditions and to formulate the path. The works of Tannenbaum and Yomdin (1987) and Yap (1987) are in the same vein. This approach requires that the obstacle locations are known exactly, and is in general conmputation intensive. Using the heuristic approach, Lozano-Perez (1983,1987)proposed the use of the configuration space in path planning for obstacel avoidance. Obstacles in the configuration space are decomposed recursively into lower dimensional slices, the one-dimensional slice being the obstacle constraint on a single robot joint. An efficient algorithm is developed based on this idea. Implementation of the algorithm is also reported. Similar results are reported by Brooks (1984), Donald (1985), Faverjon (1984), Gouzenes (1984), Laugier and Germain (1985), and Udupa (1987). A different approach is taken by Lumelsky in (1986, 1987). The obstacles are assumed to be unknown. Detection of the obstacles rely on the use of sensors. The goal of the motion planning is such that an automaton can find its way in an unstructured space filled with unknown obstacles. The path needs to be generated as the automaton goes along. An algorithm is presented for the solution of this problem for planar robot arms. Khatib (1986) proposed the potential field approach. In this approach, an artificial potential field is constructed in the robot workspace. The object location is assigned with a low potential and the obstacles are assigned with high potentials. The robot is to move in the direction of decreasing potential energy. Moving obstacles are accounted for by considering the potential field to be time-varying. Other variants of the obstacle avoiding path planning problem are also explored, such as the on-line obstacle avoidance problem treated by Freund (1984), Hogan (1985), Khatib and Le Maitre (1978), and Krogh (1983).

When the dynamic constraints of the robot arm are considered, the motion planning problem takes the form of minimum time/energy robot trajectory plannning. Hollerbach (1984) proposed the application of the dynamic time scaling property in robot motion planning. The robot dynamic equation is written as

$$T(t) = M(\theta(t)) \cdot \ddot{\theta}(t) + \dot{\theta}(t)^T \cdot C(\theta(t)) \cdot \dot{\theta}(t) + G(\theta(t))$$

where T is the vector of input torques or forces, $M(\theta) \cdot \ddot{\theta}$ is the vector of inertia terms, $\dot{\theta}^T \cdot C(\theta) \cdot \dot{\theta}$ is the vector of centrifugal and Coriolis forces, and $G(\theta)$ is the gravity loading term. If $r(t)$ is a function of time such that $\dot{r}(t) > 0$ and a new trajectory $\tilde{\theta}(t) = \theta(r)$ is defined, then the torque dependent on velocity and acceleration, T_a, for the new trajectory $\tilde{\theta}(t)$ is

$$
\begin{aligned}
\tilde{T}_a(t) &= M(\tilde{\theta}(t)) \cdot \ddot{\tilde{\theta}}(t) + \dot{\tilde{\theta}}(t)^T \cdot C(\tilde{\theta}(t)) \cdot \dot{\tilde{\theta}}(t) \\
&= (M(\theta(r)) \cdot \theta''(r) + \theta'(r)^T \cdot C(\theta(r)) \cdot \theta'(r)) \cdot \dot{r}^2(t) \\
&= \dot{r}^2 T_a(r) + \ddot{r} M(\theta(r)) \theta'(r)
\end{aligned}
$$

where θ'' and θ' denote the first and second order derivatives of θ with respect to r. This formula can be used to facilitate the computation of joint torques required by a particular trajectory and help to determine modifications needed to render an inexecutable trajectory to an executable one. Rajan (1985) and Lin and Chang (1985) both used dynamic time scaling to compute torques required for the movement along a given joint trajectory. Different joint trajectories described by a number of splines are evaluated by using Pontryagin maximum principle (Rajan, 1985) or by using nonlinear programming (Lin, Chang, 1985). A graph search is conducted to find an optimal joint trajectory combination. Sahar and Hollerbach (1985) also used dynamic time scaling to compute torques. A tesselation of joint space is carried out to provide different joint trajectories such that some of them are eliminated by applying torque limits, thus reducing computation load. A graph search then yields the optimal trajectory. Shin and McKay (1986a, 1986b) investigated the minimum cost trajectory planning problem from two viewpoints. The first one assumes that a predefined set of joint paths is given. Using dynamic programming, the authors are able to determine a set of velocity and acceleration curves along the joint paths. The joint velocity/torque limit curves are incorporated into the optimization process, thus there is no need to check the validity of the joint trajectories after they are thus obtained. The second one considered the effects of different combinations of joint trajectories on the time required to move a robot from one configuration to another, derived a lower bound on this time, and determined the form of the joint path which minimizes this lower bound. Geering, et al (1986) studied three common robot configurations and obtained the form of a time-optimal solution for the path with constrained forces and torques. Pfeiffer and Johannni (1986) described the position-torque curve as a sink in the phase space and applied Bellman's principle to obtain an algorithm for evaluating the time minimum curve from a sequence of accelerating/decelerating extremals. Lin et al. (1983) considered the minimum time formulation of cubic spline interpolated joint trajectories. Constraints on velocity, acceleration and jerk are taken into account, while the optimization cost function is simply the sum of time intervals needed by the joint to travel each segment of the trajectory described by a cubic polynomial. On the formulation of task space minimum time trajectories, Schiller and Dubowsky (1987) proved that the minimum time path is usually not a straight line in the task space and provided a computationally simple algorithm to obtain this path. They have also extended this work to the search for a global time optimal robot motion trajectory (Schiller, Dubowsky, 1989).

Works related to the study of robot kinematic capabilities also appear in the form of robot workspace analysis. Various algorithms have been proposed, both from the point of view of robot designers and from those of robot motion planning. We delay discussion on such works until Section II.

Our approach to the robot motion planning problem is different from the traditional ones. We try to take a unified look at the kinematic and dynamic capabilities of robot arms, especially those of multiple cooperating robot arms. Obstacle avoidance motion planning will be combined with robot kinematic capability analysis. Our emphasis, however, is on how these capabilities can be made the most use of through the robot kinematic and dynamic motion space analysis. Center to our approach of robot task space motion planning is the application of the nonlinear feedback controller, which allows the use of robot task commands as direct input to the robot controller, and is "intelligent" in the sense that it reacts to changes occuring in the robot task space via the task space motion planner.

This chapter is organized into five sections. In Section II, we treat the robot task space motion planning problem for single-arm robot systems. We start by presenting past relevant results on the kinematic capabilities of single robot arms. This is followed by our results on the single-arm dynamic workspace analysis problem. We define the problem of the robot dynamic workspace analysis in this section, and present our results on the solution of this problem for both the non-redundant and redundant robot arms. An algorithm is presented to solve the dynamic workspace analysis problem. In Section III, we treat the kinematic and dynamic capabilities of multiple cooperating robot arms. Our work on the motion space analysis of an object handled by two cooperating robot arms is first presented. Based on past work, we define the concept of the motion space of an object handled by multiple robot arms. We identify the various constraints on this motion space and propose an algorithm to obtain it. Our algorithm is illustrated by an example. Computer simulations are performed and the results are reported here. We next present the robot motion planning problem under the dynamic constraints. Our work on dynamic motion space analysis is discussed here. This work proposes a computationally and conceptually simple method to obtain the maximum forces and torques multiple cooperating robots can apply to the environment, constrained by the joint drive torque limits and by the internal force limits that an object being handled can endure. In Section IV, we describe how a desired robot motion path, as a geometric curve in the robot task space, can be time-parametrized into a robot motion trajectory. We also suggest a robot motion planning and control system in which our results on the analysis of the robot kinematic and dynamic capabilities can be utilized to generate robot motion trajectories that are ready to be executed by the nonlinear feedback robot controller. We illustrate our approach to the robot motion planning problem with an example in which each of the two robot arms are required to travel on a circle and the synchronization between the two arms is to be maintained. A procedure is given that realizes the robot motion planning for this particular task based on discussions of earlier sections. Our work, including future plans, is summarized in Section V.

II. KINEMATIC AND DYNAMIC MOTION SPACE ANALYSIS OF SINGLE ROBOT ARMS

Here we treat the kinematic and dynamic capabilities of single robot arms. We use the term <u>kinematic</u> to refer to the position-velocity-acceleration properties of a path arising from the geometric relations between the robot links and joints and its end effector. A path with a specified and coordinated position-velocity-acceleration profile will become a trajectory. We then define the <u>dynamic</u> motion space analysis problem to be one of finding the force/torque application capabilities of robot arms for given trajectories. Any intelligent robot motion scheme would have to include the verification of these kinematic and dynamic capabilities for robot motion trajectory.

We first summarize the results on the analysis of single arm kinematic capabilities, usually referred to as the workspace analysis of robot arms. The dynamic motion space analysis of single robot arms is discussed in subsection II.2.

II.1 Workspace Analysis of a Single Robot Arm

The kinematic capabilities of a single robot arm is represented by its workspace. For a single robot arm, its workspace is defined to be all the robot task space positions and orientations that its end effector can achieve. Research work on the analysis and evaluation of a robot arm's workspace has produced many results.

Roth pioneered work on this problem in (Roth, 1980). A set of notations are defined for different types of robot joints, such as R for revolute joints and P for primatic joints. A number of questions related to the robot workspaces are discussed, including reachable positions and orientations by the robot hand, effects of different configurations, the coupling of position and orientation. It is pointed out that the position and orientation of a robot are decoupled if the last three joints have intersecting axes. A table is given to show the number of configurations to obtain a fixed position and orientation by robot arms of different degrees of freedom and containing different types of joints. Gupta and Roth (198?) discussed in detail the shape of a robot workspace. It is pointed out that holes and voids exist within a robot workspace. Examples of the workspaces of robots with three mutually intersecting axes are examined. Gupta further presented in-depth discussions on the robot workspace theory in (Gupta, 1986a) and presented the zero reference position method for robot workspace analysis in (Gupta, 1986b). By defining a zero reference position for the robot arm in consideration, a modular approach can be obtained for the derivation of closed form robot kinematics equations, thus helping the computationally simple evaluation of a robot workspace.

Sugimoto and Duffy (1981a,1981b) studied the determination of extreme distances of a robot end effector, i.e., the maximum and minimum reaches of the robot end effector, generalizing the work of Shimano and Roth (1978). It is shown that all intermediate joint axes of a robot intersect an extreme distance line from a point in the robot end effector to an arbitrary base point and they also intersect an extreme perpendicular distance line from a point in the the robot end effector to an arbitrary line in the robot task space. Based on this, an iterative algorithm is developed to solve for the extreme distances of the robot end effector in question by searching for the extreme distance lines. This algorithm is further modified to accomodate cases where the robot arm has intersecting axes, such as in the case of the PUMA 560 robot arm. Kumar and Waldron (1981a, 1981b] classified a robot workspace into two parts: reachable workspace, in which every point can be reached by a point in the robot end effector; and the dextrous workspace, in which every point can be reached by a point in the robot end effector with any desired orientation. An algorithm is developed to evaluate the reachable workspace. It is based on the idea of applying an imaginary force to the robot end effector until it reaches its maximum extension. At this time all the joint axes of the robot will intersect with the line identifying the force vector. The algorithm then calculates the robot joint variables that satisfy the above condition. It is also shown that robot singularities cannot occur within the dextrous workspace. This approach has the problem that it does not identify the holes and voids in the robot workspace.

Tsai and Soni (1983) proposed an algorithm to determine profile of the workspace of a robot arm on an arbitrary plane. The algorithm contains the steps of locating the robot end effector on the plane, moving the end effector on the plane until the boundary is reached, and tracing the boundary. Optimization techniques are used to reduce the number of points to be eveluated. This method also has the problem that that it may not plot all the holes and voids inside a robot workspace. Cwiakala and Lee (1985) presented a similar algorithm, though the optimization cost functions are chosen differently. Kohli and Spanos proposed the use of polynomial discriminants in robot workspace analysis (Kohli, Spanos, 1985a). Viewing the robot kinematics equations as a set of polynomials and observing that the at least two real equal roots appear for each of these polynomials at the workspace boundary, this method enables one to obtain an analytical expression for the workspace boundary surfaces in the robot task space. It also helps identify the number of configurations of the robot arm at an end effector task space location and obtain the conditions on the singularity. This theory is applied to seven types of robot arms with spherical wrists (Kohli, Spanos, 1985b). By combining the robot joint limits into the solution procedure, holes and voids inside the robot workspace are identified. This method does address the reachable orientations by the robot end effector at a position within the robot workspace. Hansen et al. (1983a, Hansen, 1983) developed a noniterative algorithm to generate and evaluate a robot workspace, taking advantages of techniques to reduce the number of points of the workspace to be evaluated. Lee and Yang (1983a, 1983b) showed that, for a given robot structure, the ratio of the volume of the workspace to the cube of its total link length is a constant. It is proposed that this ratio can be used as a kinematic performance measure for robot arms. Algorithms based on the grid-scaning technique are developed to obtain the boundary of the workspace as well as detecting voids and holes in the workspace.

Once the kinematic workspace analysis is done for a given robot arm, it can be used to guide the motion planning of the robot such that the robot is not required to move to any location outside of its workspace.

As far as multiple cooperating robot arms are concerned, no work has been done on their workspace analysis. We will address this problem in Section III of this chapter.

II.2 Dynamic Motion Space Analysis of Single Robot Arms

When a robot arm is to perform force control tasks, it is necessary to know the maximum force/moment that the robot can produce at each point within its workspace. This knowledge will provide a basis for the planning of robot force control. We develop the basic technique for the dynamic motion space analysis and restrict our attention to the case of a single robot arm in this section. This restriction enables us to explain the technique in detail. First we consider a single nonredundant robot arm. The dynamic motion space analysis for a single redundant robot arm is then treated. The discussion in this section is a preparation for discussion on the multiple cooperating robot arm dynamic motion space analysis.

II.2.1 Maximum Force/Moment Capability of a Nonredundant Robot Arm

When a robot arm is required to apply a force/moment to an object that it is holding steady or moving with, we need to know whether or not the desired force/moment exceeds the capability of the robot arm. This leads to the problem of finding the maximum force/moment in any given direction that the robot arm can apply in any given configuration. Assuming that the object can produce a reaction force/moment to any force/moment applied to it by the robot, the maximum applicable force/moment of a robot arm is then decided by the robot mechanical structure, its geometric configuration, and the joint motor capabilities. The robot mechanical structure will need to support internal force/moment that are not propagated to the joints. The propagation of the force/moment applied at the robot end effector to its joints varies with the change of the robot's geometric configurations. The joint motor driving torque limits are the deciding factor on what maximum force/moment the robot can apply at its end effector. In this study, we assume that the robot mechanical structure is strong enough to withstand any practical internal forces.

In the study of force control of robot arms, the compliance frame is often defined to simplify the specifications of the task and the needed computation. A compliance frame is a 3-dimensional time-varying orthogonal coordinate system. Its origin is usually chosen to be the contact point of the robot end effector with the environment such that at each instant and along each axis that task can be expressed as a pure trajectory or force control problem. For example, if the robot end effector is holding a piece of chalk and writing on a blackboard, a compliance frame can be defined such that its origin is located at the tip of the chalk and its z-axis is normal to and points into the blackboard. Let C be a compliance frame thus defined in the object, and let

$$F = [f_x \ f_y \ f_z \ m_x \ m_y \ m_z]^T$$

be the force/moment vector that the robot applies in C. As seen, F is composed of two 3-dimensional subvectors $[f_x \ f_y \ f_z]^T$ and $[m_x \ m_y \ m_z]^T$, each well defined in the geometrical 3-dimensional frame C. Let $\tau = [\tau_1 \cdots \tau_n]$ be the robot joint driving torque vector. Then, if the robot is held steady in a given configuration $q = [q_1 \cdots q_n]$, and ignoring gravity terms, we have by the principle of virtual work,

$$J^T(q) \cdot F = \tau \,, \tag{2.2.1}$$

where $J(q)$ is the manipulator Jacobian matrix defined in C, and the superscript T stands for transpose.

Eq. (2.2.1) describes the role of the Jacobian $J(q)$ in the propagation of force/moment through the robot arm to the drive axis of the joints, i.e., $J(q)$ decides how much force/moment each robot joint must produce in order to balance the force/moment that originates from the robot end effector and propagates to it through the other joints/links successively for each configuration q.

The generalization of Eq. (2.2.1) to the case when the robot arm is moving and combining the gravity terms is easily accomplished by adding the corresponding terms to Eq. (2.2.1) to obtain

$$D(q) \cdot \ddot{q} + E(q, \dot{q}) + G(q) + J^T(q) \cdot F = \tau , \qquad (2.2.2)$$

where $D(q)$ is the 6×6 matrix of effective and coupling inertias, $E(q, \dot{q})$ is the vector of centrifugal and Coriolis forces, and $G(q)$ is the gravity loading terms.

We note that the friction and stiction terms have been ignored here in Eq. (2.2.2). This is for simplicity of discussion. Since these terms can be approximated by an additive term to Eq. (2.2.2), our analysis will still be true with the presence of these temporarily neglected terms.

Assume that we are given a directional vector $u^1 = (a_1^1 \ a_2^1 \ a_3^1)^1$ in C, along which we wish to apply a force $f = (f_x \ f_y \ f_z)^T$ with the largest possible magnitude, and a directional vector $a^2 = (a_1^2 \ a_2^2 \ a_3^2)^T$ in C, around which we wish to apply a moment $m = (m_x \ m_y \ m_z)^T$ with the largest possible magnitude. We can obtain a unit vector a in the 6-dimensional F-space,

$$a = (a_1 \ a_2 \ a_3 \ a_4 \ a_5 \ a_6)^T ,$$

which is simply the normalization of the vector $(a_1^1 \ a_2^1 \ a_3^1 \ a_1^2 \ a_2^2 \ a_3^2)^T$. The vector a is thus composed of the directions of the desired force f and the desired moment m. We define a 6-dimensional space F, called the F-space as follows,

$$F = \{(f_1 \ f_2 \ f_3 \ t_1 \ t_2 \ t_3)^T \mid (f_1 \ f_2 \ f_3)^T \text{ and } (t_1 \ t_2 \ t_3)^T \text{ are represented in } C\},$$

where $f = (f_1 \ f_2 \ f_3)^T$ is the force and $t = (t_1 \ t_2 \ t_3)^T$ is the torque applied by the robot arm.

Given a configuration q, its time derivatives $\dot{q}$ and $\ddot{q}$, the term $\tau - D(q)\ddot{q} - E(q, \dot{q}) - G(q)$ is the net torque available to produce F. We denote this term as τ_{net}. The lower and upper limits of τ_{net} is obtained by translating the limits of τ by the amount $-(D(q)\ddot{q} + E(q, \dot{q}) + G(q))$. Choosing the objective function as PI $= |F|$, where $|F|$ is the magnitude of F, the problem of solving for maximum force/moment then becomes the following optimization problem:

maximize PI with respect to τ_{net} ,

subject to

$$J^T(q) \mid F \mid a = \tau_{net},$$
$$\tau_{i(net)min} \leq \tau_{i(net)} \leq \tau_{i(net)max}, i = 1, \cdots, n.$$

where $\tau_{i(net)min}$ and $\tau_{i(net)max}$ are the minimum and maximum torques joint i can produce.

We name it the "basic problem". This is the optimization problem that we actually solve.

The above "basic problem" can be solved using a number of tools available for this type of optimization problems, such as linear programming. We used a conceptually simple method to solve this problem. We define S_τ as the set of available net joint driving torques, i.e., the τ_{net} vectors with each of its components bounded by the corresponding $\tau_{i(net)min}$ and $\tau_{i(net)max}$. S_τ is then given by

$$S_\tau = \{\tau_{net} \mid \tau_{i(net)min} \leq \tau_{i(net)} \leq \tau_{i(net)max}, i = 1, \cdots, n\}.$$

We also define a set S_F as the set of force/moment vectors that the available net joint driving torques can produce, i.e.,

$$S_F = \{F = (J^T(q))^{-1} \cdot \tau_{net} \mid \tau_{net} \epsilon S_\tau\}.$$

Clearly, S_F is the image of the rectangular parallelepiped S_τ under the mapping $(J^T(q))^{-1}$ which is linear for given values of q, hence a general parallelepiped. The F-vector of maximum length in any given direction lies on the boundary of S_F. S_F can be computed geometrically by computing the images of the 2^n vertices of S_τ under $(J^T(q))^{-1}$, and the hyperplanes that form the sides of S_F.

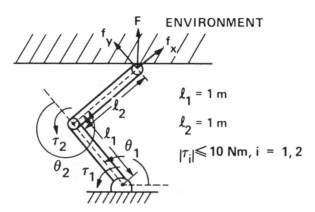

Figure II.1: A Two Link Planar Robot Arm

The following example illustrates the above method.

Example II.1. We consider a two-link planar robot arm as shown in Figure II.1 with parameters $l_1 = l_2 = 1.0$ m, $\tau_{1min} = \tau_{2min} = -10.0$ Nm., $\tau_{1max} = \tau_{2max} = 10.0$ Nm. For the sake of simplicity, we ignore the gravity terms by assuming that the two link robot is in a horizontal plane and assume that the robot is in a steady state. We define the F-vector in the robot tool frame $\{f_x \ f_y\}$ as shown in Figure II.1.

The corresponding robot Jacobian matrix is given by

$$J = \begin{bmatrix} l_2 s_2 & 0 \\ l_2 + l_1 c_2 & l_2 \end{bmatrix},$$

where $s_i = \sin\theta_i, c_i = \cos\theta_i, i = 1, 2$. Thus we have

$$\begin{bmatrix} f_x \\ f_y \end{bmatrix} = \begin{bmatrix} \frac{1}{l_1 s_2} & -\frac{l_2 + l_1 c_2}{l_1 l_2 s_2} \\ 0 & \frac{1}{l_2} \end{bmatrix} \cdot \begin{bmatrix} \tau_1 \\ \tau_2 \end{bmatrix},$$

$$-10 \le \tau_1 \le 10, -10 \le \tau_2 \le 10.$$

For a given configuration $q = (\frac{\pi}{2} \ \frac{\pi}{6})$, we have

$$\begin{bmatrix} f_x \\ f_y \end{bmatrix} = \begin{bmatrix} 2 & -(2 + \sqrt{3}) \\ 0 & 1 \end{bmatrix} \cdot \begin{bmatrix} \tau_1 \\ \tau_2 \end{bmatrix}.$$

The mapping $(J^T(\frac{\pi}{2}, \frac{\pi}{6}))^{-1}$ maps the square S_τ in the τ-space (Figure II.2(a)) to the parallelepiped S_F in the F-space (Figure II.2(b)). In Figure II.2, the points A, B, C, D are mapped to the points A', B', C', and D' respectively. S_τ in Figure II.2(a)

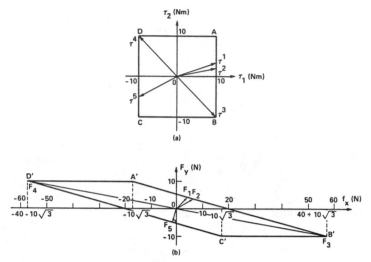

Figure II.2: (a). S_τ for the Robot in Figure II.1. (b). S_F for the Robot Arm in
Figure II.1

is a square due to the given condition that τ_1 and τ_2 have the same bounds, i.e., 10
Nm.

If it is desired to find the maximum force that this robot can apply in the
direction $a_1 = (\frac{\sqrt{2}}{2}, \frac{\sqrt{2}}{2})^T$ at $\theta = (\frac{\pi}{2}, \frac{\pi}{6})$, we immediately see that this is given
by the vector $F_1 = (\frac{20}{3+\sqrt{3}}, \frac{20}{3+\sqrt{3}})^T$ in Figure II.2(b), which has a length of 6 N.
The corresponding joint driving torques are $\tau^1 = (10, \frac{20}{3+\sqrt{3}})^T$ in Figure II.2(a). The
maximum force that the robot can apply in the direction $a_2 = (\frac{1}{2}, \frac{\sqrt{3}}{2})^T$ is found to be
$F_2 = (\frac{10\sqrt{3}}{1+\sqrt{3}}, \frac{10}{1+\sqrt{3}})^T$, with the corresponding joint driving torque $\tau^2 = (10, \frac{10}{1+\sqrt{3}})^T$.

We note that there exists some similarity between the method for solving the
maximum force/moment problem presented here and the one used by Schiller and
Dubowsky (1987) in solving for maximum acceleration in task space of a robot's
end effector. Schiller and Dubowsky considered the robot dynamic equation

$$D(q)\ddot{q} + E(q, \dot{q}) + G(q) = \tau,$$

and transformed it into the robot task space using inverse kinematics to obtain

$$\bar{D}(S)\ddot{S} + b(S)\dot{S}^2 + \bar{G}(S) = \tau.$$

Thus the maximum acceleration in the robot task space starting from a point P

with zero velocity can be obtained by inspecting the image of the rectangular parallelepiped

$$\{\tau_i \mid \tau_{i_{min}} \leq \tau_i \leq \tau_{i_{max}}, 1 \leq i \leq n\}$$

under $\bar{D}^{-1}$. Based on this information, they proposed the use of B-splines to represent a near-time-minimum trajectory between two task space points. It is obvious that the problem discussed here is different from the one in (Schiller, Dubowsky, 1987). We consider here the transformation between force/moment in task space, and torque in joint space, both are dynamic quantities, and use the kinematic transformation $J(q)$. While in (Schiller, Dubowsky, 1987) the dynamic transformation $\bar{D}(q)$ is used to relate the kinematic quantity $\ddot{S}$ in task space and the dynamic quantity τ in joint space.

We conclude this subsection by noting that our method for solving the maximum force/moment problem is valid even at singularities of $J(q)$. However, special care has to be taken at those points.

II.2.2 Maximum Force/Moment Capability of Redundant Robot Arms

For a redundant robot arm the dynamic equation takes the same form as for non-redundant robot arms

$$D(q)\ddot{q} + E(q, \dot{q}) + G(q) + J^T(q)F = \tau,$$

$$F \in R^m, \tau \in R^n, n > m.$$

The only difference is that the $J^T(q)$ is no longer a square matrix.

We assume that $J(q)$ is of full rank, i.e., $\text{rank}(J(q)) = m$. This implies that the robot actuators can produce a force/moment in any m-dimensional task space direction. Then the constraint

$$J^T(q)F = \tau_{net}, F \in R^m, \tau_{net} \in R^n, m < n, \qquad (2.2.3)$$

examplifies the nonuniqueness of the robot joint driving torques corresponding to the same end effector force/moment F. For a given value of τ_{net}, Eq. (2.2.3) may not be exactly solvable. Thus the question arises: when is Eq. (2.2.3) exactly solvable, or for which τ_{net} is Eq. (2.2.3) exactly solvable?

To answer this question, let us first multiply both sides of Eq. (2.2.3) by $J(q)$, then we have

$$J(q)J^T(q)F = J(q)\tau_{net}, \qquad (2.2.4)$$

where $J(q)J^T(q)$ is an $m \times m$ matrix and is invertible since $J(q)$ is assumed to be of full rank. Thus Eq. (2.2.4) is exactly solvable for any value of τ_{net}. This is because, by multiplying both sides of Eq. (2.2.3) by $J(q)$, the component of τ_{net} in $T_J = N(J(q))$, the null space of $J(q)$, is mapped to zero. Therefore we see that Eq. (2.2.3) is exactly solvable for a given τ_{net} if and only if $P_{T_J}(\tau_{net}) = 0$, where P_{T_J} is the projection function of τ_{net} into T_J. Physically, T_J contains those net joint driving torques that are not transmitted to the end effector to produce F, thus are not described by Eq. (2.2.3).

Based on this discussion, we can solve our "basic problem" for redundant robot arms as one of finding the image S_F of S_τ under the pseudo-inverse mapping

$$(J^T)^\dagger = (J(q)J^T(q))^{-1}J(q)$$

in F-space. Here again S_τ is as defined following the statement of the "basic problem" stated for the non-redundant case. However, S_F here may not be a parallelepiped anymore. This is so because J^T is no longer a square matrix and $(J^T)^\dagger$ is a mapping between two spaces of different dimensions. In general, S_F is a convex polygon or polyhedron. The maximum F thus obtained will satisfy Eq. (2.2.4). The corresponding net joint driving torque τ_{net} can be obtained by plugging this F into Eq. (2.2.3). It will have the additional property that $| \tau_{net} |$ is minimized by the property of the chosen pseudo-inverse of J^T (Campbell, Meyers, 1979).

Pseudo-inverses of the manipulator Jacobian matrix have been widely used in the study of redundant robot arms, e.g., in Hollerbach and Suh (1987). It is usually necessary in the solution of the inverse kinematics of redundant robot arms, i.e., solving for joint velocities $\dot{\theta}$ from desired task space velocities $\dot{x}$ by the equation

$$\dot{x} = J(\theta)\dot{\theta}.$$

Here the dimension of $\dot{\theta}$ is higher than that of $\dot{x}$, and the equation is underdetermined. A solution that minimizes $|\dot{\theta}|$ is given by

$$\dot{\theta} = J^T(\theta)(J(\theta)J^T(\theta))^{-1}\dot{x}.$$

Since Eq. (2.2.4) is overdertermined, we have to choose the pseudo-inverse differently to minimize

$$| J^T(q)F - \tau_{net} | .$$

The computation of the image S_F of S_τ by $(J(q)J^T(q))^{-1}J(q)$ is no different from the case of non-redundant robot arms. Again, the maximum force/moment that the robot arm can produce along the direction of unit vector a is obtained by solving for the intersection of an F-vector along the unit vector a and the boundary of S_F.

III. KINEMATIC AND DYNAMIC MOTION SPACE ANALYSIS OF MULTIPLE COOPERATING ROBOT ARMS

We treat the kinematic and dynamic capabilities of multiple cooperating robot arms in this section. The motion space of an object handled by two cooperating robot arms and the relevant motion space analysis problem is formulated and a solution procedure is proposed in subsection III.1. The use of this procedure is illustrated by an example in subsection III.2. Dynamic capabilities of multiple co-operating robot arms is treated in subsection III.3.

III.1 Kinematic Motion Space Analysis of Multiple Robot Arms

The applications of two or more cooperating robot arms usually involve the moving and orienting of a large object whose weight or size goes beyond the capability of a single arm. The desired motion trajectory of such an object is usually given relative to a point and a coordinate frame (the object frame) fixed in the object. The capability of the two cooperating robot arms can be described by how far the two arms can move this point and what orientations of the object frame can be achieved. Let this point be denoted by P, and the orientations of the object frame by a set of three orthogonal unit vectors, denoted by

$$S = (s_1 \ s_2 \ s_3) .$$

This situation is illustrated in Figure III.1.

When a motion trajectory of an object handled by two robot arms is planned, it is necessary to know that the positions and orientations on the planned trajectory

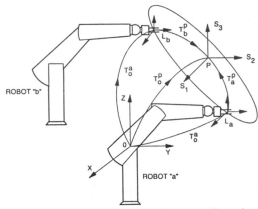

Figure III.1: Object Frame and the Related Transformations

is accessible by the object under the constraints of the two robot arms' and the object's kinematic properties. We propose to define a measure of the two robot arms' kinematic capability by the motion space of an object handled by the two robot arms. This motion space is composed of all the positions and orientations that P and S can achieve relative to the universal coordinate frame. Therefore, we define the motion space of an object handled by two cooperating robot arms to be a subset of the space $R^3 \times SO(3)$, where $SO(3)$ is the space of 3×3 orthogonal matrices. We denote it by W.

Note that this definition of the motion space of an object handled by two cooperating robot arms as a measure of the two robot arms' kinematic capability is derived by expanding the definition of the workspaces of a single robot arm. In the two arm case, P is defined away from the robot end effectors in a fixed position and orientation located within the object handled by the arms.

We point out here that the common workspace of two cooperating robot arms have traditionally been regarded as simply the intersection of the individual workspaces of the two robot arms. Therefore, the common workspace of two cooperating robot arms is empty if their individual workspaces do not intersect. This definition however does not reflect the fact that the two arms can still perform certain tasks when they are grasping an object of nonzero dimensions even when their workspaces do not intersect. Therefore it can not be used as a good measure of the two robot arms' object handling capability. We observe that the locations in the object at which the two robot arms are grasping, henceforth referred as the grasping locations, are crucial in analyzing what work can be done by the two cooperating robot arms and should be reflected in the definition of their common workspace. Our above definition is exactly based on this observation. That is, by considering a point and a coordinate system defined in the object explicitly when we define the motion space of the object handled by the two robot arms, our two arm kinematic capability measurement will be dependent on the grasping locations. In fact, as we will see later, it changes with the change of the grasping locations.

We mention that we assume that the two robot arms are grasping rigidly at the object to be moved.

Let l_a and l_b denote the grasping locations of robot arms a and b defined in the object's coordinate frame. Then l_a and l_b are limited by the object's physical dimensions. Let L be a set describing these dimensions. Then we have $L \subset R^3 \times SO(3)$. Since the motion space as we defined in this study depends on the grasping locations l_a and l_b, we write it as being parametrized by l_a and l_b and denote it by W_l. Therefore, our problem can be summarized as that of given the dimensions of an object to be handled by two robot arms, determine, for each l_a and l_b in the object, the motion space W_l based on the two robot arms individual workspaces and other kinematic constraints. We note that this definition of the motion space analysis problem makes it a geometric one. By this restriction, we do not discuss here other kinematic properties of the two cooperating robot arms such as the maximum velocity and acceleration an object can have when handled by them.

Assume that we have

$$W_a \in \{(P_a, S_a) \mid P_a \in R^3, S_a \in SO(3)\}$$

and

$$W_b \in \{(P_b, S_b) \mid P_b \in R^3, S_b \in SO(3)\}$$

defined respectively as the first and second robot arms' individual workspaces, i.e., the accessible end effector tip positions and orientations of the two robot arms, respectively. Then the motion space of an object handled by the two robot arms can be obtained as the image of a mapping from $W_a \times W_b$ to $R^3 \times SO(3)$, parametrized by the grasping locations. Let K_l be such a mapping. Then we have

$$W_l = K_l\left(W_a \times W_b\right) \subset R^3 \times SO(3),$$

under suitable kinematic constraints of the two robot arms and of the object. Let

$$F_i\left(l_a, l_b, P_a, S_a, P_b, S_b, P_{a1}, S_{a1}, \ldots, P_{an_a}, S_{an_a}, P_{b1}, S_{b1}, \ldots, P_{bn_b}, S_{bn_b}\right) \leq 0,$$

$$i = 1, \ldots, k$$

describe such constraints, where P_{ai}, S_{ai}, $i = 1, \ldots, n_a$, and P_{bi}, S_{bi}, $i = 1, \ldots, n_b$ are the link positions and orientations of the two robot arms. Then we can formulate the problem of the geometric motion space analysis of an object handled by two cooperating arms as follows:

> Given a) the dimensions of an object to be handled by two cooperating robot arms, b) the individual workspaces W_a and W_b of the two robot arms, and c) the distance between the bases of the two arms, find the motion space of the object W_l such that
>
> $$W_l = K_l\left(W_a \times W_b\right) \subset R^3 \times SO(3),$$
>
> $$F_i\left(l_a, l_b, P_a, S_a, P_b, S_b, P_{a1}, S_{a1}, \ldots, P_{an_a}, S_{an_a}, P_{b1}, S_{b1}, \ldots, P_{bn_b}, S_{bn_b}\right) \leq 0,$$
>
> $$i = 1, \ldots, k, \ \forall l_a, \ l_b \in L$$

The general solution to the geometric motion space analysis problem as formulated above will be to a large extent dependent on the mapping K_l. Here we try to analyze the general characteristics of this mapping and to develop a solution approach based on this analysis.

The individual workspaces of the two robot arms are usually described by sets of algebraic equations and inequalities of the form $f\left(P_e, S_e\right) \leq 0$, where f is a vector function of polynomials (Tannenbaum, Yomdin, 1987), P_e and S_e are the position and orientation of the robot end effector. That is, we can write W_a and W_b as

$$W_a = \left\{\left(P_a, S_a\right) \in R^3 \times SO(3) \mid f_a\left(P_a, S_a\right) \leq 0\right\} \tag{3.1.1}$$

and

$$W_b = \left\{\left(P_b, S_b\right) \in R^3 \times SO(3) \mid f_b\left(P_b, S_b\right) \leq 0\right\}, \tag{3.1.2}$$

where f_a and f_b are vector functions describing W_a and W_b respectively. We emphasize here that the functions f_a and f_b are algebraic ones.

The mapping K_l that relates the individual robot workspaces W_a and W_b to the common workspace W_l can be formulated as follows. Let T_0^a be the Denavit-Hartenberg matrix representation of the position and orientation of the end effector tip of robot a and T_0^b be that for robot b. We observe that these matrices can be equally written as functions of (P_a, S_a) and (P_b, S_b) respectively. Let the position and orientation of the object to be moved be represented similarly by T_0^p. Then we have $T_0^p = T_a^p \times T_0^a$, and $T_0^p = T_b^p \times T_0^b$, as shown in Figure III.1. Note that T_a^p and T_b^p are parametrized by the grasping locations l_a and l_b. Since

$$T_0^p = \begin{bmatrix} S & P \\ 0 \ 0 \ 0 & 1 \end{bmatrix},$$

we have, by slight change of notation,

$$(P, S) = K_l(P_a, S_a, P_b, S_b,) = T_0^p = T_a^p \times T_0^a.$$

under the constraint

$$T_b^p \times T_0^b = T_a^p \times T_0^a,$$

or

$$T_0^b = (T_b^p)^{-1} \times T_a^p \times T_0^a = T_p^b \times T_a^p \times T_0^a = T_a^b \times T_0^a.$$

Since the above equation involves only multiplications and additions, we see that K_l is also an algebraic mapping.

The constraint above simply requires that the two arms grasp the object at the locations l_a and l_b as specified by the matrices T_a^p and T_b^p, or T_a^b. Note that T_0^a and T_0^b equivalently give (P_a, S_a) and (P_b, S_b) and these are constrained by conditions (3.1.1) and (3.1.2).

Another constraint that needs to be considered is that of possible workspace occupancy conflicts between the object and the robot arms. This occurs when the positions and orientations of the object moves into those occupied by the robot arms' link bodies. Let P_{ai} and S_{ai} denote the position and orientation of the ith link of robot a, $i = 1, \ldots, n_a$, P_{bi} and S_{bi} denote the position and orientation of the

i^{th} link of robot a, $i = 1, \ldots, n_b$, where n_a and n_b are the degrees of freedom of the robots a and b respectively, Then these constraints can be written as

$$g_a\left(l_a, l_b, P_{a1}, S_{a1}, \ldots, P_{an_a}, S_{an_a}\right) \leq 0,$$

and

$$g_b\left(l_a, l_b, P_{b1}, S_{b1}, \ldots, P_{bn_b}, S_{bn_b}\right) \leq 0,$$

Again, we note that functions q_a and q_b are generally algebraic ones. An additional constraint would be the colision avoidance between the arms. We do not consider it explicitly here, and later in this chapter we will come back to this problem.

Finally, we have the constraint of the object's dimensions, described by $l_a \in L$, $l_b \in L$.

Putting the above equations and inequalities together, we have a more detailed description of the geometric motion space of an object handled by two cooperating robots, as given by the following conditions:

$$K_l(P_a, S_a, P_b, S_b) \; = \; T_a^p \times T_0^a, \tag{3.1.3}$$

$$f_a(P_a, S_a) \; \leq \; 0, \tag{3.1.4}$$

$$f_b(P_b, S_b) \; \leq \; 0, \tag{3.1.5}$$

$$T_0^b \; = \; T_a^b \times T_0^a, \tag{3.1.6}$$

$$g_a\left(l_a, l_b, P_{a1}, S_{a1}, \ldots, P_{an_a}, S_{an_a}\right) \; \leq \; 0, \tag{3.1.7}$$

$$g_b\left(l_a, l_b, P_{b1}, S_{b1}, \ldots, P_{bn_b}, S_{bn_b}\right) \; \leq \; 0, \tag{3.1.8}$$

$$l_a \in L, \qquad l_b \in L. \tag{3.1.9}$$

Based on the above discussions, we can conclude that the set of positions P and orientations S that are accessible by the object is semialgebraic. The topological structure of such sets are studied by Collins (Collins) and Schwarz and Sharir (1983a,1983b).

The solution procedure for the motion space analysis problem will necessarily be composed of solving the P and S that the object can have for given values of l_a and l_b, and then iterating over the set L that describe all the possible l_a's and l_b's.

Therefore the procedure to solve the motion space analysis problem can be given as containing the following steps:

1. Obtain the functions K_l, f_a f_b, g_a, g_b;

2. Select grasping locations l_a and l_b from the set of permissible grasping locations L;

3. For all P_a and S_a satisfying Inequality (3.1.4), find the corresponding permissible P_b and S_b via Eq. (3.1.6), disregard the P_b's and S_b's that do not satisfy Inequality (3.1.5);

4. For all the P_a, S_a, P_b, S_b obtained in last step, find the corresponding P_{ai}, S_{ai}, P_{bi}, S_{bi} via the two robot arms' kinematics equations;

5. Check that conditions (3.1.7) through (3.1.8) are satisfied. If not, disregard the corresponding P_a, S_a, P_b, S_b;

6. Plug the values of P_a, S_a, P_b, S_b obtained from the above steps into Eq. (3.1.3) to get the corresponding position and orientation of the object P and S;

7. Iterate over the set L until done.

In many cases, the descriptions of the range that the end effectors' and individual links' positions and orientations of the robot arms can obtain are not given by conditions (3.1.4), (3.1.5), and (3.1.7) through (3.1.8) in terms of P and S. This is especially true for the orientations. Instead, they have to be found through the robot arm's kinematics equations under the constraints of joint variable limits. Let θ_a and θ_b be the vectors of the joint variables of the two robot arms. Then the motion space analysis problem can be solved by use of the following equations and inequalities:

$$K_l(\theta_a, \theta_b) = T_a^p \times T_0^a(\theta_a), \qquad (3.1.10)$$

$$g_a(l_a, l_b, P_{a1}, S_{a1}, \ldots, P_{an_a}, S_{an_a}) \leq 0, \qquad (3.1.11)$$

$$g_b(l_a, l_b, P_{b1}, S_{b1}, \ldots, P_{bn_b}, S_{bn_b}) \leq 0, \qquad (3.1.12)$$

$$(P_a, S_a) = h_a(\theta_a), i = 1, \ldots, n_a, \qquad (3.1.13)$$

$$(P_b, S_b) = h_b(\theta_b), i = 1, \ldots, n_b, \qquad (3.1.14)$$

$$(P_{ai}, S_{ai}) = h_{ai}(\theta_a), \qquad (3.1.15)$$

$$(P_{bi}, S_{bi}) = h_{bi}(\theta_b), \qquad (3.1.16)$$

$$\theta_a \in \Theta_a \quad , \quad \theta_b \in \Theta_b, \tag{3.1.17}$$

$$T_0^b(\theta_b) = T_a^b \times T_0^a(\theta_a), \tag{3.1.18}$$

$$l_a \in L, \qquad l_b \in L, \tag{3.1.19}$$

where h_a, h_b, h_{ai}, h_{bi}, are the kinematic functions of the two individual robot arms, Θ_a and Θ_b are the sets of permissible values for θ_a and θ_b.

Since the kinematic functions of robot arms in general contain trigonometric operations, the equations and inequalities above are no longer algebraic. Solution procedure in this case would have to start with selection of values of θ_a and θ_b. The remaining part of the procedure is similar to the earlier one.

We mention that the constraints on the motion space of an object handled by two robot arms do not include those that are related to the time-varying aspects of the object motion trajectory. This is important, for example, in the case when the object being handled by the two robot arms has to be turned upside down and the arms are required to cross each other during the motion. Collision of the two arms would be caused by this manipulation. Thus it is prohibited for the two robot arms to attempt this manipulation. The motion space analysis problem for an object handled by two cooperating robot arms with this constraint is another form of the "piano mover's problem". The solution of it for a single robot arm has been studied in (Schwartz, Sharir, 1983a, 1983b) The same technique can be used here.

III.2 An Example

To illustrate the problem of geometric motion space analysis of an object handled by two robot arms and its solution procedure proposed above, we give an example here with some more restricted assumptions. We consider here the object being handled to be a long one. Also, we assume that the grasping locations have fixed orientations. In particular, we assume that the robot end effectors are perpendicular to the object and parallel to the floor. To envision such a task, consider two robot arms cooperate to move a rod placed on a jig on a wall to another jig below the original location of the rod on the same wall. The origin of the coordinate frame in the object is chosen to be the geometric center of the object, and the two grasping points have equal distance to the origin of the object coordinate frame. By these

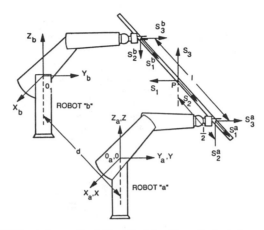

Figure III.2: A Long Rod Handled by Two Robot Arms

assumptions, the grasping locations degenerate into the grasping length l. This situation is as shown in Figure III.2. For the sake of simplicity, we assume that the object's orientations are required to be fixed such that the x, y, z axes of the object frame are aligned with the x, y, z axes of the universal frame respectively. In this section, we determine how far two Puma 560 robot arms can move the rod under the above constraints. Since the rod and the two robot arms form a closed mechanical chain, the mechanical constraints of the two arms restrict the reachable locations by the rod. As described in the last section, these constraints include the robot joint limits, the occupancy conflict condition, the grasping location constraints, and the task space obstacles.

The individual workspaces of two robot arms with rotational joints and spherical wrists, such as the PUMA 560 robot arms, are shown in Figure III.3. Notice that there are holes in the individual workspaces. The distances between the two robot arms are assumed to be fixed as d. Assume that we are given a long object of total length L. We now develop an algorithm to compute the motion space of the long object under the constraints that its orientations be fixed as described in the beginning of this section. These constraints translate into the following conditions,

$$S_1 = (1 \quad 0 \quad 0)', \; S_2 = (0 \quad 1 \quad 0)', \; S_3 = (0 \quad 0 \quad 1)',$$

while no constraints are put on P.

Correspondingly, we have the constraints on the end effector locations of robot

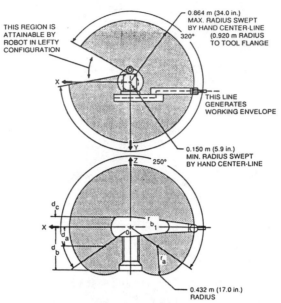

Figure III.3: PUMA 560 Robot Workspace

a and *b* respectively,

$$S_a^1 = \pm(1 \ \ 0 \ \ 0)', \ S_a^2 = \pm(0 \ \ 0 \ \ 1)', \ S_a^3 = \pm(0 \ \ 1 \ \ 0)',$$

$$P_{ax} = P_x, \ P_{ay} = P_y + l/2, \ P_{az} = P_z, \tag{3.2.1}$$

and

$$S_b^1 = \pm(1 \ \ 0 \ \ 0)', \ S_b^2 = \pm(0 \ \ 0 \ \ 1)', \ S_b^3 = \pm(0 \ \ 1 \ \ 0)',$$

$$P_{bx} = -P_x, \ P_{by} = -P_y - d - l/2, \ P_{bz} = P_z, \tag{3.2.2}$$

We also need the inverse kinematics solutions of the two robot arms in order to solve the motion space analysis problem. This will become clear when we present the solution algorithm later.

Possible occupancy conflict for this particular case occurs only between the object and the base links of the two robot arms, since the other links can always move out of the object's occupancy by changing the robot arms' configurations. This condition is given by

$$|P_x| \geq \max\{r_1, r_2\}, \text{if} P_z \leq \max\{l_1, l_2\} \tag{3.2.3}$$

where r_1, r_2 are the radii of the base links of the two robot arms, and l_1, l_2 are the heights of the base links of the two robot arms respectively. Again, collision of the two robot arms is not considered here, though collisions are possible.

Combining with the results obtained in the previous subsection, we can formulate the following algorithm to compute the motion space of the object under the orientation and occupancy conflict constraints,

1. Obtain a tessellation of L, the set of all possible grasping lengths. For a given l in this tessellation, obtain W_p;

2. Obtain a tessellation of W_p. Select a point P in this tessellation;

3. Determine if the rod will collide with the robot arms as given by condition (3.2.3). If yes, go back to 2;

4. Compute θ_a corresponding to the location of the rod from condition (3.2.1), using the inverse kinematics of robot arm a and its joint variable limits. If no solution exists, go back to 2. The inverse kinematics used in our simulations is the 6-joint inverse kinematics solution for Puma 560 robots. In practice, it may be advantageous to use the so-called differential inverse kinematics, as given by $\triangle q = J^{-1} \triangle x$, since the robot task space is already discretized in step 2. Closed form solution for the inverse Jacobian exists for some industrial robot arms such as the Puma 560;

5. Compute θ_b corresponding to the location of the rod from condition (3.2.2), using the inverse kinematics of robot arm b and its joint variable limits. If no solution exists, go back to 2. Otherwise, P belongs to the motion space of the object under the orientation and occupancy conflict constraints;

6. Select the next point in the tessellation of W_p. Go back to 3;

7. Repeat 2–6 over L. The procedure concludes when all tessellation points in L are checked.

We have implemented the above algorithm in a Tektronix 4337 Graphics workstation. Our setup is a pair of PUMA 560 robot arms as installed in the Robotics

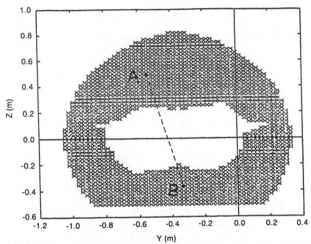

Figure III.4: A Profile of the Motion Space of the Long Rod. $x = 0.25\,\text{m}, l = 1.016\,\text{m}$

Lab at the Department of Systems Science and Mathematics, Washington University. An additional constraint in this setup is the presence of a worktable, which is located in the universal coordinate frame at $z = -0.5334\,\text{m}$. Therefore we have the constraint

$$z \geq -0.5334\,\text{m}.$$

The implementation program is written in Fortran. The results are shown in Figures III.4 and III.5. The universal coordinate frame has been chosen to be identical with the frame of robot a. The distance between the tessellation points is 0.0254 m. Figure III.4 shows the cutout profile of the motion space by the plane defined by $x = 0.25\,\text{m}$ for $l = 1.0160\,\text{m}$. The orientations of the two arms have been fixed to be constant, though the arm configuration still have all the possible choices. For this choice of the grasping length l, there is a sizable hole in the center of the motion space. Therefore, we can draw conclusions relevant to the existence of certain motion paths for this setup. For example, there exists no straight line path between the points A and B as shown in Figure III.4. The direct reason for this highly restricting motion range of the two robot arms is the fixed orientation of the object. This choice of orientation is such that joint 5 of the robot arm a is out of its permissible range most of the time. Figure III.5 is obtained by reducing the

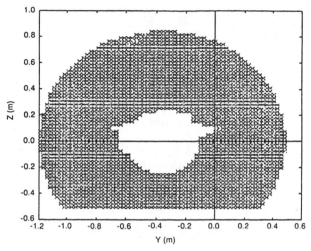

Figure III.5: A Profile of the Motion Space of the Long Rod. $x = 0.25\,\text{m}, l = 0.711\,\text{m}$

grasping length to $l = 0.7112\text{m}$. It can be seen that the hole in the motion space is reduced while the diameter of the motion space is also increased. Therefore, for this particular task, it is preferable to choose this grasping length.

In computing the inverse kinematics of the PUMA 560 robot arms, results obtained by C.S.G. Lee (1984) are applied.

III.3 Dynamic Workspace Analysis of Multiple Cooperating Robot Arms

In this subsection, we formulate and solve the dynamic workspace analysis problem for multiple cooperating robot arms.

Problems relating to the dynamic workspace analysis of multiple robot arms have been addressed by a few authors in their study of coordinated force control of two or more robot arms. A typical problem studied is the force distribution, where a force/torque is to be applied by a system of multiple robot arms to a given object, and the joint drive torques of each arm are to be determined. Since the solution to this problem is usually not unique, various optimization cost functions and techniques can be applied. Works on this problem includes Orin and Oh (1981), Orin and Cheng (1989), Luh and Zheng (1988), and Walker et al. (1988). We note that the force distribution problem is different from the problem we treated previously where we were more interested in the capabilities of multiple robot arms to handle a load, while the force distribution problem concentrates more on how a load can be handled.

When multiple robot arms are cooperating to handle a load, an internal force is generated on the object being handled. The modeling and evaluation of the internal force is discussed in Uchiyama and Dauchez (1988), Walker et al. (1989), and Nakamura (1988).

The approach that we adopt to solve the dynamic workspace analysis problem for multiple cooperating robot arms is similar to the one used to solve the dynamic workspace analysis problem for single robot arms. We will therefore emphasize the peculiarities associated with the case at hand. For more details, we refer the reader to the paper by Li et al. (1990).

We model the multiple robot arms together with the object they are in contact with as a single closed mechanical chain. Assuming that m robot arms are involved, each having n_i one DOF joints, a dynamic model for them has been developed in Tarn et al (1988b) in the form of a set of differential equations,

$$D(q)\ddot{q} + E(q,(\dot{q})) + G(q) = J'_\Theta \Gamma \tag{3.3.1}$$

where $q = (q_1 \ldots q_p)'$ is the set of generalized coordinates corresponding to the p degrees of freedom of the closed mechanical chain, p is given by the Kutzbach-Grubler criterion (Tarn et al., 1988b); $\Gamma = ((\Gamma^1)' \ldots (\Gamma^m)')$ is the vector of the joint actuator drive forces/torques of the mechanical system, $\Gamma^i = (\Gamma_1^i \ldots \Gamma_{n_i}^i)'$ is the vector of joint drive forces/torques of the ith robot arm; $J_\Theta = \frac{\partial \Theta}{\partial q}$ is the Jacobian of the function $\theta = \Theta(q)$, that relates the joint variables of the m robot arms, $\theta = ((\theta^1)' \ldots (\theta^m)')'$, $\theta^i = (\theta_1^i \ldots \theta_{n_i}^i)'$, and the genetalized coordinates, q.

The $D, E,$ and G matrices are the matrices of inertias, Coriolis and centrifugal forces and gravitational force terms, and $n = \sum_{i=1}^m n_i$. These are derived in detail in (Tarn et al., 1988b).

Eq. (3.3.1) is formulated for the use of dynamic robot motion trajectory control and does not yet describe the relation between the force/torque applied by the robot arms and the robot joint torques. We define a 6-dimension output (forward kinematics) function H in the compliance frame C. Then, applying the principle of virtual work, we have

$$D(q)\ddot{q} + E(q,(\dot{q})) + G(q) + J_H' F = J_\Theta' \Gamma \tag{3.3.2}$$

where $J_H = \frac{\partial H}{\partial q}$ is the Jacobian of H.

When multiple robot arms are rigidly connected to a single object and applying forces/torques to it, an internal force/torque is produced, The internal force/torque produces the "squeezing" effect. It will not cause any movement of the object and is withstood only by the mechanical structure of the object. The magnitude of this internal force/torque must in many cases be restricted by an upper bound in order that the object not be damaged. The dynamic workspace analysis of multiple cooperating robot arms needs to be able to deal with this constraint.

Consider the case when an egg is picked up and moved by a set of multiple cooperating robot arms/fingers, the force applied by the robot arms/fingers must not exceed a given limit so that the egg will not break. We note that this limit can be described by the maximum magnitude of the component of the force applied by a robot finger that is normal to the egge shell at the point of contact. Thus the internal force applied by each robot arm/finger that we need to put limit on has a given direction.

Assuming that $F^i, i = 1,\ldots,m$ are the 6-dimension force/torque vectors represented in the compliance frame C applied by the m cooperating robot arms, then, since F^i's are represented in the same coordinate frame as F which is the force/torque vector applied by the m robot arms, we have immediately for F :

$$F = \sum_{i=1}^{m} F^i \tag{3.3.3}$$

or,

$$F = [I_6 \ldots I_6] \begin{bmatrix} F^1 \\ \vdots \\ F^m \end{bmatrix} = W \begin{bmatrix} F^1 \\ \vdots \\ F^m \end{bmatrix}. \tag{3.3.4}$$

Let F_c^i be the component of F^i that corresponds to the internal force/torque, $W^\dagger$ be a pseudo-inverse of W, we have $(F^1 \ldots F^m)' = W^\dagger F + (F_c^1 \ldots F_c^m)$. Thus $\sum_{i=1}^{m} F_c^i = 0$. Therefore, the internal force/torque constraint can be described by

$$\sum_{i=1}^{m} F_c^i = 0, \tag{3.3.5}$$

and

$$| F_c^i \leq F_{cm}, i = 1, \ldots m \qquad\qquad (3.3.6)$$

where F_{cm} is a positive scalar.

Conditions (3.3.4), (3.3.5), (3.3.6) together relate the internal force/torque to the individual and total force/torque, F^i's and F, applied by the m robot arms, and describe a constraint on it.

When no constraint on the internal force/torque is present, the dynamic workspace analysis of multiple cooperating robot arms can be formulated as a simple optimization problem of deciding the maximum magnitude of F in a given direction in the compliance frame C, subject to the constraints of maximum joint driving torques and equation(3.3.2). Solution of this problem can be based on the same approach that we proposed in subsection II.2. Specifically, we have the following optimization problem,

Maximize $| F |$ with respect to $\Gamma_{net'}$,
Subject to

$$F = | F | a,$$
$$J_H' F = J_\Theta' \Gamma_{net},$$
$$\Gamma_{j(net)min}^i \leq \Gamma_{j(net)}^i \leq \Gamma_{j(net)max}^i,$$
$$j = 1, \ldots, n_i, \quad i = 1, \ldots, m.$$

where $\Gamma_{j(net)}^i$ is the net driving torque of the jth joint of the ith arm, $\Gamma_{j(net)min}^i$ and $\Gamma_{j(net)max}^i$ are the net joint driving torque limits.

Note that J_H and J_θ are both constant matrices for given values of q. We assume they are of full rank. Define

$$S_\Gamma = \{\Gamma_{net} \mid \Gamma_{j(net)min}^i \leq \Gamma_{j(net)}^i \leq \Gamma_{j(net)max}^i, j = 1, \ldots, n_i, \quad i = 1, \ldots, m\}.$$

S_Γ is the set of available net joint driving torques to produce F, i.e., the Γ_{net} vectors with each of its components bounded by $\Gamma_{j(net)min}^i$ and $\Gamma_{j(net)max}^i$.

Let S_F denote the set of all force/torque vectors F that the m robot arms can jointly produce, then we have

$$S_F = \{F = (J_H')^\dagger J_\Theta' \Gamma_{net} \mid \Gamma_{net} \epsilon S_\Gamma\},$$

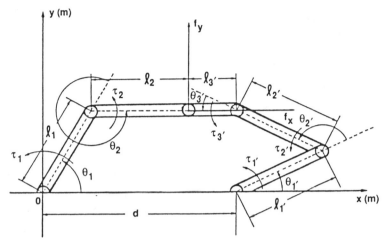

Figure III.6: (a). A Two Arm Chain Consisting of a Two Link Arm and a Three Link Arm

where $(J'_H)^\dagger$ is the pseudo-inverse of J'_H, since J_H is usually a nonsquare matrix.

The solution procedure for the dynamic workspace analysis of multiple robot arms is thus composed of the following major steps,

1. . Compute the mapping $(J'_H)^\dagger J'_\Theta$;

2. , Compute the images of the vertices of S_Γ under $(J'_H)^\dagger J'_\Theta$;

3. . Form the convex hull of the image points obtained in 2. above to obtain S_F in F;

4. . Solve for the intersection of a vector in the direction of a and the boundary of S_F.

The following is an example that illustrates the application of the above method to a simple two arm closed chain.

Example III.1. We consider a two arm closed chain consisting of a two link planar arm and a three link planar arm rigidly connected together at the end points of their last links, as shown in Figure III.6(a). The parameters of the two robots are given as $-10n.m. \leq \tau_1, \tau_2 \leq 10n.m., l_1 = l_2 = 1m.$ for the first arm, and $-10n.m. \leq \tau'_1, \tau'_2 \leq 10n.m., -5n.m. \leq \tau'_3 \leq 5n.m., l_{1'} = l_{2'} = 1m, l_{3'} = 0.5m$ for the second arm. The distance between the center points of the bases of the two arms is

$d = 2m$. We present here only numeric results because of space limits. We choose generalized variables q to be $q = (q_1 q_2)' = (\theta_1 \theta_2)'$. That is, they are chosen to be the joint variables of the two link arm. The origin of the compliance frame C is assumed to be at the connection point of the two arms, and its axes are chosen to be parallel to those of the world frame. Based on these choices and the given data, we have, at $q = (\frac{\pi}{3}, \frac{5\pi}{3})'$, $\theta_{1'} = 25.66°, \theta_{2'} = 128.68°, \theta_{3'} = 25.66°$. These values are obtained from the relation $\theta = \Theta(q)$. After we obtain the analytic forms of $J_H(q)$ and $J_\Theta(q)$, we can plug the value of q into them to obtain

$$J_H'(\frac{\pi}{3}, \frac{5\pi}{3}) = \begin{bmatrix} \frac{-\sqrt{3}}{2} & \frac{3}{2} \\ 0 & 1 \end{bmatrix},$$

$$j_\Theta'(\frac{\pi}{3}, \frac{\pi}{3}) = \begin{bmatrix} 1 & 0 & -8.465 & 18.93 & -9.465 \\ 0 & 1 & -7.100 & 14.20 & -6.100 \end{bmatrix}.$$

We ignore the gravity terms and assume that the robot chain is at a steady state. Then we have

$$S_\tau = \{\tau \mid -10n.m. \leq \tau_1, \tau_2, \tau_{1'}, \tau_{2'} \leq 10n.m., -5n.m. \leq \tau_{3'} \leq 5n.m.\}$$

which is a 5-dimensional rectangular parallelepiped. The image S_F of S_τ under $(J_H'(\frac{\pi}{3}, \frac{5\pi}{3}))^{-1} J_\Theta'(\frac{\pi}{3}, \frac{5\pi}{3}))$ is a convex decagon. The graphes of S_τ and S_F are shown in Figures III.6(b) and III.6(c). The mapping between S_τ and S_F is such that point (A,L) is mapped to P_1, (B,L) to P_2, (B,I) to P_3, (B,J) to P_4, (B,F) to P_5, (C,F) to P_6, (D,F) to P_7, (D,G) to P_8, (D,H) to P_9, and (D,L) to P_{10}. Given a direction in the compliance frame C in which we wish to apply a force as large as possible, we can immediately find an F vector in this direction on the boundary of S_F. For example, the maximum load this robot chain can lift is the maximum force it can apply along the f_y axis. This is given by the vector F_1 in Figure III.6(c), which has a length of 170.83 n.

Next, we modify the solution procedure presented above to apply to the dynamic workspace analysis of multiple robot arms with internal force constraints.

When the internal force/torque constraint is present, the optimization problem associated with the dynamic workspace analysis problem for multiple robot arms has the following form:

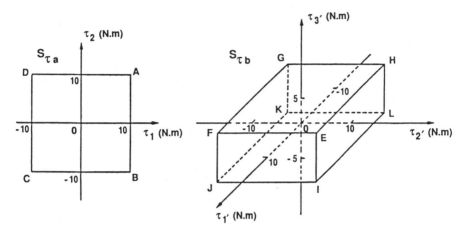

Figure III.6: (b). S_τ for the Two Arm Chain, $S_\tau = \{(\tau_a^T \; \tau_b^T)^T \; | \; \tau \in S_{\tau a}, \tau \in S_{\tau b}\}$

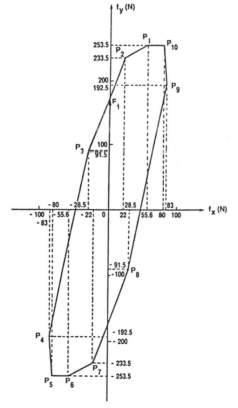

Figure III.6: (c). S_F for the Two Arm Chain at $q = (\frac{\pi}{3}, \frac{5\pi}{3})$

Maximize $|F|$ with respect to Γ_{net},

Subject to

$$F = |F|\, a,$$
$$\sum_{i=1}^{m} F_c^i = 0,$$
$$|F_c^i| \leq F_{cm}, i = 1, \ldots, m,$$
$$J_H' F = J_\Theta' \Gamma_{net},$$
$$\Gamma_{j(net)min}^i \leq \Gamma_{j(net)}^i \leq \Gamma_{j(net)max}^i,$$
$$j = 1, \ldots, n_j, \quad i = 1, \ldots, m,$$

where $F_c^i, i = 1, \ldots, m$ are as defined previously.

We define the following notations in our discussion of maximizing F while limiting the magnitude of internal force/torque:

$$\Gamma_{net}^i = (\Gamma_{1(net)}^i \ldots \Gamma_{n_i(net)}^i)',$$
$$J_\Theta^i = (\frac{\partial \Theta_i}{\partial q}), i = 1, \ldots, m,$$
$$S_\Gamma^i = \{\Gamma_{net}^i \mid \Gamma_{j(net)min}^i \leq \Gamma_{j(net)}^i \leq \Gamma_{j(net)max}^i, 1 \leq j \leq n_i\},$$
$$S_F^i = \{F^i = (J_H')^\dagger (J_\Theta^i)' \Gamma_{net}^i \mid \Gamma_{net}^i \epsilon S_\Gamma^i\},$$
$$i = 1, \ldots, m.$$

Define $\sum_{i=1}^{m} S_F^i = \{\sum_{i=1}^{m} F^i \mid F^i \epsilon S_F, i = 1, \ldots, m\}$. It is easy to see that $S_F = \sum_{i=1}^{m} S_F^i$. That is, the polygon S_F can be found by obtaining S_F^i's first and then adding them together.

Thus the optimization problem associated with the dynamic workspace analysis of multiple cooperating robot arms with internal force constraint can be written as:

Maximize $|\sum_{i=1}^{m} F^i|$ with respect to Γ_{net},

Subject to

$$F^i \epsilon S_F^i, i = 1, \ldots, m,$$
$$\sum_{i=1}^{m} F_c^i = 0,$$
$$|F_c^i| \leq F_{cm}, i = 1, \ldots, m,$$
$$\sum_{i=1}^{m} F^i = |\sum_{i=1}^{m} F^i|\, a.$$

The solution of the above problem, based on the previous discussion, can be obtained by the following procedure:

1. . Compute polygons $S_F^i, i = 1, \ldots, m$, through their own definitions;

2. . Search the boundary of each S_F^i to find F^i's such that $\sum_{i=1}^m F_c^i = 0$, $\mid F_c^i \mid \leq F_{cm}$, $i = 1, \ldots, m$, and $F = \sum_{i=1}^m F^i$ is maximal.

Obviously, the above procedure provides only a guideline to find the maximum force/torque that m cooperating robot arms can apply to the environment in a given direction without violating the internal force/torque constraint. Efficient computation algorithms are to be developed to speed up the computation. The following simple example illustrates the above approach.

Example III.2. We consider a two arm closed chain consisting of two identical two link robot arms configured as in Figure III.7(a). We find the maximum weight this two arm chain can lift (maximum force applicable in the f_y direction) under the internal force constraint $\mid F_c^i \mid \leq F_{cm}$, where the internal force direction is given as parallel to the f_x direction. The parameters of the two arm chain are as follows: $l_1 = l_2 = l_{1'} = l_{2'} = 1m.$, $\mid \tau_i \mid \leq 10n.m.$, $i = 1, 2, 1', 2'$. The configuration of the two arms are given as $\theta_1 = \frac{\pi}{2}$, $\theta_2 = -\frac{\pi}{6}$, $\theta_{1'} = \frac{\pi}{2}$, $\theta_{2'} = \frac{\pi}{6}$. Given these parameters and the configuration of the two arm chain we obtain the maximum force parallelegrams S_F^1 and S_F^2 as shown in Figure III.7(b). The maximum weight that this two arm chain can lift is obtained by adding up the components of F^1 in S_F^1 and F^2 in S_F^2 in the direction of f_y, while maintaining the components of F^1 and F^2 in the f_x direction to be of the same magnitude but with opposite direction. It can be seen that, under these requirements, the maximum weight this two arm chain can lift remains at 20 n. as long as $10\sqrt{3}n. \leq F_{cm} \leq 40 + 10\sqrt{3}n.$ The maximum weight it can lift decreases if $F_{cm} < 10\sqrt{3}n.$

IV. INTELLIGENT MACHINE MOTION PLANNING AND CONTROL

In this section, we discuss the impact of the nonlinear feedback controller on the robot motion planing and on the structure of the corresponding robot motion planner. For the structure of the nonlinear feedback robot controller, refer to the chapter "Nonlinear Control Algorithms in Robotic Systems" by T.J. Tarn and A.K. Bejczy in this volume. In subsection IV.1, we discuss the impact and requirement of

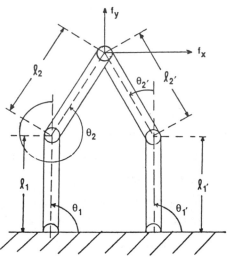

Figure III.7: (a). A Two Arm Chain Consisting of Two Identical Two Link Arms

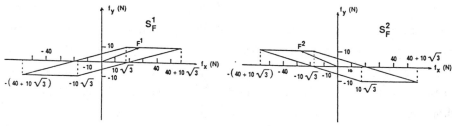

Figure III.7: (b). Maximum Force Parallelograms of the Two Arm Chain in Figure
III.7 (a).

the nonlinear feedback robot controller on the robot motion planner and illustrate
our point with an example. We introduce a description of intelligent robot motion
planning and present a structure of the robot motion planning and control system
in subsection IV.2 that is based on the discussion of the two previous sections. The
components of this robot motion planning and control structure is elaborated based
on our perceptions of intelligent robot motion planning and on the materials presented in the other sections of this chapter. An implementation example of the
proposed robot motion planner and controller is given in subsection IV.3.

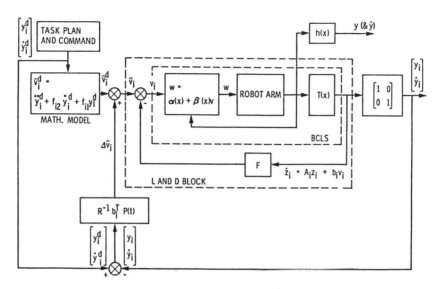

Figure IV.1: Configuration Diagram of the Nonlinear Feedback Robot Controller

IV.1 Task Planning for the Nonlinear Feedback Robot Controller

The feedback linearized and simultaneously output decoupled dynamic control technique based on nonlinear feedback and state transformation actually transforms robot arm control problems from the joint space to the task space and performs robot servoing in terms of task space variables within a linear system frame, allowing also the use of powerful techniques from optimal control of linear systems. On the joint space level, this dynamic control technique commands drive forces or torques or their equivalent quantities addressed to the joint drive system.

The "task-level" nature of the new dynamic control technique becomes apparent by examining the upper left part of Figure IV.1. As seen there, the expected output from a task planning program is a commanded trajectory "y" in the task space as a function of time. From this function and its time derivatives a linear mathematical model "v" is constructed which becomes a linear control input to the feedback linearized robot arm. Thus, the planned and commanded trajectory together with its time derivatives *directly drives* the robot arm through a linear system model.

We also note that the optimal error correction loop shown in the lower left part of

Figure IV.1 directly operates on the task level and not on the joint control level. The task level errors are then decomposed by the nonlinear gain matrix $\beta(x)$ into joint force or torque drive commands as shown in Figure IV.1. It is also important to note that the nonlinear compensating vector $\alpha(x)$ and the nonlinear gain matrix $\beta(x)$ in the nonlinear feedback equation do not change from task to task, unless the load carried by the robot hand is changing drastically relative to a nominal value assumed in the nonlinear feedback design. In that case, changes in the $\alpha(x)$ vector and the $\beta(x)$ matrix can be carried out by using straightforward algebraic formulas (Tarn et al., 1988b). Consequently, the basic control method do not need readjustment from task to task. In that sense the control method is "intelligent" since it is capable of directly responding to changing task level commands "y" embodied in the linear task control input "v".

With the development of this new robot controller, robot task planning is made much easier. Transformation from task space trajectory to joint space motion commands is no longer necessary since this transformation is now carried out by the nonlinear feedback. Typically, task planning now can be formulated as follows: given certain goal data on the six d.o.f. movement of the robot hand in the task space, find a mathematical representation of a six d.o.f. robot hand motion trajectory in the task space in the form of a function of time which can be used as input to the robot arm control system to achieve a desired robot hand motion. The given goal data can include a geometric description of a curve in the 3-D space and the robot hand's orientation along it, or points in the 3-D space that the robot hand has to pass through, specification of velocities at some points in the task space including initial and final velocities. As soon as a function of time is obtained for any desired robot hand motion trajectory, it can be used directly to formulate the control input to the nonlinear feedback robot controller.

As an example, we show below the use of polynomial approximation in the robot task planning. This is rather easy for desired end effector positions which, in general, is given by a curve γ in the task space. In the task space we fix nothing but the origin. Thus points on γ can be viewed as vectors starting from the origin, i.e., bound vectors. The motion description by polynomial approximation is independent of the choice of the task space reference frame since all coordinate systems in R^3 are related by nonsingular transformations. For orientation vectors, w, however, we do not have immediate mathematical representation since they are free vectors in R^3. They have to be transformed into bound vectors before polynomial approximation can be applied. This transformation can be done via the moving frames on γ.

Suppose that the robot end effector is required to travel along a straight line segment in the task space from point A at time $t_0 = 0$ to point B at time t_f such that the end effector is at rest at points A and B. This implies

$$\dot{\gamma}(t_0) = \dot{\gamma}(t_f) = 0 .$$

We would also like the robot motion from A to B to be smooth, i.e., the acceleration and its time derivative should be continuous along γ and

$$\ddot{\gamma}(t_0) = \ddot{\gamma}(t_f) = 0 .$$

A fifth order polynomial representation of the line segment $\overline{AB}$ would satisfy these requirements. That is, we choose

$$\tilde{\gamma}(t) = a_5 t^5 + a_4 t^4 + a_3 t^3 + a_2 t^2 + a_1 l + a_0 ,$$

with conditions

$$\tilde{\gamma}(t_0) = A, \tilde{\gamma}(t_f) = B,$$
$$\dot{\gamma}(t_0) = \tilde{\gamma}(t_f) = \ddot{\tilde{\gamma}}(t_0) = \ddot{\tilde{\gamma}}(t_f) = 0 .$$

In the task space universal frame we have:

$$\tilde{\gamma}(t) = \begin{bmatrix} x(t) \\ y(t) \\ z(t) \end{bmatrix} = \begin{bmatrix} \gamma(t)\sin\theta_B\cos\phi_B + x_A \\ \gamma(t)\sin\theta_B\sin\phi_B + y_A \\ \gamma(t)\cos\theta_B + z_A \end{bmatrix}$$

Denoting the length of $\overline{AB}$ by L, we find from six algebraic equations in t under the specified initial and final conditions that

$$\gamma(t) = \frac{6L}{t_f^5}t^5 - \frac{15L}{t_f^4}t^4 + \frac{10L}{t_f^3}t^3 .$$

We note that this $\gamma(t)$ has a very desirable velocity profile (Figure IV.2). It reaches its maximum value at $\frac{t_f}{2}$ where the acceleration becomes zero. We also have that

$$\ddot{\gamma}(t) > 0, \text{ for } t < t_f/2$$

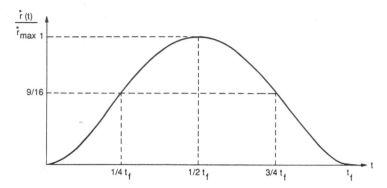

Figure IV.2: Velocity Profile of a 5th Order Polynomial

and

$$\ddot{\gamma}(t) < 0, \text{ for } t > t_f/2.$$

These facts also facilitate the handling of cases when, along with the value of L, the maximum velocity is given instead of t_f. In that case we have to find t_f by solving a system of four algebraic equations which is always solvable since the coefficients of the four equations form linearly independent vectors.

Suppose now that we have an orientation requirement for the robot end effector (e/e) while it is moving from A to B as follows:

At t_0:	e/e roll axis is parallel to x axis of task space.
For $t \le t_1$:	orientation is unchanged.
For $t \epsilon [t_1, t_2]$:	e/e roll axis is changed to become parallel to Z axis of task space, where $t_0 \le t_1 \le t_2 \le t_f$ for given t_1 and t_2.
For $t \ge t_2$:	orientation is unchanged.

This orientation requirement as stated is not representable as a closed function of time. Therefore, as we discussed before, parallel coordinate frames along $\overline{AB}$ need to be assigned so that the orientation can be written as

$$w(t) = [\tilde{\gamma}(t), \nu(t)] \text{ for } t_0 \le t \le t_f.$$

This description effectively projects the orientation vectors into one single R^3 space W.

Without loss of generality, we assume that orientation vectors are unit vectors, and we choose the path of them in W to be the arc of radius 1 in the first quadrant in the $X - Z$ plane. Then we only have to find a representation for $\nu(t)$. We have

$$
\nu = \begin{bmatrix} 1 \\ 0 \\ 0 \end{bmatrix} \quad \text{for } t_0 \leq t \leq t_1 \,,
$$

$$
\nu = \begin{bmatrix} 0 \\ 0 \\ 1 \end{bmatrix} \quad \text{for } t_2 \leq t \leq t_f \,.
$$

To provide smooth transition in orientation, we require that

$$
\dot{\nu}(t_1) = \dot{\nu}(t_2) = \ddot{\nu}(t_1) = \ddot{\nu}(t_2) = 0 \,.
$$

After some simple algebraic manipulations and having introduced again a fifth order polynomial to represent $\nu(t)$ in a way similar to the representation of $\gamma(t)$, we obtain for the orientation task:

$$
\tilde{\nu} = \begin{cases} x(t) \\ y(t) \\ z(t) \end{cases} ,
$$

where

$$
\begin{aligned}
x(t) &= \sqrt{1 - z^2(t)} && \text{for } 0 \leq t \leq t_f \,, \\
y(t) &= 0 && \text{for } 0 \leq t \leq t_f \,, \\
z(t) &= \begin{cases} 0 & \text{for } 0 \leq t \leq t_1 \,, \\ (t - t_1)^3 [p_5(t - t_1)^2 + p_4(t - t_1) + p_3] & \text{for } t_1 < t \leq t_2 \,, \\ 1 & \text{for } t_2 < t \leq t_f \,, \end{cases}
\end{aligned}
$$

with

$$
p_5 = \frac{6}{(t_2 - t_1)^5} \,, \quad p_4 = \frac{15}{(t_2 - t_1)^4} \,, \quad p_3 = \frac{10}{(t_2 - t_1)^3} \,.
$$

The above equations then can be used to obtain the orientation task control input to the nonlinear feedback robot controller. This orientation representation has the same desirable properties as does the position representation discussed earlier.

It can be seen from the above example that task planning with the nonlinear feedback robot controller is indeed very easy once the geometric path and the relevant goal data are given for the robot motion trajectory. The nonlinear feedback robot controller is intelligent in the sense that it can "understand" task space motion commands. However, this also makes possible that more "intelligent" features can be combined into robot task planning such that the robot control system becomes more autonomous, being able to make certain motion-related decisions on its own. Therefore, intelligent robot motion planning becomes more feasible with the development of the nonlinear feedback robot controller. We discuss in greater detail what we mean by intelligent robot motion planning in the next section.

IV.2 Intelligent Robot Motion Planning and its Structure

Robot motion planning, as it is now, includes the steps of inputing a set of data that generally specify the initial and final robot end effector locations, transforming them into the joint positions, and interpolating to obtain a continuous motion trajectory. The input data typically do not include specification on the dynamic aspects of the motion trajectory, e.g., the velocity and acceleration profiles. This is caused by the limits of the state-of-the-art robot controller. With the development of the nonlinear feedback robot controller, it is now possible to obtain the complete robot motion trajectory in the robot task space as a time function consistent with the robot's dynamic capabilities and use it as the input to the robot controller. The planner can concentrate on the more intelligent generation of executable motion trajectories that satisfy the robot kinematic and dynamic constraints as well as constraints of the environment.

The intelligent robot motion planning has been receiving more and more attention lately. We started our work in this direction in (Tarn et al., 1986). Brockett discussed the distribution of the computation and the organization of the communication in the motion control of multiple-degree automated machines in (Brockett, 1988). Meystel (1988) gave a detailed discussion on the subject of intelligent control.

The realization of the intelligent robot motion planning depends on the use of sensor feedback information. Literature exists on the use of sensor based control of robotic manipulators, such as in (Lee et al.,1985, Weiss et al., 1987). However, our emphasis here will be rather on the use of sensor information for the motion planning than for the servo control. The dynamic and kinematic constraints of the robots themselves must also be satisfied by any executable robot motion trajectory. Progresses are being made steadily on combining the robot dynamic constraints with motion planning. Tarn et al. (1988a) presented a computationally simple algorithm to evaluate the effect of joint actuator constraints on the maximum force/torque that a robot arm can apply in its task space. Pfeiffer and Johanni(1986) gave a new concept for the use of dynamic constraints in the planning of minimum time path for robots and Shiller and Dubowsky(1987) presented a very efficient algorithm for the same purpose. The kinematic constraints have also received their due attention in robot motion planning. Preliminary results on the generation of motion trajectories that avoids collision with the obstacles and other machines in the robot workspace are reported in (Lee, Lee, 1987, Freund, Hoyer, 1988, Nagata et al., 1988). The most prominent use of the kinematic constraints are however in the robot workspace analysis as part of robot motion planning. This has been done for various robot arms, see (Duffy, 1980). We also developed a general algorithm for the analysis of the motion space of an object handled by two robot arms (subsection III.1 and III.2 of this chapter, Li et al, 1989).

The definition of intelligent robot motion planning can indeed be various. Here we present what we view should be the necessary functions of intelligent robot motion planning:

1. Parsing of initial goal data given by the robot operator for the desired robot motion and generation of individual robot motion subtasks. We define a sub-task to be a single robot movement from start to stop. For example, when the robot is required to perform an assembly task, the task is to be decomposed into robot end effector movements between various initial and final robot task space locations. Each of these movements will be viewed as a subtask.

2. Formulation of a geometric motion path for each desired robot movement based on given goal data and sensor data on the environment. Note here that we use the term "path" to indicate a geometric curve in the robot task space, while the term "trajectory" is the path parametrized by the time, i.e., a time function. The criterion of selecting the geometric motion path, such as minimum time path or shortest path, should be given by the operator.

3. Parametrization of the geometric path into a time function continuous to the third derivative with respect to time. We suggest the use of fifth order polynomials for this purpose.

4. Verifying that the time function representation of the robot motion trajectory satisfies the robot kinematic and dynamic constraints, as well as environmental constraints by performing the kinematic and dynamic robot motion space analysis.

5. Incorporating the capability to modify a planned robot motion trajectory in the event that the robot kinematic and dynamic constraints and the environment constraints are violated. The decision as to when and how the modification should be done must be based on feedback information from lower levels of the robot motion planning and control structure. It is therefore mandatory that the lower level feedback contains enough information so that proper decisions can be made by the higher levels.

6. Selecting the proper controller design and determining the appropriate controller parameters for the task to be performed.

7. Passing the executable robot motion trajectory obtained from above to the selected robot controller.

We propose a hierarchical structure for the intelligent robot motion planning and control. This structure is shown in Figure IV.3. In the following subsections, we define and elaborate on the components of this hierarchical structure.

IV.2.1. The Interfacing/Subtask Generation Level

The top level of the hierarchy is the planner/operator interfacing and subtask generation level. The robot motion planner receives robot motion commands form the robot operator in this level. Determination of the minimal robot operator commands for a task is still an open research problem. In our design, these commands should enable the robot motion planner to generate basic motion subtasks such as moving between two locations, tracking a given task space curve, applying a force to

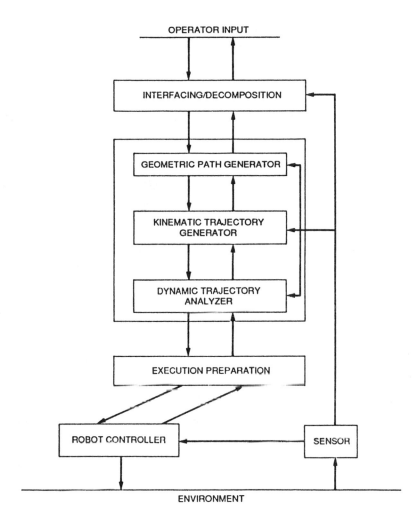

Figure IV.3: Hierarchical Structure of Intelligent Robot Motion Planner

the environment. The operator input commands will then have to contain data on the start and finish locations of each subtask and the relative time/space sequence between the subtasks. Additional information can also be provided to help the robot motion planner carry out the motion planning. This information can include the maximal total time for performing the task, the criterion to be used in motion path selection such as minimal time trajectories or shortest task space paths.

The interfacing/subtask generation level, upon receiving these commands, will analyze them, determine the number of robot arms needed for the task, and decompose the overall task into motion subtasks for each robot arm in the manufacturing workcell accordingly. The subtask generation follows a priori given rules and is programmed to work with a class of tasks. For example, a part assembly task that involves the use of two cooperating robot arms can be decomposed into the following subtasks for each of the robot arm: moving to the part pickup location; picking up the part; moving to the part assembly location; assembly of the parts. In some cases, the part assembly subtask can be further decomposed. The motion goal points for each subtask of each robot arm, such as the initial and final locations of the robot end effectors, the maximum motion time for each subtask, are generated after the subtask generation. Furthermore, this level would reject obviously inexecutable motion requirements from the operator, such as unreasonablely high motion velocity or load, thus incorporating a certain degree of decision making capability. Another function of this level is to report the status of the robot motion planning and control system to the operator. If the desired motion fails, the specific reason of the failure is passed up from lower levels of the system and reported to the operator.

IV.2.2. The Motion Trajectory Generator

The second level is the motion trajectory generator. The executable robot motion trajectory for each subtask as specified in the interfacing/subtask generation level is obtained here. The detailed description of the components in the motion trajectory generator depends to some extent on the robot motion trajectory generation strategy adopted. As an example, we describe the components of the motion trajectory generator as follows assuming that the shortest motion path is the design goal.

1. The geometric robot motion path generator.

The geometric robot motion path for each subtask is formulated here. Based on the description of the robot workspace, geometric constraints on the motion path such as the existence of obstacles obtained from vision systems, and the desired initial and final locations of robot motion, robot kinematic motion space analysis is performed and a smooth shortest robot task space motion path is formulated. In general, this path can be composed of straight line segments connected smoothly by circular arcs. The smoothness constraint is imposed because of the requirement of the nonlinear feedback controller as well as for reduction of wear of the robot hardware. All of the components of the path will have analytical representations so that the path can be readily parametrized as a time function and used as input to the nonlinear robot controller. Finally, the geometric path generator is to determine the geometric feasibility of the required robot motion. That is, e.g., if there does not exist a path between the initial and final locations of the desired motion the geometric path generator is to report to the higher levels this conclusion.

In some cases, the planned geometric path can not be executed since it may violate the robot dynamic constraints with respect to a given maximum total traveling time. This would be decided by the next two components in the motion trajectory generator. When this happens, it is possible that the geometric path can be modified so that it becomes executable. This happens when the turning radius of the path is increased.

2. The kinematic robot motion trajectory generator.

Once a satisfactory geometric path has been obtained by the geometric path generator, it needs to be parametrized into a function of time, with velocity and acceleration satisfying the time constraints specified by the operator. Furthermore, the robot motion trajectory needs to be continuous at least to the second derivative. This is mandated by the nonlinear feedback robot controller. We have discussed this problem in (Tarn et al, 1986). Our solution is to parametrize the straight line parts of the geometric robot motion path by a 5th polynomial and connect these parts together smoothly by arcs of certain radii. These radii can be expressed as functions of the total traveling time, while the velocity profile of the 5th order polynomials can be specified by the robot operator.

Thus, our task space motion planning uses polynomial approximation. This is rather easy for end effector position which, in general, is given by a curve γ in the task space. In the task space we fix nothing but the origin. Points on the curve γ can then be viewed as vectors starting from the origin, i.e., bound vectors. This motion description by polynomial approximation is independent of the choice of the task space reference frame since all coordinate systems in $\mathbb{R}^3$ are related by nonsingular transform mapping. For orientation vectors, v, however, we don't have immediate mathematical representation since they are free vectors in $\mathbb{R}^3$. Thus, we have to transform them into bound vectors like the position vectors γ that describe the curve $\gamma(t)$. This transform can be accomplished via the frame bundle, i.e., via moving frames on γ (Boothby, 1975). Instead of detailing the theory which can be found in (Tarn et al, 1986), we use later in this Chapter (see IV.3) a simple example to illustrate our task space position and orientation planning procedures.

The kinematic motion trajectory generator also needs to have the feature of modifying the velocity and acceleration profile of a previously generated robot motion trajectory based on lower level feedback information. The parameters of the robot motion trajectory that can be modified to satisfy the robot dynamic constraints can be various. For path tracking problems, the total traveling time has the most direct effect on the dynamic properties of the planned motion trajectory. The kinematic motion generator, on receiving the information that the planned trajectory cannot be executed for violation of the dynamic constraints in path tracking, will increase the total traveling time until either the trajectory becomes executable or a ceiling on the total traveling time is reached. In the latter case, it is reported to the upper level that the planned geometric path is not executable.

3. The dynamic robot motion trajectory analyzer.

Once a robot motion trajectory is generated by the above two components of the motion trajectory generator, it needs to be tested against the robot dynamic constraints to ensure that it does not violate the robot's dynamic capabilities. These capabilities are described by the robot joint drive torque/force limits and their relation to the robot task space velocity and acceleration values, as well as to the robot force output. Verification to ensure that the planned robot motion velocity and acceleration are within the maximum ones for the robot motion path is to be done by simulation to determine the corresponding robot joint drive torques required. If the joint drive torque limits are violated, the planned robot motion trajectory is rejected. The reason of rejection should also be reported to the kinematic motion generator. If it is the force to be applied that exceeds the robot dynamic capabilities, then it is simply reported that the desired force can not be applied. Otherwise, the kinematic motion generator attempts to change the total traveling time so that the joint driving torques needed to execute the planned motion trajectory can be kept within their limits.

IV.2.3. The Execution Preparation Level

The third level of the robot motion planning and control system makes the connection between the robot motion planner and the robot controller. It prepares the robot controller for the execution of the robot motion trajectory obtained from the higher levels.

Since the robot arm can be expected to perform intrinsically different tasks, implementations of different control algorithms are necessary. For trajectory tracking tasks, the nonlinear feedback robot controller is used. It is intimately compatible with the motion planner in the above level. A fine motion controller is also necessary for local maneuvering. For force control tasks, a controller based on the nonlinear feedback controller has been developed in (Yun, 1987). Decision on which one of these controller to use is made by a decision component in this level. This component accepts control algorithm selection commands from higher levels. It also makes decision on its own when such commands are absent, based on the description of the subtask to be performed. Finally, this level can incorporate the capability of tuning the control parameters such as control feedback gains according to the properties of the planned robot motion trajectory. A set of predetermined optimal feedback gains can be stored for the various applications and recalled when necessary.

After a robot motion trajectory for a subtask that satisfies the robot kinematic and dynamic constraints as well as those of the environment is obtained by the robot motion planner as outlined above, it can now be passed to the robot servo controller along with the parameters selected for the controller. The controller and the sensors interacts directly with the environment. The sensory block as shown in Figure IV.3 contains the sensors and all sensory data processing algorithms. The vision sensor system is used to detect geometric constraints in the environment and the data obtained is sent to the geometric motion path generator for robot kinematic workspace analysis. Measuring systems and force sensors are interfaced to the robot controllers for sensor guided control.

We emphasize that the robot motion planner as discussed above are used for off-line operations. Sensor guided control is realized only at the controller level. Furthermore, the robot motion planner as shown here is designed to perform time-based robot motion planning, i.e., the motion planning depends on the specification of a given set of initial and final time and the planned motion is performed accroding to the time evolution. This does not rule out the possibility of the planned time-based robot motion trajectory to become inexecutable because of uncertainties, such as the case when the task execution time has run up and the robot end effector is not yet at the desired final location. Such cases are dealt with by the so called event based robot motion planning. We do not address this type of motion planning here. However, it is worthy of pointing out that the robot motion planner as shown here does not exclude the incorporation of event based motion planning schemes.

IV.3 Two Robot Arms Traveling along Circular Arcs Synchronously

In this section, we look at the motion planning for two cooperating robot arms performing the following task: the two arms are grasping the two ends of a rigid rod and traveling along circular arcs of given center, radius, orientations, and length. Given these information, as well as a bound on the time it takes for the robot arms

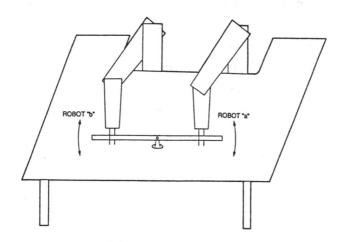

Figure IV.4: Two Robot Arms Cooperating to Turn a Big Latch

to travel the arcs, we can derive the equations needed by the motion planner to formulate motion trajectories for the two robot arms performing such tasks. This task is performed by the two PUMA 560 robot arms installed in the Robotics Lab at Washington University. An example of such tasks is the case when two robot arms are used to turn a big latch as illustrated in Figure IV.4. The equations obtained for this task will be stored in the motion planner together with equations describing other basic task space curves to be used by the robot motion planner for the automatic generation of robot motion trajectories.

The circular arcs to be traveled by the two robot arms are specified as follows: they are located in a plane parallel to the x-y plane of the PUMA 560 universal coordinate frame, defined by $z_0 = -0.4$ m. The two arcs have the same radius and arc length, denoted by r_c and l_c. The centers of the two arcs are given by $(x_{a0}\, y_{a0}\, z_0)^T$ for robot arm a and $(x_{b0}\, y_{b0}\, z_0)^T$ for robot arm b. In addition, we are given an upper bound of the time it takes for the two robot arms to travel through the arcs, T_{max}. The robot end effectors will grasp the rod vertically such that the z-axis will be aligned with the negative z direction of the universal coordinate frame. The other axes of the robot end effectors will be fixed such that the orientation of the rod does not change during the motion. Based on these information, we derive the equations of the arcs and their time-parametrizations to obtain task space motion trajectories that can be used directly by the nonlinear feedback robot controller to formulate input command.

Since we envision that robot motion trajectories in general can be given as combinations of straight line segments and circular arcs, we give a general representation of such arcs in the following. Given an arbitrary circular arc in the robot task space, its origin and radius, a coordinate frame can be fixed in the plane containing the arc such that its x-axis is normal to the plane and its y and z axis chosen according to the right hand rule, its origin chosen to be the center of the arc. We call this coordinate frame the arc frame. To further simplify the computations, we can choose the x-axis of the arc frame to be such that its projection in the x-y plane of the universal coordinate frame is parallel to the x-axis of the universal frame. Once the arc frame is fixed, it is related to a coordinate frame parallel to the universal frame by two rotational transformations: one to rotate the arc frame around its x-axis to an intermediate frame whose z-axis is parallel to the x-z plane of the universal frame, the other to rotate around the y-axis of the intermediate frame to bring it into a frame parallel to the universal frame. Let ϕ and ψ denote the positive angles of the two rotations respectively according to the right hand rule. Then we have, in general, the following equations which describe an arbitrary circle in the robot task space:

$$
\begin{aligned}
x &= r_0 \cos \psi \cos \theta - r_0 \sin \psi \sin \phi \cos \theta + x_0, \\
y &= r_0 \cos \phi \sin \theta + y_0, \\
z &= -r_0 \sin \psi \cos \theta - r_0 \cos \psi \sin \phi \cos \theta + z_0,
\end{aligned}
\tag{4.3.1}
$$

where $(x_0 \ y_0 \ z_0)^T$ is the origin of the circle, r_0 is the radius of the circle, and $0 \leq \theta < 2\pi$ is the variable of a point's position on the circle.

The time-parametrization of the above equation takes the simple form

$$
\begin{aligned}
x &= r_0 \cos \psi \cos \omega(t) - r_0 \sin \psi \sin \phi \cos \omega(t) + x_0, \\
y &= r_0 \cos \phi \sin \omega(t) + y_0, \\
z &= -r_0 \sin \psi \cos \omega(t) - r_0 \cos \psi \sin \phi \cos \omega(t) + z_0,
\end{aligned}
\tag{4.3.2}
$$

where $\omega(t)$ is the velocity at which a point, here the tip of the robot end effector, is traveling along the circle.

If it is desired that the robot end effector travels only part of the circle, i.e., an arc, defined by the condition

$$\theta_0 \leq \theta \leq \theta_f,$$

in the time interval

$$t_0 \leq t \leq t_f,$$

then we need to have

$$
\begin{aligned}
x &= r_0 \cos \psi \cos(\omega(t) + \theta_0) - r_0 \sin \psi \sin \phi \cos(\omega(t) + \theta_0) + x_0, \\
y &= r_0 \cos \phi \sin(\omega(t) + \theta_0) + y_0, \\
z &= -r_0 \sin \psi \cos(\omega(t) + \theta_0) - r_0 \cos \psi \sin \phi \cos(\omega(t) + \theta_0) + z_0, \quad (4.3.3)
\end{aligned}
$$

for

$$\omega(t_0) = 0, \text{ and } \omega(t_f) = \theta_f - \theta_0.$$

This is in consideration of the fact that $\omega(t)$ is the velocity along the arc and we require $\omega(t)$ to be zero when the motion of the robot end effector starts at θ_0.

With Eq. (4.3.3) available, we can parametrize the equations describing an arbitrary arc in the robot task space. Specifically, we let $\omega(t)$ be a 5th order polynomial, then we have,

$$\omega(t) = a_0 + a_1 t + a_2 t^2 + a_3 t^3 + a_4 t^4 + a_5 t^5, \quad (4.3.4)$$

with

$$
\begin{aligned}
\omega(t_0) &= 0, & \omega(t_f) &= \theta_f - \theta_0; \\
\dot{\omega}(t_0) &= 0, & \dot{\omega}(t_f) &= \omega_f; \\
\ddot{\omega}(t_0) &= & \ddot{\omega}(t_f) &= 0,
\end{aligned}
$$

assuming that the robot end effector has velocity ω_f at the position θ_f and maintains smooth acceleration and deceleration along the arc. Assuming the orientation of the arc, represented by the angles ϕ and ψ, as well as its radius and center, are given as operator input to the robot motion planner, then a robot motion trajectory can be immediately formulated from Eq. (4.3.4) and from the coefficients given above.

For our example, the motion trajectories for the two robot arms are obtained by plugging into Eqs. (4.3.1), (4.3.2), (4.3.3), and (4.3.4) the radii and centers of the two circles that the two robot arms are tracking respectively. The orientations of the circles are fixed at $\phi = \psi = 0$. Let $(x_a \ y_a \ z_a)^T$ and $(x_b \ y_b \ z_b)^T$ denote the end effector positions of the two robot arms a and b, and $(n_a \ o_a \ a_a)$ and $(n_b \ o_b \ a_b)$ the end effector orientation matrices of the arms, all defined in the universal coordinate frame coincident with the coordinate frame of Robot a. Then we have the following steps that the robot motion planner has to perform:

1. Transform the given data and the circle equations from the universal coordinate frame to the individual coordinate frames of the two robot arms. The variables of Eq. (4.3.1) are also transformed accordingly,

2. Perform kinematic motion space analysis for the two cooperating robot arms in the plane containing the circles, using the procedure of Section III. We obtain, for this example, that the circles are within the kinematic capability of the two arms if the radii and centers of the two circles are chosen to be $r_0 = 10$ cm, $x_{a0} = x_{b0} = 50$ cm, $y_{a0} = 0$ cm, $y_{b0} = 110$ cm, $z_{a0} = z_{b0} = -40$ cm, for the desired orientation of robot end effectors, all parameters are expressed in the universal coordinate frame, which is chosen to be coincident with the coordinate frame of Robot a. We also include the constraint on possible collision between the two arms and between the rod and the two robot arms in the kinematic motion space analysis,

3. Parametrize the circle equations as given by Eqs. (4.3.3) and (4.3.4). Here the motion planner is to determine the acceleration, constant motion, and deceleration phases of the motion. The time span for the acceleration and deceleration phases can be determined by operator input or predetermined values. In this example, we assume that they are given in the operator input. The $\omega(t)$ of the motion trajectory will be divided into three parts accordingly. This division is also done automatically by the robot motion planner. In our case, we assume that $\theta_0 = 0$ and $\theta_f = 2\pi$. We require that the robot end effectors accelerate to a given maximum velocity ω_f in the first quadrant of the circle, maintain this constant velocity until $\theta = \frac{3}{4}\pi$, and decelerate from thereon to zero velocity. The time periods for the the different motion segments are denoted by $[0, \ t_1]$ for acceleration, $[t_1 \ t_2]$ for constant velocity motion, and $[t_1, \ t_f]$ for deceleration. Plugging these data into Eq. (4.3.4), we can obtain

the velocity functions for each of the acceleration, constant velocity motion, and deceleration periods. Once these circle equations are obtained, the only thing remaining is to determine the parameters in the expression of $\omega(t)$. This is the task of the next step,

4. Perform the dynamic analysis on the circle equations by verifying that the desired velocity and acceleration profiles are realizable by the two robot arms. This includes the following steps:

   a). Select a total traveling time t_f and a maximum velocity ω_f. Set counter to 1.

   b). Add 1 to current counter content. The time instants t_1, t_2 can be decided by

   $$(t_2 - t_1)\omega_f = \pi,$$
   $$t_f - t_2 = t_1;$$

   c). Plug the values of t_1, t_2, t_f, ω_f into the expression of $\omega(t)$ to obtain numeric representation of $\omega(t)$;

   d). Plug the expression of $\omega(t)$ into the circle equations to obtain the numeric form of these equations;

   e). Simulate the robot nonlinear feedback controller using the circle equations to formulate desired input to the controller. Record the joint torques required by robot controller to travel the desired circles and verify that they are within the robot joint torque limits. If they are within the joint torque limits, go to h). Otherwise, carry out f);

   f). Determine where the maximum torque is needed on the circle trajectory from the information passed back from d). If it happens on the constant velocity motion part of the circle trajectory, change the value of ω_f. An appropriate optimal search algorithm can be implemented for the selection of next ω_f, e.g., assuming that the joint accelerations are small, velocity can be viewed to be linearly related to the term $\tau - G(q)$, where $G(q)$ is the gravity term. Therefore the next ω_f can be selected by using

   $$\omega_f^n = \frac{\tau_{max} - G(q)}{\tau^m - G(q)}\omega_f,$$

where ω_f^n is the value of ω_f to be used in next simulation, τ_{max} is the robot joint torque limit, and τ^m is the maximum torque required in the constant velocity motion part of the circle trajectory. Go back to b) and continue. If the maximum torque occurs in the acceleration or deceleration part of the circle trajectory, go to next step;

g).If counter content is equal to 1, attempt to increase ω_f by using the equation

$$\omega_f^n = \frac{\tau^m - G(q)}{\tau_{max} - G(q)}\omega_f.$$

Go to b). If counter content is large than 1, increase t_f by using the dynamic time scaling formula (Hollerbach,1984). If t_f is larger than T_{max}, the maximum allowable time for traveling the circles, quit the procedure and report that the task can not be performed within the given traveling time limit. Otherwise go back to b);

h). Select the nonlinear feedback robot controller for trajectory tracking control. Select the corresponding control gains. Load the executable circle trajectory obtained from above into the controller.

After the above steps are performed, an executable circle trajectory is obtained and the robot is ready to perform the task.

We point out here that the above procedure, especially the part for dynamic analysis, is not only limited to the particular example we are considering. Since we envision that most robot motion trajectories are composed of straight line segments connected smoothly by arcs and their velocity profiles contain acceleration, constant velocity motion, and deceleration parts, the above procedure will be applicable to most types of trajectories with minimum modifications.

V. SUMMARY AND CONCLUSION

The formulation of intelligent robot motion planner based on the nonlinear feedback robot controller, the robot kinematic and dynamic capabilities, and the implementation of the intelligent robot motion planner for a experimental task are outlined in this chapter. By the term intelligent robot motion planner, we are not implying that it will be a fully autonomous motion planner with learning and reasoning capabilities. Instead, we view the generation of robot motion trajectories based on a suitable task description by the operator and based on satisfying the robot kinematic and dynamic constraints as an action of intelligence. The proposed robot motion planner is intelligent in this sense.

The concept of kinematic motion space as a measure of the kinematic capability of two cooperating robot arms is proposed. This is defined based on the traditional concept of a single robot arm's workspace. We noted that for two cooperating robot arms, their kinematic capability is closely related to the geometric properties of the object they are handling. Various constraints on the kinematic motion space are identified and formulated into algebraic equations and inequalities in the case of two cooperating PUMA 560 robot arms. An algorithm is further proposed to solve for the kinematic motion space. Its use is illustrated via an example of two PUMA robot arms cooperating to move a long object. The kinematic motion space for this setup is computed and represented graphically.

The dynamic motion space of single and multiple cooperating robot arms is defined. This measures the maximum force/moment application capability of the robot arms. A simple algorithm is proposed to obtain the dynamic motion space of a single robot arm, based on the concept of geometric projection. This algorithm is modified to be applied to the case of multiple cooperating robot arms. The dynamics of the multiple cooperating robot arms is modeled via the closed mechanical chain approach. The dynamic motion space analysis problem and the proposed solution approach are illustrated in two examples, in which we find the maximum applicable force of a single robot and two cooperating robot arms. We also discussed the evaluation of a planned robot motion trajectory against the robot joint torque/force constraints and its modification in case these constriants are violated.

It is shown that, with the introduction of the nonlinear feedback robot controller, task motion planning can be done at a higher level than before, i.e., in the robot task space. This leads naturally to the development of robot motion planner that plans robot motion trajectories in the task space.

A hierarchical structure for the organization of the intelligent robot motion planner is proposed. This structure contains the levels of user interface and subtask generation, motion trajectory generation, and setting up of the robot controller for the execution of the task. The descriptions of each level are given in detail, as well as the inter-relations between them.

The intelligent robot motion planer is further explained, its operation illuminated via its application in the motion planning for a two robot arm cooperation task. A step-by-step procedure is given to show the details of the implementation of the robot motion planner proposed.

The results reported in this chapter provide a framework under which intelligent robot motion planners can be designed. Several critical procedures in this motion planner has been developed. Since the intelligent robot motion planning is a vast territory to be explored, especially considering the development of more autonomous robot arms, the results presented here are by no means complete and final. Many opportunities exist in this research area and many problems remain to be solved. We indicate some of them in the following, together with some observations on how they could be attacked.

In the study of the kinematic motion space analysis, we have obtained a procedure to generate the kinematic motion space of two cooperating robot arms. The use of the graphic representation of the kinematic motion space certainly helps the robot operator to determine possible motion paths for the given task. On the other hand, the robot motion planner can perform an exhaustive search over the part of the kinematic motion space that contains the initial and final locations of the desired motion and determine if a path consisting of prespecified path segments, such as straight lines and arcs, are executable as far as the kinematic constraints are concerned. However, a more efficient method needs to be developed such that the knowledge on the kinematic motion space can be directly utilized by the robot motion planner in a more intelligent manner. The solution of this problem depends on the development of an efficient computer representation of the kinematic motion space, as well as algorithms that facilitate the search over it for suitable path segments.

In the generation of the kinematic motion space for the example given here, we have used a tesselation of the robot task space to compute the kinematic motion space. More efficient algorithms may be developed to speed up the computation.

One way to accelerate the generation of the kinematic motion space may be to trace along the boundaries of the kinematic motion space and the holes/voids inside it. If the number of holes/voids inside the kinematic motion space are known, such algorithms can be easily developed. Therefore, more theoretical study needs to be done on the topological structure of the kinematic motion space. One possible direction is that relations may exist between the number of polynomial constraints together with their degrees and the number of holes/voids in the kinematic motion space. This will be indeed a very challenging problem to solve.

In our proposed intelligent robot motion planner structure, we have assumed that geomatric path planning and the dynamic motion planning are two separate steps in the robot motion planning, and treated the motion planning accordingly. However, these two steps are intrinsically related, as in the case when the radius of a circular section of a geometric robot path affects the dynamic qualities of the motion trajectory on this section intimately. It is conceivably that these two steps in the robot motion planning can be combined so that the motion planning can be more efficient. This will also provide greater insight into robot motion. How this can be done remains an open question.

References

Bejczy, A.K. (1974) Robot Arm Dynamics and Control, JPL, TM 33-669

Boothby, W. (1975) An Introduction to Differential Manifolds and Riemanian Geometry, Academic Press, New York, pp. 294-297

Brockett, R. (1988) "On the Computer Control of Movement," Proceedings of 1988 IEEE International Conference on Robotics and Automation, Philadelphia, PA, pp. 534-540

Brooks, B. (1984) "Planning Collision Free Motion for Pick and Place Operations," in Robotics Research, M.Brady, R.Paul Eds., The MIT Press, Cambridge, MA

Campbell, S.L., Meyers, Jr., C.D. (1979) Generalized Inverse of Linear Transformations, Pitman, London

Chand, S., Dotty, K. (1985) "On-line Polynomial Trajectories for Robot Manipulators," The International Journal of Robotics Research, Vol.4, No.2

Collins, G. "Quantifier Elimination for Real Closed Fields by Cylindrical Algebraic Decomposition," in Second GI Conference on Automata Theory and Formal Languages, Lecture Notes in Computer Science, Vol. 33, Springer-Verlag, Berlin, pp.134–183.

Cwiakala, M., Lee, T.W. (1985) "Generation and Evaluation of a Manipulator Workspace Based on Optimum Path Search," ASME Journal of Mechanisms, Transmissions, and Automation in Design, Vol. 107, pp.245–255

Donald, B. (1985) "On Motion Planning with Six Degree of Freedom: Solving the Intersection Problems in Configuration Space," Proceedings of 1985 IEEE International Conference on Robotics and Automation, IEEE, St. Louis, MO

Faverjon, B. (1984) "Obstacle Avoidance Using an Octree in the Configuration Space of a Manipulator," Proceedings of IEEE International Conference on Robotics, Atlanta, GA

Duffy, J. (1980) Analysis of Mechanisms and Robot Manipulators, John Wiley and Sons, New York

Freund, (1987) "Collision Avoidance in Multi-Robot Systems," Proceedings of the Second International Symposium on Robotics Research, Kyoto, Japan, MIT Press, Cambridge, MA

Freund, E., Hoyer, H. (1988) "Real-Time Pathfinding in Multirobot Systems Including Obstacle Avoidance," The International Journal of Robotics Research, Vol.7,

No.1, pp.42–70

Geering,H., Guzzella, H., Hepner, S., Onder, C. (1986) " Time-Optimal Motion of Robots in Assemply Tasks," IEEE Transactions on Automatic Control, Vol.AC-31, No.6

Gouzenes, L. (1984) "Strategies for Solving Collision-Free Trajectory Problems for Mobile and Manipulator Robots," International Journal of Robotics Research, Vol.3, No.4

Gupta, K.C., Roth, B. (1982) "Design Considerations for Manipulator Workspace," Transactions of the ASME, Journal of Mechanical Design, Vol.104, pp.704–711

Gupta, K.C. (1986a) "Kinematic Analysis of Manipulators Using the Zero Reference Position Description," The International Journal of Robotics Research, Vol.5, No.2, pp.5–13

Gupta, K.C. (1986b) "On the Nature of Robot Workspace," The International Journal of Robotics Research, Vol.5, No.2, pp.112–121

Hanson, J.A., Gupta, K.C., Kazerounian, S.M.K. (1983) "Generation and Evaluation of the Workspace of a Manipulator," The International Journal of Robotics Research, Vol.2, No.3, pp.22–31

Hansen, J.A. (1983) Generation and Evaluation of the Workspace of a Manipulator, M.S. Thesis, University of Illinois, Chicago

Hayati, H. (1986) "Hybrid Position/Force Control of Multi-Arm Cooperating Robots," Proceedings of the 1986 IEEE International Conference on Robotics and Automation, San Francisco, CA, pp.82–89

Hogan, N. (1985) "Impedance Control: An Approach to Manipulation – Theory,

Implementation, and Applications," Transactions of ASME, Journal of Dynamic Systems, Measurement, and Control, Vol.107, pp.1–24

Hollerbach, J. (1984) "Dynamic Scaling of Manipulator Trajectories," Transactions of the ASME, Journal of Dynamic Systems, Measurements, and Control, Vol.106

Hollerbach, J.M., Suh, K.C. (1987) "Redundancy Resolution of Manipulators through Torque Optimization," IEEE Journal of Robotics and Automation, Vol.RA-3, No.4, pp.308–316

Khatib, O. (1984) "Real-Time Obstacle Avoidance for Manipulators and Mobile Robots," The International Journal of Robotics Research, Vol.5, No.1, 1986

Khatib, O., Le Maitre, J.F. (1978) "Dynamic Control of Manipulators Operating in a Complex Environment," Proceedings of the Third CISM-IFToMM International Symposium on Theory and Practice of Robots and Manipulators, Udine, Italy

Kohli, D., Spanos, J. (1985a) "Workspace Analysis of Mechanical Manipulators using Polynomial Discriminants," ASME Journal of Mechanisms, Transmissions, and Automation in Design, Vol. 107, pp.209–215

Krogh, B.H. (1983) "Feedback Obstacle Avoidance Control," Proceedings of the 21st Allerton Conference, University of Illinois

Kumar, A., Waldron, K.J. (1981) "The Workspaces of a Mechanical Manipulator," ASME Journal of Mechanisms, Transmissions, and Automation in Design, Vol. 103, pp.665–672

Kumar, A., Waldron, K.J. (1980) "The Dextrous Workspaces," ASME Paper No.80-DET-108

Laugier, C., Germain, F. (1985) "An Adaptive Colision-Free Trajectory Planner,"

Proceedings of International Conference on Advanced Robotics, Tokyo, Japan

Lee, C.S.G. (1984) "Geometric Approach in Solving Inverse Kinematics of PUMA Robots," IEEE Transactions on Aerospace and Electronics Systems, Vol. AES-20, No. 6, pp.695–705.

Lee, S., Bekey, G., Bejczy, A.K. (1985) "Computer Control of Space-Bourne Teleoperators with Sensor Feedback," Proceedings of the 1985 IEEE Conference on Robotics and Automation, St. Louis, MO, pp.205–214

Lee, B.H., Lee, C.S.G. (1987) "Collision-Free Planning of Two Robots," IEEE Transactions on Systems, Man, and Cybernetics, Vol.SMC-17, No.1, pp.21–32

Lee, T.W., Yang, D.C.H. (1983a) "On the Evaluation of Manipulator Workspace," Transactions of the ASME, Journal of Mechanisms, Transmissions, and Automation in Design, Vol.105, pp.70–77

Li, Z., Tarn, T.J., Bejczy, A K. (1990) "Dynamic Workspace Analysis of Multiple Cooperating Robot Arms," submitted to the IEEE Transactions on Robotics and Automation.

Li, Z., Tarn, T.J., Bejczy, A.K., Ghosh, B. (1989) "Motion Space Analysis of an Object Handled by Two Robot Arms," Proceedings of the 28th IEEE Conference on Decision and Control, Tempa, FL

Lin, C., Chang, P., Luh, J.Y.S. (1983) "Formulation and Optimization of Cubic Polynomial Joint Trajectories for Industrial Robots," IEEE Transactions on Automatic Control, Vol.AC-28, No.12

Lin, C., Chang, P. (1985) "Appropriate Optimum Paths of Robot Manipulators inder Realistic Physical Constraints," Proceedings of 1985 IEEE International Conference on Robotics and Automation, IEEE, St. Louis, MO

Lozano-Perez, T. (1983) "Spatial Planning: A Configuration Space Approach," IEEE Transactions on Computers, Vol.C-32, No.2

Lozano-Perez, T. (1987) "A Simple Motion-Planning Algorithm for General Robot Manipulators," IEEE Journal of Robotics and Automation, Vol.3, No.3, pp.224–238

Luh, J.Y.S., Zheng, Y.F. (1988) "Load Distribution between Two Coordinating Robots by Nonlinear Programming," Proceedings of 1988 ACC, Atlanta, GA

Lumelsky, V.L. (1986) "Continuous Robot Motion Planning in Unknown Environment," in Adaptive Control and Learning Systems: Theory and Applications, Proceedings of the Fourth Workshop on Adaptive Systems Control Theory, K.S.Narendra, ed., Plenum Press, New York

Lumelsky, V.L. (1987)"Effect of Kinematics on Motion Planning for Planar Robot Arms Moving Amidst Unknown Obstacles," IEEE Journal of Robotics and Automation, Vol.RA-3, No.3

Meystel, A. (1988) "Intelligent Control in Robotics," Journal of Robotic Systems, Vol.5, No.4, pp.269–308

Nagata, T., Honda, K., Teramoto, Y. (1988) "Multirobot Plan Generation in a Continuous Domain: Planning by Use of Plan Graph and Avoiding Collision among Robots," IEEE Journal of Robotics and Automation, Vol.RA-4, No.1, pp.2–13

Nakamura, Y. (1988) "Minimizing Object Strain Energy for Coordination of Multiple Robotic Mechanisms," Proceedings of the 1988 American Control Conference, Atlanta, GA, pp.499-504

Orin, D.E., Oh, S.Y. (1981) "Control of Force Distribution in Robot Mechanisms Containing Closed Kinematic Chains," Transactions of the ASME, Journal of Dynamic Systems, Measurement, and Control, Vol.102, pp.134–141

Orin, D.E., Cheng, F.T. (1989) "General Dynamic Formulation for the Force Distribution Equations," Proceedings of the Fourth International Conference on Advanced Robotics, Columbus, OH

Paul, R. (1975) "Manipulator Path Control," Proceedings of the 1975 International Conference on Cybernetics and Society, IEEE, New York,

Paul, R. (1979) "Manipulator Cartesian Path Control," IEEE Transactions on Systems, Man and Cybernetics, Vol.SMC-9, No.11

Pfeiffer, F., Johanni, R. (1986) "A Concept for manipulator Trajectory Planning," Proceedings of 1986 IEEE International Conference on Robotics and Automation, IEEE, San Francisco, CA

Rajan, V. (1985) "Minimum Time Trajectory Planning," Proceedings of 1985 IEEE International Conference on Robotics and Automation, IEEE, St. Louis, MO

Roth, B. "Performance Evaluation of Manipulators from a Kinematic Veiwpoint," In Performance Evaluation of Programmable Robots and Manipulators, NBS #459, U.S. Govt. Printing Office, pp.39–61.

Sahar, G., Hollerbach, J. (1985) "Planning of Minimum Time Trajectories for Robot Arms," Proceedings of 1985 IEEE International Conference on Robotics and Automation, IEEE, St. Louis, MO

Schiller, Z., Dubowsky, S. (1987) "Time Optimal Paths and Acceleration Lines of Robotic Manipulators," Proceedings of the 26th Conference on Decision and Control, Los Angeles, CA, pp.199–204

Schiller, Z., Dubowsky, S. (1989) "On Computing the Global Time Optimal Motions of Robotic Manipulators in the Presence of Obstacles," submitted to IEEE Transactions of Robotics and Automation, No. B89313, 1989

Schwartz, J.T., Sharir, M. (1983a) "On the Piano Mover's Problem: I," Comm. Pure and Applied Math. Vol.36, pp.345–398

Schwartz, J.T., Sharir, M. (1983b) "On the Piano Mover's Problem: II," Advances in Applied Math. Vol.4, pp.298–351

Shimano, B.E., Roth, B. (1978) "Ranges of Motion of Manipulators," The Third CISM–IFToMM International Symposium on Theory and Practice of Robots and Manipulators, Udine, Italy

Shin, K., McKay, N. (1986a) "A Dynamic Programming Approach to Trajectory Planning of Robotic Manipulators," IEEE Transactions on Automatic Control, Vol.AC-31, No.6

Shin, K., McKay, N. (1986b) "Selection of Near-Minimum Time Geometric Paths for Robotic Manipulators," IEEE Transactions on Automatic Control, Vol.AC-31, No.6

Spanos, J., Kohli, D. (1985b) "Workspace Analysis of Regional Structure of Manipulators," ASME Journal of Mechanisms, Transmissions, and Automation in Design, Vol. 107, pp.216–222

Sugimoto, K., Duffy, J. (1981a) "Determination of Extreme Distances of a Robot Hand— Part 1: A General Theory," Transactions of the ASME, Journal of Mechanical Design, Vol.103, pp.631–636

Sugimoto, K., Duffy, J. (1981b) "Determination of Extreme Distances of a Robot Hand— Part 2: Robot Arms with Special Geometry," Transactions of the ASME, Journal of Mechanical Design, Vol.103, pp.776–783

Tannenbaum, A., Yomdin, Y. (1987) "Robotic Manipulators and the Geometry of Real Semialgebraic Sets," IEEE Journal of Robotics and Automation, Vol. RA-3, No. 4, pp.301–307.

Tarn, T.J., Bejczy, A.K., Isidori, A., Chen, Y. (1984) "Nonlinear Feedback in Robot Arm Control," Proceedings of the 23th Conference on Decision and Control, Las Vegas, Nevada

Tarn, T.J., Bejczy, A.K., Li, Z. (1986) "Task Driven Feedback Control: A Step towards Intelligent Control, " Proceedings of the 25th IEEE Conference on Decision and Control, Athens, Greece

Tarn, T.J., Bejczy, A.K., Li, Z. (1988a) "Dynamic Workspace Analysis of Two Cooperating Robot Arms," Proceedings of 1988 American Control Conference, Atlanta, GA, pp.489–498

Tarn, T.J., Bejczy, A.K., Yun, X. (1988b) "New Nonlinear Control Algorithms for Multiple Robot Arms," IEEE Transactions on Aerospace and Electronic Systems, Vol.AES-24, No.5, pp.571–583

Taylor, R. (1979) "Planning and Execution of Straight Line Manipulator Trajectories," IBM Journal of Research and Development, Vol.3, No.4

Tsai, Y.C., Soni, A.H. (1983) "An Algorithm for the Workspace of a General n-R Robot," ASME Journal of Mechanisms, Transmissions, and Automation in Design, Vol. 105, pp.52–57

Uchiyama, M., Dauchez, P. (1988) "Statics and Kinematics of a Two Arm Robot for Coordinated Position/Force Control," submitted to IEEE Journal of Robotics and Automation

Udupa, S. (1977) "Collision Detection and Avoidance in Computer Controlled Manipulators," Proceedings of the Fifth International Conference on AI, Cambridge, MA

Walker, I.D., Freeman, R.A., Marcus, S.I. (1988) "Dynamic Task Distribution for Multiple Cooperating Robot Manipulators," Proceedings of 1988 IEEE International Conference on Robotics and Automation, Philadelphia, PA

Walker, I.D., Freeman,R.A., Marcus, S.I. (1989) "Internal Object Loading for Multiple Cooperating Robot Manipulators," Proceedings of 1989 IEEE International Conference on Robotics and Automation, Scottsdale, AZ

Weiss, L., Sanderson, A., Neuman, C. (1987) "Dynamic Control of Robots with Visual Feedback," IEEE Journal of Robotics and Automation, Vol.3, No.5, pp.404–417

Yang, D.C.H., Lee, T.W. (1983b) "On the Workspace of Mechanical Manipulators," Transactions of the ASME, Journal of Mechanisms, Transmissions, and Automation in Design, Vol.105, pp.62–69

Yap, C.K. (1987) "How to Move a Chair Through a Door," IEEE Journal of Robotics and Automation, Vol.RA-3, No.3, pp.172–181

Yun, X. (1987) Coordinated Control of Two Robot Arms by Nonlinear Feedback, D.Sc. Dissertation, Department of Systems Science and Mathematics, Washington University, St. Louis, MO

Yun, X. (1988) "Dynamic State Feedback Control of Constrained Robot Manipulators," Proceedings of the 1988 IEEE Conference on Decision and Control, Austin, TX, pp.622–626

DISCRETE KINEMATIC MODELING TECHNIQUES IN CARTESIAN SPACE FOR ROBOTIC SYSTEM

WITOLD JACAK

Institute of Technical Cybernetics
Technical University of Wroclaw
Wroclaw, Poland

1. INTRODUCTION

The integrated robotic systems play the main role in the modern technological processes. Robots can be applied to a large variety of tasks without major redesign of their manipulator structure. The robot´s versatility derives from the generality of a manipulator´s mechanical structure and sensory capabilities. However such versatility can be exploited only if a robot can be easily programmed. The automatic generation of robot´s motion trajectories in the presence of obstacles is one of the main subproblems in the synthesis of task-oriented languages for robot programming. For the purpose of automatic translation of the language commands such as "PICK-and-PLACE" or "MOVE TO" we need adequate models of a robot kinematics and work-scene.

CONTROL AND DYNAMIC SYSTEMS, VOL. 39

There are two requirements which a robot´s kinematics model should satisfy. Firstly, in order to apply methods of geometrical path of robot motion planning, a robot´s kinematics model should afford possibilities for computer simulation of a manipulator motion. Secondly, a robot´s kinematics model should be convenient to the 3D graphic simulation of robot movements in the base Cartesian frame. The computer graphic simulation is a fundamental tool for the off-line testing of correctness of the robot program.

Hence, a robot´s kinematics model should be a mathematical system endowed with an ability to produce a new configuration of the manipulator on the basis of an old configuration and a desired input signal.

A natural way of describing the configuration of a robot´s arm and its environment is to introduce an external Cartesian coordinate frame, called base Cartesian space [1], [17]. This description is objective, i.e. it does not depend on the type of robot or its environment [26]. The positions of the manipulator and the objects within its operating range are described independly of one another. The inner state of the robot´s manipulator is described by a vector of variables denoting angles of successive joints of the robot. Such a description of the configuration of the manipulator produces a space called the Joint space [1], [6], [26]. The mechanical structure of a manipulator generates the type of Joint space assumed, and its technical realization renders the Joint spaces to be different and dependent on this realization. The description of the robot in an assumed Joint space is then subjective. It does not coincide with the environment description in the Cartesian base frame. The relationship between the manipulator joint angles, present in the Joint space and the external Cartesian space is usually involved, because of the basic non-uniqueness of solutions of kinematic equations [1], [4], [5], [26].

To obtain a uniform representation of the robot and its environment, a transformation of the geometrical model of the work-scene from the base Cartesian frame into the Joint space is often performed [3], [17], [19]. The transformations, called the Configuration Space methods, are complex and inefficient, particularly for the redundant robots. (Because the determination of an appropriate effector path is not equivalent to determining the trajectory of the whole manipulator's body for such a robot) [17], [18], [20]. The Configuration Space method is also inconvenient in case of frequent changes of object locations in the work scene. Such changes require repeated, time-consuming transformations of the work-scene model into the Joint space. This situation occurs in CAD/CAM systems for robotized work-cell design [13], [19]. Furthermore, CAD/CAM systems are equipped with a task-oriented robot programming language, in order for the kinematic model used by the interpreter of such language to be relatively insensitive to such work-scene changes [13].

Due to these reasons, we try to construct a model of manipulator kinematics which would:

- be convenient for the computer simulation of robot's motion,
- facilitate direct analysis of a robot location with respect to objects in its environment,
- be not too complex computationally.

The most suitable model of robot kinematics is a discrete dynamic system M defined as follows [6], [10], [12]:

$$M = (X,U,Y,f,g) \tag{1}$$

The set X denote a set of state of manipulator. The state x should determine univocally the position of the manipulator, called the manipulator's configuration, in the real space in which the physical robot operates. U denotes a set of input signals causing a change of the configuration. The set Y denotes output signals of M. An output should ensure a possibility of the graphic representation of the robot's body in a 3D-base frame. For this purpose it is convenient to use a skeleton model of the manipulator. By the skeleton model of the manipulator, we denote a broken line in 3D-space, which represents a current configuration of the manipulator's arm.

The function $f : X \times U \to X$ is a one-step transition function of the form:

$$x(k+1) = f(x(k), u(k)) \tag{2}$$

and $g : X \to Y$ is an output function of the form:

$$g(x(k)) = y(k) \tag{3}$$

The function g(x) forms the skeleton model of the robot´s manipulator and can be used to compute the volume in the base Cartesian space occupied by different bodies of the manipulator.

The properties of such defined model of the robot kinematics depend on the method of specification of its components, specifically on the state-set X and input-set U. There could be two ways to construct such a model. The first one involves changing the state in the Joint space and then translating it into the base Cartesian space. The second one involves constructing a model which operates on the manipulator´s states directly in the base frame. These two ways lead to different specifications of model states and involve different computational complexities.

2. DISCRETE MODELS OF ROBOT KINEMATICS IN THE CARTESIAN SPACE

The description of a discrete model of a robot´s kinematics depends on the specification of model components. The most frequently used description of a state of the manipulator with n-DOF is its representation as a vector of joint angles:

$$q = (q_i \mid i=1,...,n) \tag{4}$$

where $q_i \in Q_i = [q_i^{min}, q_i^{max}]$. Q_i is a range of changes of the i-th joint angle.

In this case the state set X of system M is denoted by:

$$Q = Q_1 \times Q_2 \times ... \times Q_n \tag{5}$$

and is also called the Joint Space [1], [5] or the Configuration Space [6], [17]. Such a specification of a manipulator´s state is said to be the Joint-state [11].

Alternatively, the second possibility of description of the manipulator´s configuration is to take as a state x a vector of points in the Cartesion base frame, which represents positions of every joint of the manipulator. Hence, the state of system M is a vector

$$x = (P_i \mid i=1,...,n+1) \tag{6}$$

where $P_i = (x_i, y_i, z_i) \in E_0$ is a point in the base Cartesian coordinate frame E_0 describing the current position of the i-th joint and P_{n+1} is a position of the effector-end.

The state x is called the Cartesian-state [11]. Both state descriptions are equivalent. It is easy to observe that using the Denavit-Hartenberg method [26], one can construct the function

$$t_i : Q \to E_0 \tag{7}$$

determining the location of the i-th joint in Cartesian space i.e. $t_i(q) = P_i$.

The vector function

$$t(q) = (t_i(q) \mid i=1,...,n+1) \tag{8}$$

constitutes a transformation of the Joint-State-Space into the Cartesian-State-Space. Similarly, one can construct the inverse function Conf : $X \to Q$ by using the vector dot and cross products of the vectors $[P_{i+1} - P_i]$ and $[P_{i-1} - P_i]$ [8].

The mechanical structure of a n-degree of freedom (n-DOF) manipulator can be represented by a sequence of line segments (skeleton model). The lengths of the manipulator links are denoted by l_i. One of the most commonly used kind of the manipulator is the planar manipulator. The planarity of a manipulator means that links $l_1,...,l_n$ (n-1 DOF) lie in a plane rotating around the vertical axis Z_0 of the base frame E_0 while angle q_1 changes (the n-th DOF).

Due to planarity of the manipulator, its configuration (6) can be described in a 2-dimensional space. By introducing a coordinate frame E_A of a link such that the origin of E_A lies at point P_1, Z_A axis coincides with the Z_0 axis and X_A axis lies in the plane of the manipulator link, it is easy to find a univocal representation of the manipulator state x in the plane $X_A \times Z_A$, as follows

$$s = (p_i \mid i=1,...,n+1) \tag{9}$$

where:

$$p_i = H^{-1} \cdot P_i = (\| P_i \|_{xy}, 0, z_i) = (x'_i, o, z'_i) \tag{10}$$

is the position of the i-th joint in the plane $X_A \times Z_A$.

The form $\| P_i \|_{xy}$ denotes the Euclidean length of the projection of P_i onto the plane $X_0 \times Y_0$, while

$$P_i = H \cdot p_i = (x'_i \cdot \cos q_1,\ x'_i \cdot \sin q_1,\ z'_i) \tag{11}$$

where q_1 is the rotation angle of the manipulator link around $Z_0 = Z_A$ axis and transformation $H = Rot(Z_0, q_1)$. The coordinate frames E_0, E_A and the skeleton model of the manipulator are shown in Fig. 1. Given a state s, $c^i_j = \|p_i - p_j\|$ denotes the Euclidean distance between joints i and j. By the relative maximum (minimum) reach between joints i and j, we mean the maximum (minimum) size of the kinematic chain connecting the joints, which is denoted c^i_{jmax} (c^i_{jmin}) respectively.

Let p_i and p_j denote positions of two joints in the plane $X_A \times Z_A$. By $Rot\ \beta(p_i, p_j, \xi)$ we shall mean the new position of point p_j after a rotation around point p_i by a constant angle β toward the ξ direction ($\xi = +1 \rightarrow$ clockwise rotation, $\xi = -1 \rightarrow$ counter clockwise rotation).

Then, the new position of p_j is defined as follows:

$$p'_j = p_i + R(\xi) \cdot [p_j - p_i] \tag{12}$$

where:

$$R\ (\xi) = \begin{bmatrix} \cos \beta, & \xi \cdot \sin \beta \\ -\xi \cdot \sin \beta, & \cos \beta \end{bmatrix} \tag{13}$$

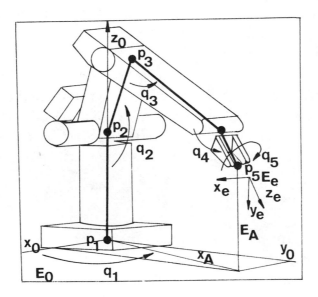

Fig. 1. The skeleton-model of robot configuration

2.1.   THE JOINT-STATE SPECIFICATION OF MANIPULATOR´S CONFIGURATION

First, we consider how to construct a model of a robot kinematics with the set of states treated as Joint-states.

One of the ways to construct of such a model is based on an arbitrary discretization of angle increments of the manipulator joints [6], [25], [27]. Let δq_i denote the smallest possible change of the i-th joint angle, so-called the discretization increment. Then any angle of the i-th joint can be expressed by:

$$q_i = q_i^{min} + j \cdot \delta q_i \tag{14}$$

where $j \in J_i$ and the set

$$J_i = \left\{ 0, 1, \ldots E\left(\frac{q_i^{max}}{\delta q_i}\right) = N_i \right\}$$

This discretization decides the accuracy with which the given effector position can be reached, as well as the exactness of the relation "manipulator-objects" of the scene. Some analysis of the choice of the discretization of successive joint angles steps has been performed in [6], [7] ensuring a suitable accuracy.

Using the fact that all the angles can change only by a define increment, we define a set U of inputs of the model M as:

$$U = X \{u_i \mid i=1,...,n\} \tag{15}$$

where: $u_i = \{-\delta q_i, 0, +\delta q_i\}$ is the set of possible (admissible) directions of changes of the i-th joint angle.

In general, there are $3^n - 1$ admissible directions, where n is the number of degrees of freedom of the robot. Having defined the set U, it is possible to describe the changes of successive configurations of the robot´s link as the quasilinear discrete system of the form:

$$q(k+1) = q(k) + \Lambda(q(k)) \cdot u(k) \tag{16}$$

where $q(k) \in Q$ is the state of manipulator at time k; $u(k) \in U$ is the vector of increments of angles of the joint at time k, (being actually the input of the system) and

$$\Lambda(q(k)) = \text{diag } [\lambda_i (q(k)) \mid i=1,...,n]$$

is a diagonal $n \times n$-matrix describing the length of the angle's step changes at each joint. Λ is dependent on the actual value of vector q in order to render it possible to change the step length when a given position of the effector-end is being reached or to impose preferences on movements of certain joints. Clearly, the dependence of Λ on q is generally nonlinear (piecewise constant). Therefore, the one-step transition function f (16) on the model is, in this case, quasi-linear.

Such a specification of the kinematics model causes the system M to provide values of the states expressed in the Joint space: In order to make it possible to check the position of the manipulator with respect to obstacles, locations it is necessary to create an ouput function g which describes the actual position of the manipulator in the Cartesian space.

As we have stated previously, the manipulator's position in Cartesian space is given as the vector:

$$x = (P_i \mid i=1,...,n+1)$$

called the skeleton model of the robot's manipulator or a Cartesian state.

Recall that the i-th joint's position and orientation in the Cartesian base frame, assuming that all the joint angles q_i are known, is described by the Denavit-Hartenberg matrix [1], [4], [26]:

$$T_i = \begin{bmatrix} \bar{n}_i, \bar{o}_i, \bar{a}_i & \mid & P_i \\ \text{---} & \mid & \text{---} \\ 0 \quad 0 \quad 0 & \mid & 1 \end{bmatrix} = A_{0,1} \cdot A_{1,2} \cdots A_{i-1,i} \tag{17}$$

where $A_{i-1,i}$ is the transformation matrix between the coordination frame E_i of the i-th joint and the coordination frame E_{i-1} of the i-1th joint. The last column of the matrix T_i determines the i-th joint position in the base frame E_0 as a function of the angles $q_1,...,q_i$. Hence, let us introduce an output function

$$g : Q \rightarrow Y \tag{18}$$

defined as:

$$g(q(k)) = (t_i(q(k)) \mid i=1,...,n+1) \tag{19}$$

where: $t_i(q) = P_i$ is the last column of the matrix T_i.

The components of the output function g are a highly nonlinear trigono-metric function.

To sum up, the robot kinematics can be modelled as a quasi-linear discrete system M_Q whose transition function f is described by formula (16) and the output function g is given by formula (19).

For example, let us consider a model of the ASEA IRb-6 robot kinematics. Its structure is shown in Fig. 2. In this case, the transition function is trivial, but the output function g has the following complex form:

$$P_1 = (0,0,0), \qquad P_2 = (0,0,l_1) \tag{20}$$

$$P_3 = (l_2 \cos q_1 \sin q_2, l_2 \sin q_1 \sin q_2, l_1 - l_2 \cos q_2)$$

$$P_4 = (\cos q_1 (l_2 \sin q_2 - l_3 \sin(q_2 + q_3)),$$
$$\sin q_1(l_2 \sin q_2 - l_3 \sin(q_2 + q_3)),$$
$$l_1 - l_2 \cos q_2 + l_3 \cos(q_2 + q_3))$$

$$P_5 = (\cos q_1(l_2 \sin q_2 - l_3 \sin(q_2 + q_3) + l_4 \sin(q_2 + q_3 + q_4)),$$
$$\sin q_1(l_2 \sin q_2 - l_3 \sin(q_2 + q_3) + l_4 \sin(q_2 + q_3 + q_4)),$$
$$l_1 - l_2 \cos q_2 + l_3 \cos(q_2 + q_3) - l_4 \sin(q_2 + q_3 + q_4))$$

The above presented model is based on an arbitrary discretization of the Joint space.

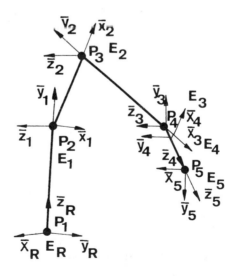

Fig.2.The kinematic structure of the ASEA IRb-6 robot

Due to the discretization of the Joint space, the number of possible states of the manipulator is finite but too large to be determined explicity. The state space of the manipulator contains

$$\overline{\overline{Q}} = \prod_{i=1}^{n} N_i$$

elements, and for a given state $q \in Q$ the state transition function f generates $3^n\text{-}1$ successors of q. Thus, it is very important to find a method to reduce the number of possible states of the manipulator. This can be done by performing the discretization of the base Cartesian Space. Additionally, we should recall that the input-data obtained from a motion task description are in the form of geometrical paths of the effector-end motion in the base Cartesian frame. From this follows that our model of the robot kinematics must be able to generate a sequence of manipulator configurations which realize the motion along the desired effector path.

This leads to the assumption that an input signal of M has to be a generalized vector of displacement of the effector-end. A position and orientation of the effector is generally described as a vector

$$\mathbf{p} = (x, y, z, \alpha, \beta, \gamma) \tag{21}$$

where: (x, y, z) is the position of the effector-end, (i.e. $P_{n+1} = (x, y, z)$) and α, β, γ are the Euler angles representing an orientation of the effector [1], [26].

By the generalized vector of the effector's displacement in base frame we mean a vector

$$\delta \mathbf{p} = (\delta x, \delta y, \delta z, \delta \alpha, \delta \beta, \delta \gamma) , \tag{22}$$

which consist of a displacement vector of the work-point $(\delta x, \delta y, \delta z)$ and of the orientation change vector $(\delta \alpha, \delta \beta, \delta \gamma)$.

Assuming that the effector position and orientation can change only by a defined increment $\delta \mathbf{p}$, we can construct the set of inputs of the model M as:

$$U \subset V_x \times V_y \times V_z \times V_\alpha \times V_\beta \times V_\gamma \tag{23}$$

where $V_x = \{-\delta x, 0, +\delta x\}$ is the set of possible increments of the effector displacement along the X axis. Similary, we define the sets $V_y, V_z, V_\alpha, V_\beta, V_\gamma$. The number of elements of the set U depends on the type of discrete geometry used to discretisize the base frame.

Having defined the set of inputs of the model, we show how to construct a one-step transistion function f and an output function g for the model M whose states are treated as Joint-states.

To determine a one-step transition function, we shall use the direct relationship between the joint velocities and the Cartesian velocities. The direct kinematic equation (17) allows us to establish the following relationship between a joint vector q and a generalized position of the effector-end [4], [26]:

$$\mathbf{p} = t_{n+1}(q) \tag{24}$$

where t_{n+1} is a continuous nonlinear trigonometric function [26]. This equation can be transformed into an equation describing the relationship between joint velocities and Cartesian velocities, i.e., we have

$$d\mathbf{p} = J(q)dq \tag{25}$$

where $J(q)$ is the Jacobian matrix $[\partial t_{n+1} / \partial q]$.

In turn, the above equation can be inverted using the pseudoinverse of the Jacobian matrix. As the result of this operation, we obtain the following equation:

$$dq = J^+ \, d\mathbf{p} + [\mathbf{1} - J^+J] \cdot h \tag{26}$$

where J^+ denotes the Moore-Penrose pseudoinverse [4] given by:

$$J^+ = J^T(JJ^T)^{-1} \tag{27}$$

and h is an arbitrary joint velocity vector, $[\mathbf{1} - J^+J] \cdot h$ is the projection into the null space of J corresponding to a self motion of the linkage that does not move the effector-end.

Liegeois [16] developed a formulation of redundancy such that a scalar criterion function $z(q)$ may be minimized by assigning to the vector h the gradient vector of the criterion function. The null space vector has been used in singularity avoidance. In turn, Yoshikawa [16] proposed minimizing a dexterity measure given by:

$$z(q) = - \sqrt{\det (JJ^T)} \tag{28}$$

From this equation it follows that

$$h(q) = \operatorname*{grad}_{q} z(q) \ . \tag{29}$$

It is obvious that, if dim q = dim $\mathbf{p}$ then

$$dq = (J(q))^{-1} \cdot dp \qquad (30)$$

The equation (26) can be disretized and as a result we obtain a one-step transition function f as:

$$q(k+1) = q(k) + [1-J^+J] \cdot h(q(k)) + J^+ \cdot \delta p(k) \qquad (31)$$

The output function g is the same as in the previous model M_Q, i.e.,

$$g(q(k)) = (t_i(q(k)) \mid i=1,...,n+1) \qquad (32)$$

This completes the full description of model M with respect to the Joint-State-Space specification. This model of the robot kinematics is a strong nonlinear discrete system with the transition function f given by formula (31) and the output function g given by formula (32). The model of kinematics obtained on this way is denoted by M_J. The model M_J possesses a large computational complexity because the function f and g are strongly nonlinear For example, the invers of Jacobian matrix of kinematics transformation of ASEA IRb-6 robot is given by:

$$J^{-1}(q) = \frac{1}{l_2 l_3 s_3} \cdot$$

$$
\begin{bmatrix}
\dfrac{-l_2 l_3 s_3 s_1}{l_4 s_{234}-l_3 s_{23}+l_2 s_2} & \dfrac{l_2 l_3 s_3 c_1}{l_4 s_{234}-l_3 s_{23}+l_2 s_2} & 0 & 0 & 0 \\[2ex]
l_3 c_1 s_{23} & l_3 s_1 s_{23} & -l_3 c_{23} & l_3 l_4 s_4 & 0 \\[1ex]
c_1(-l_3 s_{23}+l_2 s_2) & s_1(s_2 l_2-s_{23} l_3) & l_3 c_{23}-l_2 c_2 & l_2 l_4 s_{34}-l_3 l_4 s_4 & 0 \\[1ex]
-l_2 c_1 s_2 & -s_1 s_2 l_2 & l_2 c_2 & -l_2 l_3 s_3-l_2 l_4 s_{34} & 0 \\[1ex]
0 & 0 & 0 & 0 & l_2 l_3 s_3
\end{bmatrix}
$$

We now list advantages and dis-advantages of the kinematics model just presented. The advantages are:

- because the model uses the Jacobian matrix, it is well defined for general manipulator kinematics, expect for numerical problems near kinematic singularities,

- the model has a symbolic solution of components of the transition function, which can be obtained automatically.

However, the model M_J has also some weak points, namely:

- this model has an intrinsic inaccuracy due to the approximate character of the Jacobian matrix at every step, thus it accumulates errors which become larger as the velocity increases. The path tracing error increases step by step, during a motion along a given path,

- the model does not give directly the joint value for a given location of the effector-end,

- the model does not preserve the repeatability of joint values of repeated effector-end motion,

- the model possesses a large computational complexity because the functions f and g are nonlinear with many trigonometric components.

The application of above presented models to the motion planning of robot manipulator, causes that the planning is accomplished at two levels of description: changes in the manipulator configuration are defined in the Joint space (function f), while the manipulator position relative to some objects in its environment is determined in the Cartesian space (function g). The disadvantages of the models in the question, show clearly the importantance of the task of finding some methods of robot's kinematic modelling capable of handling the manipulator configuration in the Cartesian space, without restoring to the Joint space at any stage. It means, that this model should be able to supply us with successive configuration calculated immediately in the Cartesian Space.

We now present two models based on the Cartesian-state specification. As before, we construct one model by discretization of the Joint space and the other one by discretization of the Cartesian space.

2.2.   THE CARTESIAN-STATE SPECIFICATION OF MANIPULATOR'S CONFIGURATION

We begin with the assumption that every joint angle q_i has already been discretized at the increment δq_i. Hence, the value of q_i is defined by formula (14).

The model of kinematics will be synthetized in three steps related to the decomposition of the manipulator into an arm (a 2-dimensional problem) and individual joints. The first step will be concerned with synthetizing the model of a single joint. At the second step, we shall produce the model of the manipulator's arm kinematics in the arm plane $X_A \times Z_A$. At the third step, we shall describe the kinematics model of the whole manipulator in the 3-dimensional space E_0.

Assume that a single change of angle q_i amounts only to $\pm \delta q_i$ or 0. Such an assumption yield that changes of q_i can be described by means of the input alphabet $U_i = \{+1,0,-1\}$, where +1 means a change by $+\delta q_i$, -1 means a change by $-\delta q_i$ and 0 means no change. For the i-th joint, the set of states equals the set of feasible angle positions of the joint, defined as:

$$J_i = \{0,1,...,N_i\} \tag{33}$$

A change of the joint state causes that the link l_i rotates by the angle q_i in the direction defined by the input letter u_i, what in the composed kinematic chain amounts to a rotation of point p_{i+1} around the point p_i by a constant angle δq_i. For this reason, it is natural to take the output alphabet Y_i as a set of transformations defining the rotation of a point in the plane, related to input letters. Thus

$$Y_i = \{R_i(+1), R_i(-1), R_i(0)\} \tag{34}$$

where R_i are constant 2×2 matrices described by formula (13) for $\xi = \pm 1$.

The model of changes of the i-th joint kinematics is then representable as FSM (Finite State Machine) in the form:

$$M_i = (U_i, J_i, Y_i, \lambda_i, \delta_i) \tag{35}$$

where:

$$\lambda_i : J_i \times U_i \rightarrow J_i \qquad \text{is the one-step state transition function}$$

and

$$\lambda_i(j,u) = \begin{cases} \max(0,j+u) & \text{for } u = -1 \\ \min(N_i,j+u) & \text{for } u = +1 \\ j & \text{for } u = 0 \end{cases} \tag{36}$$

$$\delta_i : J_i \times U_i \rightarrow Y_i \qquad \text{is the output function and}$$

$$\delta_i\ (j,u) = \begin{cases} R_i(u) & \text{if } j \neq 0 \text{ and } j \neq J_i \\ \\ R_i(0) & \text{if } (j = 0 \text{ and } u \neq +1) \text{ or } (j=J_i \text{ and } u \neq +1) \end{cases} \tag{37}$$

Observe that the transition function λ_i has been designed in such a way that whatever the input sequence might be, the feasible range of the angle q_i change will not be violated.

The position of individual joints of the manipulator within the arm plane $X_A \times Z_A$ is described by the vector s (9) related univocally to the configuration x w.r.t. the basic Cartesian frame E_0. The discretization of the Joint space Q yields that the set S_A of feasible vectors s of the manipulator joint positions in the plane $X_A \times Z_A$ is finite and contains exactly

$$\overline{\overline{S}}_A = \prod_{i=2}^{n} N_i$$

elements. Let the set S_A be the state space of a FSM modelling changes in the arm kinematics. The input alphabet U_A of the FSM will consists of vectors $\mathbf{u} = (u_i \mid i = 2,...,n)$ whose components take values $u_i \in U_i$, i.e.

$$U_A = X \{U_i \mid i = 2,...,n\} \tag{38}$$

The output alphabet Y_A of the FSM is set equal to S_A.

Then, the model of changes of the arm kinematics within the plane $X_A \times Z_A$ can be represented as a FSM of the form

$$M_A = (U_A, S_A, Y_A, \lambda_A, \delta_A) \tag{39}$$

where $\lambda_A : S_A \times U_A \rightarrow S_A$ denotes the one-step state transition function defined as follows.

Let $s' = \lambda_A(s,\mathbf{u})$ where $s'= (p'_1,...,p'_{n+1})$, $s = (p_1,...,p_{n+1})$ and $\mathbf{u} = (u_2,...,u_n)$.

Then

$$p'_i = Rot^{\delta q_{i-1}}(p_{i-1}, Rot^{\delta q_{i-2}}(p_{i-2},...,Rot^{\delta q_2}(p_2,p_i,u_2)u_3)...u_{i-2})u_{i-1}) \tag{40}$$

for $i = 2,...,n+1$ and $p'_1 = p_1$.

The function $\delta_A : S_A \rightarrow Y_A$, being the response function of M_A, is the identity function, i.e., $\delta_A = Id$.

Now, using expression (13) which defines operation Rot, and models of kinematics for individual joints M_i , it is easy to prove the following fact:

If the Joint-FSM M_i assumes state j_i, and the input is supplied with the letter u_i for $i = 2,...,n$, then the transition function of the Arm-FSM, λ_A is defined recursively as follows:

$$p'_1 = p_1$$

$$p'_i = p'_{i-1} + (\prod_{k=2}^{i-1} \delta_{i-k} (j_{i-k}, u_{i-k})) \cdot [p_i - p_{i-1}] \qquad (41)$$
$$\text{for } i = 2,...,n+1$$

where δ_k is the output function of the k-th Joint-FSM.

Taking into account output matrices $R_k(u_k)$ of the Joint-FSM, and substituting $T_{k-1} = R_{k-1} \cdot R_{k-2} \cdot ... \cdot R_2$, one can express (41) as:

$$p'_i = p'_{i-1} + T_{i-1} \cdot [p_i - p_{i-1}] \quad \text{for } i = 2,...,n+1 \qquad (42)$$

Also observe that the recursive dependence (42) describing the function λ_A is easy to handle numerically because, while generating the new position of the i-th joint within the arm plane, one has just to multiply the matrix T_{i-2} (found for the preceding joint), by a constant matrix R_{i-1} (having 4 entries and being the value of the response function δ_{i-1} of automaton M_{i-1} simulated by the input letter u_{i-1}). To calculate the function λ_A, if suffices to perform some algebraic operations, without resorting to trigonometric functions at all.

The manipulator's position w.r.t. the base Cartesian reference frame is described by its configuration x (6). The discretization of values taken on by the joint angles produces the finite set X_M of feasible manipulator configurations. Clearly, the cardinality of $X_M, \overline{\overline{X}}_M = N_1 \cdot \overline{\overline{S}}_A$. Now, the set of fesible configurations X_M will be taken as the state space of the FSM modelling changes of the manipulator kinematics. The input alphabet U_M consists of vectors $\mathbf{u} = (u_i \mid i = 1,...,n)$ with valueas ± 1 or 0 i.e.,

$$U_M = X \{U_i \mid i = 1,...,n\} \qquad (43)$$

The output alphabet Y_M is the same as X_M.

Thus the model of changes in the manipulator kinematics w.r.t. the basic coordinate frame E_0 can be represented as an FSM of the following form

$$M_M = (U_M, X_M, Y_M, \lambda_M, \delta_M) \qquad (44)$$

where $\lambda_M : X_M \times U_M \rightarrow X_M$ is the one-step state transition function and $\delta_M : X_M \rightarrow Y_M$ is the output function.

Before synthetizing the function λ_M, recall that configuration x corresponds univocally to the state s of the Arm-FSM, determined by transformation H (10). Using this fact, we define the new-configuration $x' = \lambda_M (x, \mathbf{u})$ in 4 steps as follows:

Step 1: Taking into account configuration x, calculate the state s of the Arm-FSM.

Step 2: Calculate a new state s´, of the Arm-FSM for a vector
$$\mathbf{u}_A = (u_i \mid i = 2,...,n) \text{ as } s´ = \lambda_A (s, \mathbf{u}_A).$$

Step 3: Transform the vector s´ into an intermediate configuration $x´_p$ by using transformation H (configuration $x´_p$ lies in the "old" arm plane).

Step 4: Find new state x´ by rotating the intermediate configuration $x´_p$ around point P_1 in the plane $X_0 \times Y_0$, according to the action of the 1-Joint-FSM, M_1, for the input letter u_1.

By extending the transformation H defined in (10) to sets X_M and S_A, we obtain $\mathcal{H} : S_A \rightarrow X_M$,

$$\mathcal{H} (s) = (H(p_i) \mid i = 1,...,n+1) \qquad \text{and} \qquad (45)$$
$$\mathcal{H}^{-1}(x) = (H^{-1}(P_i) \mid i = 1,...,n+1)$$

On the basis of what we have said above, we now define the function λ_M. Let $x´ = (P´_i \mid i = 1,...,n+1) = \lambda_M(x,\mathbf{u})$, $\mathbf{u} = (u_1,\mathbf{u}_A)$ and the 1-Joint-FSM assume the state j_1. Then

$$P´_i = \begin{bmatrix} \delta_i(j_1, u_1) & \vdots & 0 \\ \text{-----} & \text{-----} & \text{-----} \\ 0 & \vdots & 1 \end{bmatrix} \text{crd}^i \mathcal{H} (\lambda_A (\mathcal{H}^{-1}(x), \mathbf{u}_A))$$

$$\text{for } i=1,..,n+1 \qquad (46)$$

where $\text{crd}^i x = P_i$ denotes the i-th coordinate of vector x.

Notice that the values of trigonometric functions sin and cos appearing in the transformation $\mathcal{H}$ are calculated for a fixed position of the arm plane. Hence they can be determined at an arbitrary point from the configuration x (e.g., $\cos q_1 = x_i / \| P_i \|_{xy}$.)

Thus the transition function λ_M of the Manipulator-FSM is a composition of the transition function λ_A of the Arm-FSM and the response function of the 1-Joint FSM responsible for the rotation within $X_0 \times Y_0$ plane. The structure of the Manipulator FSM is shown in Fig. 3. The output function δ_M, is the identity function.

The construction of the transition function λ_M of the manipulator FSM allows us to obtain, by simple computations, successive configurations of the manipulator w.r.t. the basic Cartesian space E_0. The only thing required in order to find the value of λ_M is to calculate recurrent function λ_A by successive multiplying constant rotation matrices and to perform the square root operation and the multiplication by constant matrix R_1.

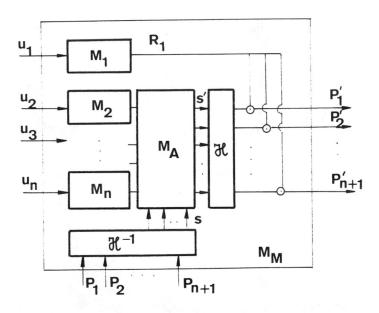

Fig. 3. The structure of the FSM model of robot kinematics

The model does not need large memory for the computations. The data which should be stored are following the actual vector state of the automata M_i, the configuration x and the values of sin and cos of the discretization increment δq_i (all together 4n variables). The FSM is a convenient exemplification of a production system. Successive states of the system are the states of the automaton M_M, and production rules are given by the transition function λ_M.

The state space X_M of the model M_M contains

$$\overline{\overline{X}}_M = \prod_{i=1}^{n} N_i$$

elements and for a given state (configuration) x, the state transition function λ_M generates 3^n-1 sucessors of x.

We now should consider the other specification of the input set of model M which is based defining U as the set of discretization increments of the effector-end displacement defined by formula (22).

This specification of U creates a model of the robot kinematics in the form of a nonlinear discrete dynamical system.

$$M_S = (X,U,Y,f,g) \tag{47}$$

which functions f and g are defined as follows:

$$x(k+1) = f(x(k), \delta p(k)) \qquad \text{and} \tag{48}$$

$$g(x(k)) = x(k) \quad \text{is the identity.} \tag{49}$$

The vector x(k) is the manipulator's state at time k in the base Cartesian frame E_o (6) and $\delta p(k) = [P_{n+1}(k+1) - P_{n+1}(k)]$ is the displacement vector of the effector-end. We can observe that $P_1(k+1) = P_1(k)$ and $P_2(k+1) = P_2(k)$ for every k.

The effector translation δp can be decomposed into a rotation around Z_o-axis and a translation of the effector-end in the arm-plan E_A. Now, using a transformation of basic frame E_o into frame E_A (10), we can introduce an adequate kinematics model in the frame E_A of the following form:

$$s(k+1) = h(s(k), d(k)) \tag{50}$$

where s(k) is the manipulator state in E_A at time k (9),

$$d(k) = [p_{n+1}(k+1) - p_{n+1}(k)]$$
is the effector displacement in the plane $X_A \times Z_A$ and

h(·) is the next state function.

It is easy to observe that both systems (47) and (50) are isomorphic [8].

The main problem faced while dealing with the model of manipulator kinematics is how to find the next state function f. According to what we have said previously, the function should be found by solving under-determined kinematics equations of the manipulators [5], [26].

The next state function will be found by employing the following scheme: hypothesis - verification - modification. Thus, at the first we shall determine a hypothetic configuration of the manipulator at time k+1 which will subsequently be verified with respect to its attainability. If the configuration appears to be non-attainable, it should be modified accordingly.

Procedure for Determining Hypothetic Configuration

In order to choose one configuration from among many possibilities, we introduce two parameters into the synthesis procedure of a hypothetic state.

Let $D = \{p,...,n-1\}$ denote an arbitrary set of initial joints of a submanipulator (i.e., the subsequence of the kinematic chain from P_p to P_{n-1}), and let $G = \{g,...,n+1\}$ where $g > p+2$ and $p_i(k) + d(k) \in E^f_A$ for $i = g,...,n+1$ stand for the set of terminal joints of the submanipulator. From sets D and G we shall choose active joints of the manipulator's motion.

Step 1: Selection of strategy of a changing configuration. The choice of the
        strategy of motion is based upon determining joints p,g according to
the   formula

$$(p,g) = \max_{i \in D \; j \in G} \; (\min \; \{(i,j) \mid c^i_{jmin} \leq c^i_j \leq c^i_{jmax} \text{ and } i < j\text{-}1\}) \quad (51)$$

where $c^i_j = \| p_i(k) - (p_j(k) + d(k)) \|$.

Step 2: Synthesis of a hypothetic configuration-state s(k+1).
        For selected joints p and g, there are three possibilities:
        a) Joint $p = g\text{-}2$
        b) Joint $p < g\text{-}2$ and $c^{p+1}_g > c^{p+1}_{gmax}$ (52)
        c) Joint $p < g\text{-}2$ and $c^{p+1}_g < c^{p+1}_{gmin}$

We assume that the joint $p+1$ lies at the maximum (minimum) distance from the joint g. It follows from the assumption that the distances between joints $p+1$ and g are given and constitute the configuration of maximum (minimum) reach of the joint g w.r.t. the joint $p+1$. The positions of those joints can be found on the basis of known maximum (minimum) distances, by recurrent runs of Joint Position-Procedure subsequently, for joints $p+1, p+2,...,g-1$. The Joint Position Procedure determines in a simple way the position of point p_i, taking into account the positions of neighbouring points p_{i-1}, p_{i+1} and distances $\rho_{i-1,i}$, $\rho_{i,i+1}$ and $\rho_{i-1,i+1} = \| p_{i-1} - p_{i+1} \|$ from these points to p_i [8].

Thus the hypothetic configuration $s(k+1)$ is determined as:

$$p_i(k+1) = \begin{cases} p_i(k) + d(k) & \text{for } i = g,...,n+1 \\ p_i(k) & \text{for } i = 1,...,p \\ i\text{-th Joint Position Procedure} & \\ JPP(p_{i-1}(k+1), p_g(k+1),\ \rho_{ig}, l_{i-1}) & \text{for } i = p+1,...,g-1 \end{cases} \quad (53)$$

where $\rho_{ig} = c^i_{gmax}$ or $\rho_{ig} = c^i_{gmin}$.

The geometric shape of the hypothetic configuration $s(k+1)$ is mainly influenced by the choice of the set D and G of initial and terminal joints, respectively. By means of the sets we can produce configurations $s(k+1)$ with regard to diverse criteria.

Here, we consider some of them:

(i) Reach Motion Strategy - minimization of the number of active terminal joints. Let $D = \{2,...,n-1\}$ and $G = \{n+1\}$. Then, due to (51) the number of terminal joints g of the configuration change is equal to n and the number of initial joints p is determined as

$p = \max \{i \le n-1 \mid c^i_{n+1\ min} < c^i_{n+1} < c^i_{n+1\ max}\}$

Such a choice of p ensures that to achieve a desirable shift of the effector-end, a minimum number of links will be set in motion, counting from the end of the manipulator. The configuration of the submanipulator from joint 1 to joint p remains unchanged [8].

(ii) Gross Motion Strategy - minimization of the number of active initial joints. Let $D = \{2,...,n-1\}$ and $G = \{4,...,n+1\}$. Thus, the required shift of the effector-end is realized by moving a minimum number of joints counting from the base of the manipulator. The submanipulator from joint g to joint n+1 preserves its geometrical shape but the positions of its joints are shifted by the vector d(k). A sequence of those configurations would correspond to the gross motion aimed at placing the effector-end in the neighbourhood of the terminal point [6], [8].

Verification of Hypothetic Configuration s(k+1)

A basic requirement imposed by physical realizability of a given configuration is that all the values of joint angles fall into their allowable ranges.

It can be demonstrated that there exists a bijective map taking state s of the manipulator to vector q of joint angles. This holds true since for the configuration constructed previously, we can immediately calculate the vector of sin and cos functions of joint angles by using the vector dot and cross products of the vectors $[p_{i+1} - p_i]$ and $[p_{i-1} - p_i]$ [8]. If we know the vectors, the realizability conditions can be verified in a straight forward manner without resorting to the calculation of functions arcsin and arccos.

Based on the realizability of a configuration at time k and the algorithm of hypothetic configuration synthesis, it can be proved that at most three joint angles, namely q_p, q_{p+1}, q_g are subject to the verification. The other joint angles already lie within their allowable ranges.

Modification of Hypothetic Configuration

The modification of a configuration with respect to joint angle i means a rotation of the appropriate link (usually l_i or l_{i-1}) by a fixed angle β followed by forming a new configuration for such modified joint position. The new configuration should be reached in such a way that the number of angles subject to a change be minimum. This can be done by using the JPP procedure for certain segments of the kinematic chain of the manipulator.

Alternatively, a new configuration can also be produced based upon joints preceding or following joint i. If so, the rotation of links at joint i will be followed by some changes at other joints. Notice that in the former case, the range of possible movements of each joint is bounded, which may make the modification simply unrealizable. To the contrary, in the later case, the range of possible movements of joint n is unbounded. The modification of subsequent angles is performed in accordance with Link Rotation Procedure (LRP).

We get $q_i < q^{min}_i$. Link Rotation Procedure for the i-th joint.

Step 1: The determination of a modyfying joint

$$m = \min_{j}\{j = 2,...,i-2, i+2,...,n+1 \mid q_j << q^{max}_j\} \tag{54}$$

The procedure makes possible to check which joint angle allows us to perform the modification by rotation.

Step 2: The determination of new configuration. If $m < i$ then:

Time 1: Link l_i is rotated around joint i+1 (we assume that in the submanipulator from joint m to joint i-1 angles are kept fixed) counter clockwise by a given modification angle β. The new position is calculated as (12)

$$p'_i = Rot^\beta (p_{i+1}, p_i, -1) \tag{55}$$

Time 2: Position p'_{i-1} is calculated using the JPP procedure for given $p_m, p'_i, l_{i-1}, c^m_{i-1}$.

Time 3: Angle increment $q_m = \gamma$ is determined. If $q_m + \gamma > q^{max}_m$, go back to step 1 with $j \neq m$. Otherwise calculate positions p'_j for $j = m+1,...,i-2$ as a result of rotations in joint m by angle γ i.e.,

$$p'_j = Rot^\gamma (p_m, p_j, -1) \tag{56}$$

If $m > i$, then we assume that in the manipulator segment from joint i+1 to m all angles are kept constant, and as Time 1 and Time 2 we have:

Time 1′: $p'_i = Rot^\beta (p_{i-1}, p_i, +1)$ \hfill (57)

Time 2′: Position p'_{i+1} is calculated using JPP procedure for given $p_m, p'_i, l_i, c^m_{i+1}$.

Step 3: Check if $q_i \geq q^{min}_i$. If not go to step 1, else stop.

The modified configuration for the case $q_i > q^{max}_i$ is found analogously. Observe that m exists as a pseudojoint n+1 which satisfies conditions of being modifying joint. After applying the procedure a finite number of times the unattainable angle eventually enters the allowable range. Detailed formulas describing such a model of robot kinematics can be found in [8]. The algorithm of synthesis of s(k+1) operates according to the illustration in Fig.4.

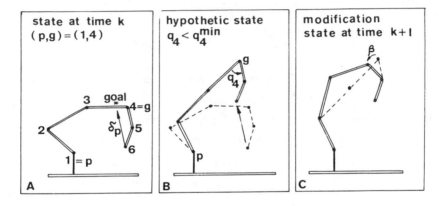

Fig. 4. The state-transition function of the sequential model M_s

The model presented above has the following disadvantages:
- the model does not preserve the repeatability of joint values of repeated effector-end motion,
- the model requires additional iterations for the transition function computation if the modification is required.

However, it has the following important advantages:
- the model gives directly the joint values for a given location of the effector-end,
- the desired geometric shape of a configuration can be obtained by a choice of active joints,

- the model has an acceptable computational complexity because only
algebraic formulas appear in the transition function f,
- the model has a trivial output function.

Let r denote the number of selected strategies of changing configuration, and let the displacement vector $\delta \mathbf{p}$ have one of the forms: $(\pm \delta x,0,0),(0,\pm \delta y,0)$ or $(0,0,\pm \delta z)$ where $\delta x, \delta y, \delta z$ denote the vectors of the effector displacement along X,Y,Z axes of the base frame E_0, respectively.

Then, for a given state x, the state transition function f generates $6 \cdot r$ successors of x, independent of the DOF-number of the manipulator.

The different models of robot kinematics are shown in Fig. 5.

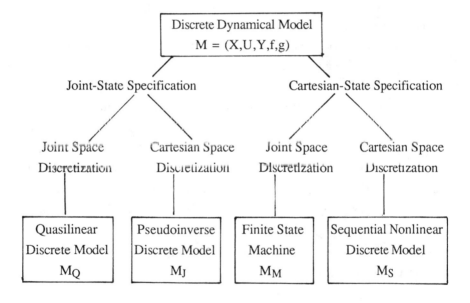

Fig. 5. Models of robot kinematics for different model´s components
specification

3. THE APPLICATION OF KINEMATIC MODELS TO COLLISION-FREE ROBOT MOTION PLANNING

The models of the manipulator kinematics dealt with in this paper are appplied in a system for planning collision-free paths of robot movements.

The automatic generation of robot motion trajectories in the presence of obstacles is one of the main subproblems in the synthesis of task-oriented languages for robot programming. The system solving such a problem should be able to determine (on the basis of a geometric and kinematic description of the robot, a geometric description of its environment and the initial and final positions of the end of effector), the collision-free track of the manipulator motion (an envelope for the manipulator trajectory) from initial to goal effector locations.

This research problem has been treated in various ways by Pieper [27], Nielsson [24], Lozano-Perez [20], Khatib [15], Ozaki [25], Brooks [2], Gouzenes [6], Lumelsky [21], Herman [7]. The problem becomes particularly hard for redundant robots for which the determination of an appropriate effector path is not equivalent to determining the trajectory of the whole manipulator. The global methods based on the notion of the configuration space [6], [17], [18] face considerable difficulties related with transforming obstacles to the configuration space, with the appearance of pseudoobstacles [12] as well as with the construction of the visibility graphs. This being so, one is left with local methods, starting from an initial position of the manipulator and transforming the position successively according to selected motion strategies.

3.1 COLLISION-FREE MOTION PLANNING BASED ON FSM-MODEL Of ROBOT KINEMATICS

The problem of finding a collision-freee trajectory of a manipulator, from an initial configuration x_0 to a terminal point $G \in E_0$, can be formulated as follows [9], [10]: Let $X_G = \{x \mid crd^{n+1}x = G\}$ be the set of terminal configurations.

The problem amounts to finding such a sequence $u^*_p = (u_0,...,u_p)$ of input letters of the FSM M_M that the terminal configuration reaches point G i.e.

$$\lambda^*_M(x_0,u^*_p) \in X_G, \tag{58}$$

where $\lambda^*_M(x_0,u^*_p) = \lambda_M(\lambda^*_M(x_0,u^*_{p-1})u_p)$, and that every configuration corresponding to sequence u^*_p is feasible i.e.

$$\lambda^*_M(x,u^*_j) \in X^f_M \qquad \text{for } j = 0,...,p \tag{59}$$

where $X^f_M \subset X_M$ denotes the set of collision-free configurations.

Additionally it is often required that sequence u^*_p minimize a cost function $V : U^* \rightarrow R$ (expressing for example the length of the trajectory).

The problem of finding the collision-free trajectory is then a problem of "reachability" of the set of states X_G from the state x_0 of automaton M_M for which not all of its states are feasible.

In order to solve the problem, we employ procedures of graph searching used in AI [24]. For this purpose we exploit the state transition graph of automaton M_M generated implicitly by applying the function λ_M as production rules. The development of the search graph starts from the node x_0, by the action λ_M for alle possible letters of alphabet U_M. Except for the letter **0**, which does not change the state, every node of the graph has 3^n-1 successors, i.e., $X(x) = \{\lambda_M(x,u) / u \in U_M \setminus \{0\}\}$. Thus, it becomes essential to quickly check for non-repeatability of the nodes generated and their feasibility, i.e., to test successive nodes.

A configuration x is said to be feasible if it does not collide with any obstacle in the work-scene.

Let O_i E_0, for $i = 1,...,l$, denote obstacles in the scene. The space occupied by the robot manipulator at configuration x can be approximated by rotary cylinders whose axes are individual arms l_i of the skeleton model [18]. In order to check if the manipulator moves in a collision-free way only on the basis of its skeleton model, let us extend the obstacles in each direction by the value of the maximal radius of cylinders which approximate arms from 1 to n. The extended objects are denoted by 0^e_i. The extending algorithm clearly depends on the type of approximation and the description of the volume occupied by the i-th obstacle. Due to the cylindric-like model of the manipulator, the growing procedure is simplified considerably. Taking into account the above facts, we are able to define the set of feasible configurations as

$$X^f_M = \{x \mid Seg(x) \cap \bigcup_{i=1}^{1} 0^e_i = \varnothing\} \tag{60}$$

where $Seg(x)$ is a broken line joining points $(P_1,...,P_{n+1})$. Now checking for non-intersection of segments $[P_i,P_{i+1}]$ with solids O_j in R^3 can be replaced by more effective algorithms in R^2, by using the joint space discretization. The discretization of the q_1 joint angle results in a finite number of admissible positions of the arm E_A plane w.r.t. the basic frame E_0.

Therefore, checking for the collision-freeness of the configuration x can be reduced to the "broken line-polygon" intersection detection problem in the plane. This problem can be solved by a few efficient algorithms based on the plane-sweep techniques [23], [28]. The intersection detection of a broken line $Seg(s)$, composed of n segments with polygon P of m vertices is a problem of computational complexity $0(n+m)$ for convex P or $0((n+m)lg(n+m))$ for non-convex P [23], [28].

The robot kinematics model as a FSM not only facilitates us a simple computation of successive configurations (simple production rules), but also offers efficient-test-algorithms of configuration feasibility.

In order to solve the reachability problem of terminal configurations, we apply the graph searching procedure [10], [24] to the state-transition graph. This procedure generates (makes explicit) a part of an implicity specified graph. This implicit specification is given by the start node, representing the initial configuration x_0 and the rules (one-step state transition function λ_M) that alter configurations. The process of applying the transition function λ_M to a node is called expanding the node. Expanding x_0, successors of x_0 etc., ad infinitum, makes explicit the graph that is implicitly defined by x_0 and λ_M. Each node x in X_G is a goal node. A graph search control strategy is a process of making explicit a portion of an implicit graph, sufficient to include a goal node.

The way of expanding the graph depends on the form of the cost function of the evaluation function and on the way in which nodes are expanded.

Below we use Nilsson's [24] notation $k(x,x')$ to denote the cost of directed arc from node x to node x'. The cost of a path between two nodes in the sum of the cost of all the arcs connecting the nodes along the path. Let the function g(x) give the actual cost of the path in the search tree from start node x_0 to node x, for all x accessible from x_0, and the function h(x) give a cost estimate of the path from node x to a goal node. We call h the heuristic function. The evaluation function f(x) at any node x gives the sum of the cost of the path from x_0 to x plus the cost of the path from x to a goal node, i.e., $f(x) = g(x) + h(x)$. General Graph Searching Procedure [24] relies on expanding the search graph GR, beginning with the start node x_0 set initially in the list of nodes OPEN. The list OPEN is a set of nodes ordered with the help of the evaluation function f(x). After the best node from the list OPEN has been chosen, it is transferred from this list to the list CLOSED (which is initially empty). If the node is not a goal node, the process of expanding nodes continues. From the set of node successors, feasible vertices are selected, and indicators are found establishing univocally their predecessors is such a way that the expanded graph is a tree. Next, the feasible nodes are added to OPEN, are ordered, and the procedure is repeated.

Let $x' = \lambda_M(x,u)$. For the cost function between two nodes x, x' joined by an arc, we assume the Euclidian distance travelled by the effector while passing from configuration x to x', i.e.,

$$k(x,x') = \| crd^{n+1}x' - crd^{n+1}x \| \tag{61}$$

The function which estimates the distance from the actual configuration x to the set X_G is defined as the recti-linear distance between the effector position in configuration x and terminal position G, i.e.,

$$h(x) = \| crd^{n+1}x - G \| \tag{62}$$

It can easily be seen that the heuristic function h(x) is a lower bound of the function h* calculated for the optimal path between x and a goal node.

Let configuration x be chosen as the best node from the list OPEN. (X is not a goal node.) For x let us find the set of successors X(x). Remove from X(x) the configurations x' which either are ancestors of x, or do not cause any effector translation, i.e., for which $k(x,x') = 0$, or are colliding, i.e., they do not satisfy formula (22). In order to simplify the procedure of collision-freeness testing, we can additionally introduce a list of forbidden nodes FORBID, initially empty, and add to it there successive colliding configurations. Then the testing configuration x may begin with checking if $x \in$ FORBID. If $x \notin$ FORBID, the collision-freeness testing procedure can be turned on.

From the reduced set of successors x, we choose those configurations x´, which have not been present until now on lists OPEN or CLOSED, and introduce the configurations into OPEN, establishing x as their predecessor. For other configurations x´, we compare the values of the cost path function g calculated first for the path passing through x, and then for the path going through the node which has been denoted previously, in the search tree, as their predecessor.

For all the new and newly marked nodes of the list OPEN, we calculate the evaluation function f, and order the list according to increasing values of f.

After the list OPEN has been ordered, it may happen that there exist several configurations with the same minimal value of the evaluation function. In order that a univocal choice of the minimal element be possible in the next step, we introduce the following arrangement index. For every "best" node x, there is known its predecessor x´ and the input letter **u**. We define the "energy" function, required for a transition to the node x

$$V(x) = \sum_{i=0}^{n-1} d_i \, | \, u_i \, | \tag{63}$$

where $d_i > 0$ denotes the cost of launching the i-th joint and $d_i > d_{i+1}$ is satisfied for $i=1,...,n-1$. Inequality means that launching initial joints of the manipulator is more energy consuming than for the terminal joints.

Employing function V, we choose from configurations with a minimal value of the function f the one for which V takes its minimum.

An illustration of this method of planning a collision-free path of the manipulator is shown in Fig. 6 (a,b) where simulation results are presented for a robot with 5-DOF in macro-movements. A scene has been assumed with obstacles approximated by a polyhedra. In order to simplify procedures of an arm collision testing, successive goal points G_0, G_1, G_2 of the effector-end position have been placed in the arm plane E_A. Fig. 6a visualizes the effector path from G_0 to G_1. Fig. 6b shows the path from G_1 to G_2 and the successively selected configurations of the entire manipulators.

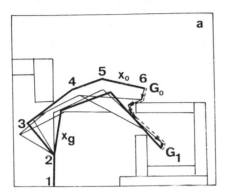 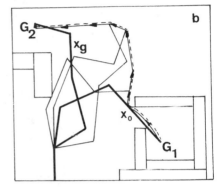

Fig. 6. Collision-free paths of robot motion obtained by using FSM model of robot kinematics

3.2 COLLISION-FREE ROBOT MOTION PLANNING BASED ON SEQUENTIAL, NONLINEAR MODEL OF THE MANIPULATOR KINEMATICS

The model of manipulator´s kinematics described by fromulas (48), (49) is now used in order to determine a robot´s trajectory in the presence of obstacles.

An algorithm of searching for a collision-free manipulator trajectory concists of two main procedures. The first accomplishes planning of the path of the end of a manipulator by using the potential method of the Find Path problem solving [14], [15]. The other procedure defines configurations of the manipulator for successive positions of the effector, checks if they collide, and if so, modifies the configurations.

Effector-Path Planning Procedure (EPP)

In order to find a path of the effector, the potential method has been applied. The method relies on the transformation of the problem of the point $P_{n+1} \in E_0$ motion planning to the problem of the motion modelling of material point charges with negative electricity within a potential field generated by obstacles 0_i (negative charges) and terminal point G (a positive charge). The potential field generated by individual obstacles 0_i is defined as

$$F_i = a_i \cdot \exp \left(-e_i \left(\rho \left(P_{n+1}, 0_i\right) - \varepsilon_i \right)\right) \qquad i=1,...,l \qquad (64)$$

where P_{n+1} denotes the position of the end of manipulator, ρ is the distance of point from set, ε_i means feasible approach of the effector toward obstacles, which takes into account dimensions of the manipulator, e_i, a_i are constants which determine the field magnitude and the range of its activity. The field generated by the terminal point G is expressed as

$$F_0 = a_0 \cdot \exp \left(-e_0 \left(\| P_{n+1} - G \| - \varepsilon_0 \right)\right) \qquad (65)$$

where ε_0 stands for the accuracy of reaching point G. Using the above listed relationships which describe the value of the potential field at point P_{n+1}, we deduce the following equation of motion of a free negative electric charge placed at P_{n+1}

$$\ddot{P}_{n+1} = \sum_{i=1}^{M} f_i \left(P_{n+1}\right) + f_0 \left(P_{n+1}\right) \qquad (66)$$

where $f_i \left(P_{n+1}\right) = - \operatorname*{grad}_{P_{n+1}} F_i$ is the force with which the i-th obstacle effects the test charge and $f_0 = \operatorname*{grad}_{P_{n+1}} F_0$ is the force with which terminal point G acts upon the charge.

By numerically solving the equation of motion of the point P_{n+1}, we obtain successive positions $P_{n+1}(k)$ of the point $k = 0,1,...$. The accuracy of the solution depends on the size of the time discretization increment and the type of obstacle's approximation.

Collision Avoidance Procedure (CAP)

The EPP procedure results in successive positions of the end of manipulator $P_{n+1}(k)$. On the basis of positions $P_{n+1}(k)$ and $P_{n+1}(k+1)$, and the configuration $x(k)$ of the manipulator, it is possible to find new configuration $x(k+1)$ using the model of the manipulator's kinematics. Clearly, the new configuration is selected depending on the choice of parameters D and G of the model. In what follows, we are going to restrict our attention only to two strategies of selecting configuration $x(k+1)$, namely to the reach motion strategy and the gross motion strategy. One of the strategies is employed with respect to the positions of the actual point P_{n+1} $(k+1)$ and terminal point G, and to the distance between them. More specifically, if $\| P_{n+1}(k+1) - G \| \leq \delta_{gr}$, then we use the reach motion strategy. Otherwise the gross motion strategy is employed. Due to this assumption, it is possible to uniquely determine configuration $x(k+1)$ corresponding to the position of the end of manipulator at point $P_{n+1}(k+1)$. The position also determines the plane $E_A(k+1)$ within which the manipulator arm lies. If configuration $x(k+1)$ is collision-free, then

$$\delta ([P_{i-1},P_i],0_j) \geq \varepsilon \qquad \text{for } i = 2,...,n+1 \text{ and } j - 1,...,l \qquad (67)$$

and $[P_{i-1},P_i]$ denotes a line segment joining the points $P_{i-1}(k+1)$ and $P_i(k+1)$ of the configuration $x(k+1)$. The computational complexity of checking the above condition depends on the type of obstacles' approximation. We can assert that the condition is easily testable in the case of the approximation by spheres and ellipsoids, whereas some difficulties are met in the case of polyhedral approximation. Using the arm plane $E_A(k+1)$ determined by the point $P_{n+1}(k+1)$, the task of collision testing can be reduced to a much simpler task of testing the intersection between a broken line and a polygon. The collision-freeness of segment $[P_{i-1},P_i]$ will be written down as a predicate Free (P_{i-1},P_i).

If a collision has appeared, we define the joints of the manipulator which determine non-colliding links of the kinematic chain. More specifically,

$$P_d = \max_{j=1,...,n+1} \{ P_j \mid \forall i=1,...,j) (\text{Free}(P_{i-1},P_i)) \} \qquad (68)$$

determines the initial, collision-freee submanipulator, while

$$P_u= \min_{j=1,...,n+1} \{ P_j \mid \forall\ i= j+1,...,n+1)\ (Free(P_{i-1},P_i\ )\} \qquad (69)$$

gives the terminal, collision-free manipulator.

Each of these points divides the manipulator into two parts: one is collision-free and the other one is not. A proceudre of removing collisions is realized through the following steps:

Step 1: Rotate the arm $[P_{u-1}, P_u]$ around P_u in the arm plane $E_A(k+1)$ toward the previous position of point P_{u-1} (k) by an angle ensuring collision-freeness of this arm. Observe that the above action is equivalent to an action of a force pushing away the joint P_{u-1}, which is proportional to the depth of penetrating the obstacles by the arm $[P_{u-1}, P_u]$. The rotation causes a "tearing down" of the manipulator into two submanipulators, namely, the initial one containing joints [1,u-1] and the other one, collision free, with joint [u-1,n+1].

Step 2: If new point P_{u-1} lies within the range of initial submanipulator, then determine a new configuration of the submanipulator able to reach point P_{u-1} , applying to joints [1,u-1] the reduced model of the manipulator´s kinematics and the strategy opposite to that used for determining x(k+1). The above procedure will lead again to a fusion of both submanipulators. The configuration of the complete manipulator obtained in that way is subject to a standard procedure of verification and, perhaps, of modification. The new configuration contains the collision-free part of the kinematic chain with at least one new link added. Eventually, if the point P_{u-1} does not lie within the range of initial submanipulator, then go to Step 4.

Step 3: Check if the new configuration is collision-free. If not, repeat the procedure.

Step 4: Rotate link $[P_d\ P_{d+1}]$, around point P_d in the arm plane toward the previous position of point P_d (k+1) by an angle which guarantees collision-freeness of this link. The rotation results in a "tear" of the manipulator into two parts, one concerning joints [1,d+1] and the other with joints [d+1,n+1].

Step 5: If new point P_{d+1} lies within the range of the submanipulator [d+1,n+1] , then find its new configuration applying to joints [n+1,d+1] the reduced model of the manipulator´s kinematics, and the strategy opposity to the one used before. If P_{d+1} does not lie within the range of the terminal submanipulator, then shift the whole submanipulator [d+1, n+1] in such a way that its joint d+1 assumes new position P_{d+1} . Go to 3.

The procedure described above causes a successive pushing of indiviual colliding arms of the manipulator out of the obstacle, while keeping fixed, as long as possible, the position $P_{n+1}(k+1)$ of the end of the manipulator obtained from procedure EPP. An illustration of this method of planning a collision-free trajectory of a manipulator is shown in Fig. 7 (a,b) where simulation results are presented for a robot with 5 D.O.F. in macromovements. The scene is identical to the scene in the previous example.

The potential method of finding the effector path was adopted to approximate the obstacles by polyhedra. Fig. 7a illustrates the path from G_0 to G_1 and then from G_1 to G_2. It also shows successivley selected configurations of the whole manipulator.

Compaiing the two examples of motion planning using Cartesian-state models, one can notice the simularities of the paths from point G_0 to G_1. Different final configurations result from the redundancy of the manipulator and model properties. The second phase of motion from G_1 to G_2 has widely different paths for the two models. This is because the A* algorithm additionally uses the minimal energy criterion defined by formula (63). It prefers bigger changes in the joint angles closer to the effector-end. This makes the effector path longer but a total motion cost is smaller.

In order to compare the efficiency of both models and motion planning methods, let us estimate the complexity of one step. By a step of the method, we mean a calculation of the next configuration x(k+1) belonging to the collision-free path of motion. The first method based on the FSM model and A* algorithm has the computational complexity

$$0(M_M) = 0(n(3^n-1)) \; 0 \; (test)$$

where 0(test) is the complexity of the configuration feasibility testing.

The next method based on the MS model and potential field-FP algorithm has the complexity in the form

$$O(M_S) = O(\text{Effector Path Proc.}) +$$
$$+ \ O \ (\text{Collision Avoidance Proc.})$$

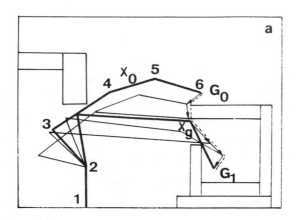

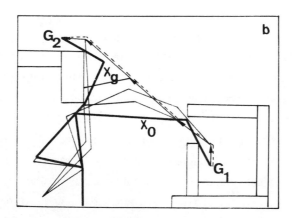

Fig. 7. Collision-free paths of manipulator motion obtained by using the sequential model of robot kinematics

Assuming the worst case, that the all joints will collide with the obstacles and that each joint will be modified, the complexity of this method can be estimated as:

$$0(M_S) = 0(n(n-2)^2) \, 0 \text{ (test)}$$

Therefore, the method employing the M_S model is more efficient. However, being a heuristic method, it does not guarantee to always find the optimal motion trajectory. Also, it does not always terminate successfully. The sequential model M_S can also be used to calculate a collision-free manipulator motion based on a discretization of the Cartesian base frame E_0 and searching for the effector path along successive edges of such obtained raster. In that case, the task of motion planning becomes equivalent to finding a path in the appropriate state-transition graph of the model [10].

In summary, we presented the kinematics models of a redundant manipulator in the form of a discrete dynamical system whose states directly express the position of subsequent joints in the Cartesian space. These models can be good tools for searching for the robot motion trajectory in the presence of obstacles and for graphical simulation of the manipulator's movements.

REFERENCES

1. M. Brady, "Trajectory Planning", *in* "Robot Motion: Planning and Control", (M. Brady, ed.), Cambridge, MA, M.I.T., 221-244 (1983)

2. R. A. Brooks, "Solving the Find Path Problem by Good Representation of Free Space", *IEEE Trans. Syst. Man. Cybern.*, Vol. SMC-13, 190-197 (1983)

3. R. A. Brooks, "Planning collision-free motions for pick-and-place operations", *Int. Journ. Robotics Res.*, Vol.2, 19-44 (1983)

4. H. Chang, "Closed form solution for inverse kinematics of robot manipulators with redundancy", *IEEE Int. Journ. Robotics and Automation*, Vol. 3, 394-403 (1987)

5. A. A. Goldenberg, B. Benhabib, and R. G. Fenton, "A Complete Generalized Solution to the Inverse Kinematics of Robots", *IEEE Int. Journ. Robotics and Automation*, Vol. RA-1, 14-20 (1985)

6. L. Gouzenes, "Strategies for solving collision-free trajectories problem for mobile and manipulator robots", *Int. Journ. Robotics Res.*, 3, 51-64 (1984)

7. M. Herman, "Fast three dimensional collision-free motion planning", *IEEE Conf. on Robotics and Automation*, Vol. 2, 1057-1063 (1986)

8. W. Jacak, "A discrete kinematic model of robot in Cartesian Space", *IEEE Trans. on Robotics and Automation*, RA-5, 435-444 (1989)

9. W. Jacak, I. Sierocki, "Planning collision-free movements of a robot: A system theory approach", *Robotica*, Vol. 6, 289-296 (1988)

10. W. Jacak, "Strategies of searching for collision-free manipulator motions", *Robotica*, Vol. 7, 129-138 (1989)

11. W. Jacak, I. Sierocki, "The system theory instrumented solver for problem of planning collision-free robot motion", *in* "Cybernetics and Systems 86", (R. Trappl, ed.), Kluwer, Acad. Pub., 751-758 (1988)

12. W. Jacak, "Modelling and Simulation of Robot Motions", *Lecture Notes in Computer Science*, Nr. 410, Springer Verlag, 307-318 (1990)

13. W. Jacak, I. Sierocki, "Software structure for design of automated work cell", *in* "Cybernetics and Systems 90", (R. Trappl, ed.), 1990 (in print)

14. B. H. Krogh, "A Generalized Potential Field Approach to Obstacle Avoidance Control", *Proc. Conf. Robotics Research*, Bethlehem, Pennsylvania (1984)

15. O. Khatib, "Commande Dynamique dans l'Escape Operationnel des Robots Manipulateurs en Presence d'Obstacles", Ph.D. Thesis, Toulouse (1980)

16. A. Liegeois, "Automatic supervisory control of the configuration of multibody", *IEEE Trans. on SMC*, Vol. SMC-13, 245-250 (1983)

17. T. Lozano-Perez, "Task Planning", in "Robot Motion: Planning and Control", (M. Brady, ed.), Cambridge, M.I.T., 463-489 (1983)

18. T. Lozano-Perez, "Automatic Planning of Manipulator Transfer Movements", *IEEE Trans. Syst. Man. Cybern.*, SMC-11, 681-689 (1981)

19. T. Lozano-Perez, "A Simple Motion-Planning Algorithm for General Robot Manipulators", *IEEE Journ. of Robotics and Automation*, RA-3, 224-238 (1987)

20. T. Lozano-Perez, "Task-level Planning of Pick-and-Place Robot Motions", *IEEE Trans. on Computers*, Nr 3, 21-29 (1989)

21. V. J. Lumelsky, "Effect of Kinematics on Motion Planning for Planar Robot Arms Moving Amidst Unknown Obstacles", *IEEE Journ. of Robotics and Automation*, RA-3, 207-223 (1987)

22. J. Y. Luh, C. Campbell, "Minimum Distance Collision-Free Path Planning for Industrial Robots with a Prismatic Joint", *IEEE Trans. Autom. Control* 29(8), 675-680 (1984)

23. D. T. Lee, F. P. Preparata, "Computational Geometry - a Survey", *IEEE Trans. on Computers* 33(12), 1072-1101 (1984)

24. N. J. Nilsson, "Principles of Artificial Intelligence", Tioga Publ. Comp. California (1980)

25. H. Ozaki, A. Mohri and M. Takata, "On the Collision-Free Movement of a Manipulator", *in* "Advanced Software in Robotics", (A. Danthine and M. Geradin, eds.), North-Holland, NY, 189-200 (1984)

26. R. P. Paul, "Robot Manipulators: Mathematics, Programming, and Control", Cambridge, MA, M.I.T. (1981)

27. D. L. Pieper, "The Kinematics of Manipulator under Computer Control", Ph.D. Thesis, Stanford University (1968)

28. M. I. Shamos, D. Hoey, "Geometric intersection problems", Proc. 17th *IEEE Ann. Symp. Found. Comp. Sci.*, 208-215 (1976)

FORCE DISTRIBUTION ALGORITHMS
FOR MULTIFINGERED GRIPPERS

JUNG-HA KIM   VIJAY R. KUMAR

Department of Mechanical Engineering
and Applied Mechanics
University of Pennsylvania
Philadelphia, PA 19104

KENNETH J. WALDRON

Department of Mechanical Engineering
The Ohio State University
Columbus, OH 43210

Abstract

The work described in this paper addresses the problem of determination of the appropriate distribution of forces between the fingers of a multifingered gripper grasping an object. The finger-object interactions are modeled as point contacts. The system is statically indeterminate and an optimal solution for the this problem is desired for force control. A fast and efficient method for computing the grasping forces is presented. Some simple grasps are used to evaluate the proposed algorithms.

CONTROL AND DYNAMIC SYSTEMS, VOL. 39

I. Introduction

Dexterous, multifingered grippers have been the object of considerable research [1], [5], [11], [19]. The kinematics and force control problems engendered by these devices have been analyzed in [2], [7], [15], [20], [21], [22]. Force control of such a system requires the specification of contact forces between the fingers and the gripped object. The gripper-object system has a high degree of static indeterminacy [3], [7], [22], [25] and the interaction between the gripper and the object is similar to the interaction between a legged locomotion system and the ground [9], [10], [16], [23]. This paper outlines a method of computing the grasping forces for a general object.

In this paper, finger-object contacts are modeled as point contacts [22] (the point contact model is accurate when the finger tips are small compared to the object being held) which means that a finger can apply any three force components but no moments. A quasi-static approach to the problem has been adopted. That is, the load wrench, which is the resultant of the inertial forces and moments on the object and all external forces and moments excluding the finger forces, is always balanced by the finger contact forces. It is assumed that the object is stationary with respect to the gripper - manipulation issues are not in context here.

It is convenient to decompose the grasping forces into *equilibrating forces* and *interaction forces* [12], [13]. Broadly speaking, the equilibrating forces are forces required to maintain the object in equilibrium without squeezing it, while the interaction forces are the forces that squeeze the object with a zero resultant. Formally, the interaction force between two fingers is defined as the component of the difference of the finger contact forces projected along the line joining the two contact points. The equilibrating forces have no interaction force components. For example, in a walking machine it is essential to compute the support forces required at the feet to maintain equilibrium with the force of gravity and the inertial forces [10], [23]. These support forces are chosen to be equilibrating forces [14], [23]. In the gripper-object system, in addition to computing the contact forces required to maintain the object in equilibrium (equilibrating forces), it is necessary to determine the finger interaction forces to ensure that the friction angle at each contact point is within allowable limits.

This problem lends itself to optimization of the contact forces through linear programming [7] or by using Generalized Inverses [18], but such techniques are very expensive in terms of computational time and are consequently unsuitable for real-time operation with currently available hardware. Instead, a different approach to the problem is discussed in this paper. The basic objective is to minimize the contact forces and at the same time keep the friction angle within the allowable limit. This method is attractive in its speed and its efficiency. It is shown to be optimal for two and three fingered planar grasps, and suboptimal for more complicated grasps.

II. Formulation

Let X_E-Y_E-Z_E be a reference frame fixed with respect to the earth. Consider a body fixed reference frame X_B-Y_B-Z_B with the origin at the grasp centroid, the centroid of the support/contact points. $ is the axis of the load wrench [4], which is the resultant of the external forces acting on the object and the inertial forces and moments. From this point on, unless otherwise specified, all quantities are described in the body fixed reference frame. In Fig. 1, we define

r_i position of the ith contact point,

n   number of contact points,

F_i contact force at r_i,

O   centroid of the n contact points,

(f, c) force-couple dyname [4] associated with the wrench about $.

It has been assumed that the contact interaction is such that moments cannot be transmitted, which means that there is a total of three force components at each contact. The equilibrium equations for the grasped object may be written in the form

$$G\ q = w \qquad\qquad (1)$$

where w is the 6×1 external load vector consisting of the inertial forces and torques, and the weight of the object, q is the unknown $3n \times 1$ force vector, and n is the number of contact interactions. Here, $q = [\ F_1^T\ F_2^T\ ...\ F_n^T\ ]$. G is the $6 \times 3n$ coefficient matrix which is analogous to the Jacobian matrix encountered in the kinematics of serial chain manipulators. Each 6×1 column vector is a zero-pitch screw through a point contact in screw (in this case, line) coordinates (see Hunt [4] for a definition of screw coordinates). If $\$_{ix}$, $\$_{iy}$, and $\$_{iz}$ are the zero-pitch screw axes parallel to the x, y, and z axes (of any convenient coordinate system) passing through the ith contact point,

$$G = [\ \$_{1x}\ \$_{1y}\ \$_{1z}\ \$_{2x}\ \$_{2y}\ \$_{2z}\ \cdots\ \$_{nx}\ \$_{ny}\ \$_{nz}\ ]$$

Alternatively, the same expression may be written as

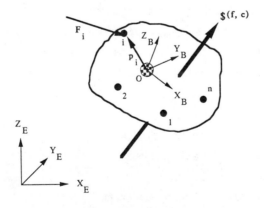

Fig. 1. Nomenclature for the gripper-object system.

$$G = \begin{bmatrix} I_3 & I_3 & \cdots & I_3 \\ R_1 & R_2 & \cdots & R_n \end{bmatrix} \tag{2}$$

where I_3 is the 3×3 identity matrix and R_i is a skew-symmetric 3×3 matrix derived from the position vector, r_i, Such that

$$R_i = \begin{bmatrix} 0 & -z_i & y_i \\ z_i & 0 & -x_i \\ -y_i & x_i & 0 \end{bmatrix}$$

and (x_i, y_i, z_i) are the components of r_i.

The problem of determining the contact forces (q) may be decomposed into two subproblems:

(a) Determination of the equilibrating forces F_{iE} required to maintain the equilibrium of the gripped body assuming that the finger interaction forces are absent.

(b) Determination of the interaction forces F_{iI} needed to produce the finger forces computed in step (a) without violating the constraints. If μ is the coefficient of friction and F_{max} is the maximum finger force that is allowable (F_{max} depends on the material of each body), the two sets of constraints are

$$(i) \quad \frac{F_i \cdot n_i}{|F_i|} \geq \cos(\tan^{-1}(\mu)) \qquad i = 1,..., n$$

$$(ii) \quad F_{max} \geq F_i \cdot n_i > 0 \qquad i = 1,..., n$$

where $F_i = F_{iE} + F_{iI}$.

The following sections elaborate on procedures for steps (a) and (b).

III. Decomposition of the contact force field

We now discuss the decomposition of the force field consisting of the contact forces, into an equilibrating force field and an interaction force field. The interaction force between any two contact points, F_{ij}, is defined as the component of the difference of the contact forces along the line joining the two contact points. In other words,

$$F_{ij} = (\mathbf{F}_i - \mathbf{F}_j) \cdot (\mathbf{r}_i - \mathbf{r}_j) \tag{3}$$

where $\mathbf{F}_i$ and $\mathbf{F}_j$ are the contact forces, and $\mathbf{r}_i$ and $\mathbf{r}_j$ are the position vectors at the ith and jth contacts (in any convenient earth-fixed or object-fixed reference frame), respectively. The equilibrating force field consists of equilibrating forces, which are the forces required to maintain equilibrium against an external load and satisfy the condition $F_{ij} = 0$ for all i and j. This zero interaction force condition is illustrated through examples for a two and a three point contact case in Fig. 2. The interaction force field consists of interaction forces which satisfy the condition $\mathbf{Gq} = 0$.

It has been shown [12], [14] that the pseudo-inverse solution for the force system belongs to the equilibrating force field. Further, the interaction force field was shown to be the set of forces belonging to the null space and the equivalence between internal forces and interaction forces was established in [12]. In this section, the relationship between the generalized inverse solution and decomposition of the force field is analyzed, and the nature of the two force fields is described. In particular, it is shown that the equilibrating force field is mathematically isomorphic to the velocity field in a rigid body. A computationally efficient, analytical method to obtain this solution is also presented.

The Pseudo-Inverse Solution

In general, the Moore-Penrose Generalized Inverse or the pseudo-inverse of G, G^+, yields the minimum norm, least squares solution [18] for the force vector q. In this problem, if the screw system defined by the 3n zero-pitch wrenches is a sixth-order screw system or a six-system, w always belongs to the column space of G. It is assumed that this is the case here. The pseudo-inverse, then, is a right inverse which yields a minimum norm solution.

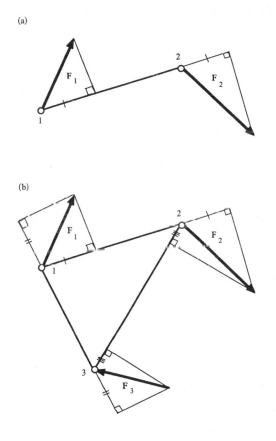

Fig. 2. The zero interaction force condition for (a) two and (b) three contacts

(F_i is the contact force at the ith contact).

$$q = G^+ w \qquad (4)$$

Since the solution vector must belong to the row space of G [18]

$$q = G^T c \qquad (5)$$

where c is a 6×1 constant vector. If F_i is the force at the ith contact point and c_0 and c_1 are two 3×1 vectors, such that $c = [\, c_0, c_i \,]^T$, then

$$F_i = c_0 + R_i c_i = c_0 - c_1 \times r_i \qquad (6)$$

It can be easily shown that this force system has no interaction forces [14]:

$$\begin{aligned}
(F_i - F_j) \cdot (r_i - r_j) &= [(c_0 - c_1 \times r_i) - (c_0 - c_1 \times r_j)] \cdot (r_i - r_j) \\
&= (r_i - r_j) \times c_1 \cdot (r_i - r_j) \\
&= 0.
\end{aligned}$$

Thus the minimum norm condition implies that the solution vector must belong to the row space of G and hence to the equilibrating force field. As a corollary, all force vectors in the interaction force field must be represented in the null space. The following statement can now be made:

Theorem 1

If a body is subjected to multiple frictional contacts modeled by point contact, and if the system of zero-pitch contact wrenches span a six-dimensional space, the Moore-Penrose Generalized Inverse (or the pseudo-inverse) solution to the equilibrium equations yields a solution vector which lies in the equilibrating force field and has no interaction force components.

The Helicoidal Vector Field

Consider a system of ∞^2 coaxial helices, each of a constant pitch h. For an infinitesimal twist of a rigid body about the common axis of the helices, the velocity of any point on the body is tangential to that helix, which passes through the point, at that point. Such a system of ∞^3 tangents has been called a helicoidal velocity field by Hunt [4] and the common axis is the instantaneous screw axis.

A helicoidal vector field (or an axial field Ψ) is defined to be a system of ∞^3 vectors associated with an instantaneous screw axis S. In Fig. 3, the vector at any point i is given by

$$\mathbf{w}_i = L\mathbf{u} \times (\mathbf{r}_i - \rho) + hL\mathbf{u} \tag{7}$$

If the force field is a helicoidal vector field, then $\mathbf{F}_i$ is of the same form as $\mathbf{w}_i$ in Eq. (7).

$$\mathbf{F}_i = L\mathbf{u} \times (\mathbf{r}_i - \rho) + hL\mathbf{u}$$

and

$$(\mathbf{F}_i - \mathbf{F}_j) \cdot (\mathbf{r}_i - \mathbf{r}_j) = L\mathbf{u} \times (\mathbf{r}_i - \mathbf{r}_j) \cdot (\mathbf{r}_i - \mathbf{r}_j) = 0$$

Hence, an important conclusion can be reached:

Theorem 2a

The force field given by a helicoidal vector field has no interaction force components and hence belongs to the equilibrating force field.

The converse of Theorem 2a is stated without proof below (see [12] for proof).

Theorem 2b

> A force field satisfying the zero interaction force components must be a helicoidal vector field.

From Theorems 2a and 2b, the equivalence of the equilibrating force field and the helicoidal vector field is evident and the following theorem may be stated:

Theorem 3

> The equilibrating force field is a helicoidal field.

From these theorems it may be concluded that a system of wrenches arising through multiple frictional contacts which spans the six-dimensional space and satisfies the equilibrium equations, belongs to the equilibrating force field, if and only if it belongs to a helicoidal field, and if and only if it is the minimum norm solution to the equilibrium equations.

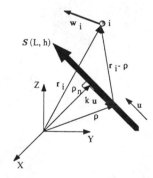

Fig. 3. The helicoidal vector field; Ψ. (S is the (screw) axis central to the vector field; h is the pitch, and L is the intensity of the field; u is a unit vector along S; ρ is the position vector of an arbitrary point on the axis; r_i is the position vector, and w_i is the vector at the ith point.)

IV. Equilibrating Forces

The best approach to finding the equilibrating force field or the minimum norm solution, is to find the axis, which is central to the helicoidal force field. Let the equilibrating forces F_{iE} be given by Eq. (7). The equations of equilibrium (1) can be written in the form

$$\sum_{i=1}^{n} F_{iE} = Q \tag{8}$$

and

$$\sum_{i=1}^{n} (r_i \times F_{iE}) = T \tag{9}$$

where Q and T are the net external force and moment components of the load vector w. Substituting the expression in Eq. (7) in Eqs. (8) and (9)

$$hL\,u + L\,u \times \bar{r} - L\,u \times \rho = \frac{1}{n}Q \tag{10}$$

and

$$\bar{r} \times hL\,u - \bar{r} \times (L\,u \times \rho) + \frac{L}{n} \sum_{i}^{n}(r_i \times (u \times r_i)) = \frac{1}{n}T \tag{11}$$

where $\bar{r}$ is the position vector of the centroid. If I is the centroidal moment of inertia tensor, given by

$$I = \begin{bmatrix} Y^2 + Z^2 & -XY & -XZ \\ -XY & X^2 + Z^2 & -YZ \\ -XZ & -YZ & X^2 + Y^2 \end{bmatrix}$$

in which

$$X^2 = \frac{1}{n} \sum_i^n (x_i - \bar{x})^2 \qquad\qquad YZ = \frac{1}{n} \sum_i^n (y_i - \bar{y})(z_i - \bar{z})$$

$$Y^2 = \frac{1}{n} \sum_i^n (y_i - \bar{y})^2 \qquad\qquad XZ = \frac{1}{n} \sum_i^n (x_i - \bar{x})(z_i - \bar{z})$$

$$Z^2 = \frac{1}{n} \sum_i^n (z_i - \bar{z})^2 \qquad\qquad XY = \frac{1}{n} \sum_i^n (x_i - \bar{x})(y_i - \bar{y})$$

where

$$\bar{x} = \frac{1}{n} \sum_i^n x_i \qquad\qquad \bar{y} = \frac{1}{n} \sum_i^n y_i \qquad\qquad \bar{z} = \frac{1}{n} \sum_i^n z_i$$

then Eq. (11) can be rewritten as

$$\bar{r} \times hL\,\mathbf{u} - \bar{r} \times (L\,\mathbf{u} \times \rho) + L(\mathbf{Iu} + \bar{r} \times (\mathbf{u} \times \bar{r})) = \frac{1}{n}\mathbf{T}$$

Substituting from Eq. (10) in this expression

$$\bar{r} \times \frac{1}{n}\mathbf{Q} + L\,\mathbf{Iu} = \frac{1}{n}\mathbf{T} \qquad\qquad\qquad (12)$$

Now, expressions for the parameters describing the field axis may be computed

$$\mathbf{u} = \frac{\mathbf{I}^{-1}}{nL}(\mathbf{T} - \bar{r} \times \mathbf{Q}) \qquad\qquad\qquad (13)$$

$$L = \left| \frac{1}{n}\mathbf{I}^{-1}(\mathbf{T} - \bar{r} \times \mathbf{Q}) \right| \qquad\qquad\qquad (14)$$

$$h = \frac{1}{nL}(\mathbf{u} \cdot \mathbf{Q}) \tag{15}$$

Finally, if $\rho = \rho_n + k\mathbf{u}$, as shown in Fig. 3, substituting into Eq. (10) and cross-multiplying by $\mathbf{u}$ yields

$$\rho_n = \frac{1}{nL}(\mathbf{u} \times \mathbf{Q}) - (\mathbf{u} \cdot \bar{\mathbf{r}})\mathbf{u} + \bar{\mathbf{r}} \tag{16}$$

In a centroidal reference frame, as shown in Fig. 1, Eqs. (12)-(15) may be written more compactly as

$$\mathbf{u} = \frac{\mathbf{I}^{-1}}{nL}\mathbf{T}^* \tag{17}$$

$$L = \left| \frac{\mathbf{I}^{-1}\mathbf{T}^*}{n} \right| \tag{18}$$

$$h = \frac{1}{nL}(\mathbf{u} \cdot \mathbf{Q}) \tag{19}$$

$$\rho_n = \frac{1}{nL}(\mathbf{u} \times \mathbf{Q}) \tag{20}$$

where $\mathbf{T}^*$ is related to $\mathbf{T}$ by the transformation: $\mathbf{T}^* = \mathbf{T} + \mathbf{Q} \times \mathbf{r}$. Now the force field can be obtained from Eq. (7). From the point of view of programming, computing the force distribution, Eqs. (13)-(16) followed by Eq. (7), involves a total $12n + 87$ multiplications and $16n + 43$ additions. In addition, It is easy to program as singularities in the algorithm can be easily detected and alternative steps can be followed for such special cases.

V. Interaction Forces

The interaction force field may be characterized using screw system theory [4]. If the interaction force field consists of n wrenches (which must have a zero resultant), the screws corresponding to the n wrenches must, in general, belong to a screw system of order n - 1 [24]. In special cases, they belong to a screw system of order less than n - 1. Further, since the contact wrenches are pure forces, the screw system must allow zero-pitch wrenches.

The interaction force field does not exist for the trivial case of a single contact. If the number of contact points is equal to 2, the interaction force field is represented by a first-order screw system. Further, the defining screw must be of zero pitch. Thus the interaction forces must lie along the line joining the two contact points and be equal and opposite. The magnitude is not uniquely determined.

If the number of contacts is equal to 3, the interaction force field corresponds to a special two-system [4], which consists of coplanar zero-pitch screws whose axes are either parallel or concurrent. Such a system of coplanar, zero-pitch wrenches has been used for three-fingered grasps [6], [13]. Clearly, there are ∞^2 choices for the point of concurrence, and a third degree of freedom corresponding to the intensity of the field. The problem of determining interaction forces becomes more intractable when the number of contacts exceeds 3. However, we discuss this problem later. We first present a method to determine optimal interaction forces for 2 and 3 fingered grasps.

Two fingered grasp

Consider the two-fingered grasp in Fig. 4. The direction of the interaction forces is determined by the geometry. They are equal and opposite forces along the vector $\mathbf{e}$. However the magnitude is not uniquely determined. We define α, the *interaction force factor*, to denote this magnitude in accordance with [6] so that $F_{1I} = \alpha\mathbf{e}$ and $F_{2I} = -\alpha\mathbf{e}$. The dimension of the interaction force field is one, which is evident from the freedom in choosing the magnitude of α.

In Figure 4a, the interaction force lies outside the friction cone (for both contact points) and hence, it is not possible to form a stable grasp [17]. On the other hand, in Figure 4b, the interaction force lies within the friction cone at each point. In this case, the interaction forces may be increased so that the net force lies within the friction cone for both contact points. Let $\cos \psi_1 = -\mathbf{n}_1 \cdot \mathbf{e}$ and $\cos \psi_2 = +\mathbf{n}_2 \cdot \mathbf{e}$. The necessary and sufficient conditions for producing a stable grip are

a)

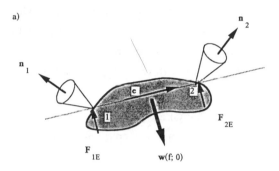

b)

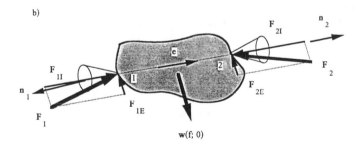

Fig. 4. Example of Interaction Forces in a Two Fingered Grasp.

(F_{iI} and F_{iE} are the interaction and equilibrating forces and

n_i is the normal to the object at the ith contact respectively.

w is the load wrench which is a pure force in this example.)

$$\psi_i \leq \tan^{-1}(\mu) \qquad\qquad i = 1, 2 \qquad\qquad\qquad (21)$$

where μ is friction coefficient at the contact points. In order to obtain a stable grasp, α must be so chosen that $F_i = F_{iE} + F_{iI}$ satisfies this inequality.

Three fingered grasp

For three fingered grasps, the contact points form a triangle which is referred as the *grasp triangle* in this discussion [6]. The interaction forces must lie on the plane of the grasp triangle and they must be concurrent [13]. This also includes the case in which all three forces are parallel when the point of concurrence (POC) lies at ∞. The interaction force field is three dimensional. The three parameters that completely specify the interaction forces are the two coordinates of the POC on the grasp triangle and the scaling parameter α.

The interaction forces F_{iI} can be resolved into normal and tangential components ($F_{iI,n}$ and $F_{iI,t}$) in this plane.

$$\begin{bmatrix} F_{iI,x} \\ F_{iI,y} \end{bmatrix} = \begin{bmatrix} n_{ix} & n_{iy} \\ n_{iy} & -n_{ix} \end{bmatrix} \begin{bmatrix} F_{iI,n} \\ F_{iI,t} \end{bmatrix} = \begin{bmatrix} n_i & t_i \end{bmatrix} \begin{bmatrix} F_{iI,n} \\ F_{iI,t} \end{bmatrix}$$

where $i = 1, 2, 3$ and n_i and t_i are unit normal and tangential vectors at each contact, where $F_{iI,x}$ and $F_{iI,y}$ now refer to the x and y components of interaction forces.

The zero resultant condition for interaction forces can be written in the following manner:

$$[A + B \Phi] X = 0 \qquad\qquad\qquad (22)$$

where

$$A = \begin{bmatrix} n_{1x} & n_{2x} & n_{3x} \\ n_{1y} & n_{2y} & n_{3y} \\ (r_{1x}n_{1y} - r_{1y}n_{1x}) & (r_{2x}n_{2y} - r_{2y}n_{2x}) & (r_{3x}n_{3y} - r_{3y}n_{3x}) \end{bmatrix}$$

$$B = \begin{bmatrix} n_{1y} & n_{2y} & n_{3y} \\ -n_{1x} & -n_{2x} & -n_{3x} \\ -(r_{1x}n_{1x} + r_{1y}n_{1y}) & -(r_{2x}n_{2x} + r_{2y}n_{2y}) & -(r_{3x}n_{3x} + r_{3y}n_{3y}) \end{bmatrix}$$

$$X = \begin{bmatrix} F_{1I,n} \\ F_{2I,n} \\ F_{3I,n} \end{bmatrix} \qquad \Phi = \begin{bmatrix} \phi_1 & 0 & 0 \\ 0 & \phi_2 & 0 \\ 0 & 0 & \phi_3 \end{bmatrix} \qquad \phi_i = \frac{F_{iI,t}}{F_{iI,n}}$$

In order to prevent slippage, the interaction forces should be chosen so that the net force lies within the friction cone at each contact point. However, it is desirable to decouple the computation of interaction forces from that of equilibrating forces for two reasons. Firstly, parallel computation enhances the efficiency of the algorithm. Secondly, in such situation, the interaction forces do not depend on the exact computation of equilibrating forces. Therefore, we pursue a technique which automatically optimizes the POC of the interaction forces so that the force field is specified except for the scaling factor, α.

In order to minimize α and the magnitude of the required interaction forces, it is beneficial to have the interaction forces to coincide with the respective contact normals, that is , $\phi_1 = \phi_2 = \phi_3 = 0$. Since this is not, in general, possible, it is meaningful to minimize the largest of ϕ_1, ϕ_2, ϕ_3. It has been shown by [6] that this is achieved when $|\phi_1| = |\phi_2| = |\phi_3|$. Therefore Φ can only be of the form:

$$\Phi = \phi U \qquad\qquad (23)$$

where **U** is one of the four matrices:

$$\begin{bmatrix} 1 & 0 & 0 \\ 0 & 1 & 0 \\ 0 & 0 & 1 \end{bmatrix}, \quad \begin{bmatrix} 1 & 0 & 0 \\ 0 & -1 & 0 \\ 0 & 0 & 1 \end{bmatrix}, \quad \begin{bmatrix} 1 & 0 & 0 \\ 0 & -1 & 0 \\ 0 & 0 & -1 \end{bmatrix} \quad \text{or} \quad \begin{bmatrix} 1 & 0 & 0 \\ 0 & 1 & 0 \\ 0 & 0 & -1 \end{bmatrix}$$

Substituting Eq. (23) in Eq. (22), we can rewrite Eq. (22) to get the eigenvalue problem:

$$\mathbf{C}\,\mathbf{X} = \phi\,\mathbf{X} \qquad \text{where } \mathbf{C} = -\,\mathbf{U}^{-1}\,\mathbf{B}^{-1}\,\mathbf{A} \qquad (24)$$

If **B** is a full rank matrix, $\mathbf{B}^{-1}$ always exist. Thus we can get 3 eigenvalues (ϕ) and the corresponding eigenvectors (**X**) for each of the four **U** matrices. That is we can get a total 12 eigenvalues and eigenvectors. Now we can choose the solution which is real and with the smallest $|\phi|$ (this is the optimal choice of ϕ) provided the elements of **X** are all positive. Once ϕ is obtained, the vector **X** yields the interaction forces except for the scaling factor, α. It is clear from the discussion on Fig. 4 that ϕ must be less than ϕ_{max}, the maximum allowable ratio of the net tangential force to the net normal force. If **B** is not of full rank, we can still get a solution by manipulating Eq. (22) as shown in [8].

n fingered grasps (n > 3)

Although it is possible to qualitatively describe the interaction force field for multiple contact grasps through screw system theory [12], a practical and productive characterization for the cases of four, five, or six contact points is through the decomposition of such grasps into multiple two or three fingered grasps. In order to facilitate this decomposition of the interaction force field, several new terms are defined next.

Dyads are defined to be pairs of equal and opposite forces acting along the lines joining the contact points. In general, the number of degrees of freedom of the interaction force field is 3n-6. Thus, 3n-6 equal and opposite force pairs are needed to specify the field. Since there exist $^nC_2$ (or $n(n-1)/2$) such pairs and $3n-6 \leq {}^nC_2$ for $n > 3$, the set of all dyads span the interaction force field. Similarly, it is also possible to decompose the interaction into *triads* of forces, which are solutions to the three point contact problem. It can be shown [8] that the number of dyads, n_d, the number of triads, n_t, the number of linearly independent dyads, l_d, the number of linearly independent triads, l_t, and the number of contact points, n, by the following equations for $n > 2$:

$$l_d = 2\, n^2 - 4\, n_d - 3, \text{ for planar (two dimensional) grasps} \qquad (25)$$

$$l_d = 3\, n^2 - 6\, n_d - 6, \text{ for spatial (three dimensional) grasps} \qquad (26)$$

$$l_t = l_d - n + 1 \qquad (27)$$

Stable dyads are defined as dyads which satisfy the inequality in Eq. (21). Similarly, *stable triads* are triads satisfying Eqs. (22)-(24), such that the angle ϕ is less than ϕ_{max}. A linear combination of stable dyads and/or triads will result in an interaction force field which will satisfy the friction angle constraints. The objective then is to find enough stable dyads and/or triads, so that the interaction force at each contact point can be controlled. Now by choosing an appropriate linear combination of these dyads and/or triads, can be made to the net force at each contact point satisfy the friction angle requirements.

This method works in most cases. But a failure of the method does not indicate that a stable grasp is not possible. For example, a planar four contact grasp can be decomposed into six dyads and five linearly independent dyads according to Eq. (25). Clearly if enough stable dyads can be found, a stable grasp can be synthesized. On the other hand, even if none of the six possible dyads are stable, it is quite possible that a stable grasp can be found, although this method will not work. An example of such an exceptional case is shown in Fig. 5, in which no dyad or triad is stable (because the maximum allowable friction angle is deliberately chosen to be very small) and yet an interaction force field which satisfies all the frictional angle constraints can be found. In such a case, a different method such as the one reported in [13] may be used.

This method is clearly suboptimal for grasps with more than three contact points. It is also suboptimal if the net force is used in the criterion for optimization. For example, a possible criterion is:

$$\mathbf{Minimize}\left\{\mathbf{Maximum}\left[\left(\frac{\mathbf{F}_i \cdot \mathbf{n}_i}{|\mathbf{F}_i|}\right), i = 1,..., n\right]\right\}$$

However, such criteria do not lend themselves to decoupled equilibrating and interaction force subproblems.

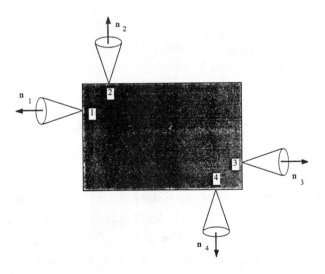

Fig. 5. Example of a planar 4 point contact grasp in which all dyads are unstable.

VII. Examples and Discussion

Example 1) Screwdriver - Three fingered grasp (see Fig. 6(a))

    Number of fingers: 3

    Coordinates of contact points (m):

                1. (0.8666, -0.5000, 0.0000)

                2. (0.0000, 1.0000, 0.0000)

                3. (-0.8666, -0.5000, 0.0000)

Load force (N):	$-1\mathbf{i} - 10\mathbf{k}$ applied at (0, 0, 10)
Load couple (N-m):	$-2\mathbf{k}$
Equilibrating forces (N):	$F_{1E} = 0.6664\mathbf{i} + 0.5774\mathbf{j} - 2.4363\mathbf{k}$
	$F_{2E} = -0.3329\mathbf{i} + 0.0000\mathbf{j} + 3.3333\mathbf{k}$
	$F_{3E} = 0.6664\mathbf{i} - 0.5774\mathbf{j} + 9.1030\mathbf{k}$
Interaction forces (N):	$F_{1I} = -31.1110\mathbf{i} + 17.9501\mathbf{j} + 0.0000\mathbf{k}$
	$F_{2I} = 0.0000\mathbf{i} - 35.9001\mathbf{j} + 0.0000\mathbf{k}$
	$F_{3I} = 31.1110\mathbf{i} + 17.9501\mathbf{j} + 0.0000\mathbf{k}$
Coefficient of friction :	$\mu = 0.25$

Friction angle: $F_{1t}/F_{1n} - 0.0633$, $F_{2t}/F_{2n} - 0.0930$, $F_{3t}/F_{3n} = 0.25$

Example 2) Planar polygon - Three fingered grasp (see Fig. 6(b))

    Number of fingers: 3

    Coordinates of contact points (m):

                1. (0, 4, 0)

                2. (-4, -5, 0)

                3. (4, -6, 0)

Load force (N):	$-10\mathbf{k}$ applied at (0, 0, 0)
Load couple (N-m):	$-1\mathbf{k}$
Equilibrating forces (N):	$F_{1E} = -0.0683\mathbf{i} + 0.0000\mathbf{j} + 5.7895\mathbf{k}$
	$F_{2E} = 0.0288\mathbf{i} - 0.0432\mathbf{j} - 2.1053\mathbf{k}$
	$F_{3E} = 0.0396\mathbf{i} + 0.0432\mathbf{j} + 2.1053\mathbf{k}$
Interaction forces (N):	$F_{1I} = -3.7981\mathbf{i} - 27.8455\mathbf{j} + 0.0000\mathbf{k}$
	$F_{2I} = 21.0958\mathbf{i} + 16.0334\mathbf{j} + 0.0000\mathbf{k}$

$$F_{3I} = -17.2976i + 11.8121j + 0.0000k$$

Coefficient of friction : $\mu = 0.25$

Friction angle: $F_{1t}/F_{1n} = 0.25$, $F_{2t}/F_{2n} = 0.1603$, $F_{3t}/F_{3n} = 0.1725$

Example 3) Sphere - Four fingered grasp (see Fig. 6(c))

　　　Number of fingers: 4

　　　Coordinates of contact points (m):

　　　　　　　　1. (1, 0, 0)

　　　　　　　　2. (0, 1, 0)

　　　　　　　　3. (0, 0, 1)

　　　　　　　　4. (-0.5774, -0.5774, -0.5774)

Load force (N):　　　　- (1i + 1j +10k) applied at (0.5, 1.0, 0.5)

Load couple (N.m):　　- 1k

Equilibrating forces (N): $F_{1E} =$ $0.2684i + 0.4748j + 0.1812k$

　　　　　　　　　　　$F_{2E} =$ $0.0436i + 0.2500j + 0.4564k$

　　　　　　　　　　　$F_{3E} =$ $0.3188i + 0.0252j + 0.2316k$

　　　　　　　　　　　$F_{4E} =$ $0.3692i + 0.2500j + 0.1308k$

Interaction forces (N):　$F_{1I} = - 1.3049i - 0.4777j - 0.4777k$

　　　　　　　　　　　$F_{2I} = - 0.4978i - 1.3598j - 0.4978k$

　　　　　　　　　　　$F_{3I} = - 0.4473i - 0.4473j - 1.2219k$

　　　　　　　　　　　$F_{4I} =$ $2.2500i + 2.2848j + 2.1974k$

Coefficient of friction : $\mu = 0.5774$

Friction Angle:　　　　$F_{1t}/F_{1n} = F_{2t}/F_{2n} = F_{3t}/F_{3n} = 0.5774$,

　　　　　　　　　　　$F_{4t}/F_{4n} = 0.051$

In this section some examples of application of the proposed method have been presented. The first example in Fig. 6(a) involves a long screwdriver with a cylindrical handle with a symmetric, three-fingered grasp. Because of the symmetry, it is easy to find a stable triad with $\phi_1 = \phi_2 = \phi_3 = 0$ (the POC is at the grasp centroid) result in large interaction forces. As the results show, the decomposition of finger contact forces into equilibrating and interaction force components works satisfactorily.

In Fig. 6(b), a three fingered grasp of a planar irregular polygon is shown. The optimal interaction forces are characterized by $\phi_1 = \phi_2 = \phi_3 = 0.1364$. These angles are nonzero due to the asymmetry in the problem.

A three dimensional four fingered grasp is considered in Fig. 6(c). In this case the interaction force field was obtained by superposing three stable dyads. These dyads are obtained by combining points 1 and 2, 1 and 3, and 1 and 4. These dyads are stable only because of a large friction angle ($\mu = 0.5774$). However, this example illustrates how the interaction forces at each contact point can be easily controlled if a sufficient number of stable dyads are available.

In all three examples, the decoupling of the equilibrating and interaction force subproblems facilitates parallel computation. Although the equilibrating force solution is clearly optimal, the interaction force solution is suboptimal. More work needs to be done in this area.

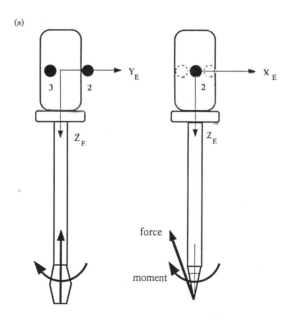

Fig. 6(a). A screwdriver with a symmetric, three-fingered grasp.

(b)

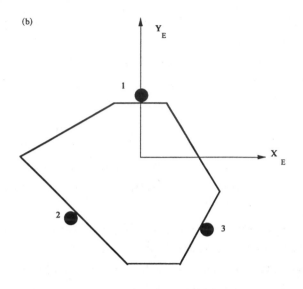

	x y		n_x	n_y
1.	(0, 4)		(0,	-1)
2.	(-4, -5)		(0.7071,	0.7071)
3.	(4, -6)		(-0.8944,	0.4472)

Fig. 6(b). A planar irregular polygon with an asymmetric, three-fingered grasp.

(c)

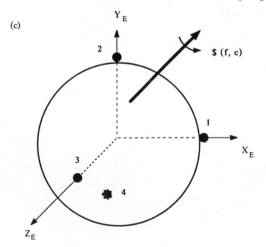

Fig. 6(c). A sphere with four-fingered grasp.

VIII. Conclusion

A fast and efficient method for computing finger forces in a multifingered grasp is presented. The method involves the decomposition of the force field consisting of the finger forces into equilibrating forces and interaction forces. The optimization problem is decoupled in the sense that the equilibrating forces and the interaction forces are obtained independently except for a scaling factor multiplying the interaction forces. This method is shown to be optimal for two and three fingered grasps and is suboptimal for more complicated grasps.

References

1. M.R. Cutkosky, Robotic Grasping and Fine Manipulation, Kluwer Academic Publishers, Boston 1985.

2. J. Hollerbach and S. Narasimhan, "Finger Force Computation without the Grip Jacobian," Proceedings of IEEE Int. Conf. on Robotics and Automation, San Fransisco, April 1986, pp. 871-875.

3. W. Holzmann and J.M. McCarthy, "Computing the Friction Forces Associated with a Three Fingered Grasp," IEEE J. of Robotics and Automation, Vol. RA-1, No. 4, December 1985, pp. 206-210.

4. K.H. Hunt, Kinematic Geometry of Mechanisms, Clarendon Press, Oxford, 1978.

5. S.C. Jacobsen, E.K. Iversen, D.F. Knutti, R.T. Johnson, and K.B. Biggers, "Design of the Utah/M.I.T. Dextrous Hand," Proceedings of IEEE Int. Conf. on Robotics and Automation, San Fransisco, April 1986, pp. 1520-1532.

6. Z. Ji and B. Roth, "Direct Computation of Grasping Force for Three-Finger Tip-Prehension Grasps," ASME J. of Mechanisms, Transmissions, and Automation in Design, Vol. 110, No. 4, December 1988, pp. 405-413.

7. J. Kerr and B. Roth, "Analysis of Multifingered Hands," Int. J. of Robotics Research, Vol. 4, No. 4, 1986, pp. 3-17.

8. J.H. Kim, Kinematics and Statics of Multifingered Grasping. Ph.D dissertation, University of Pennsylvania, Philadelphia, PA, December 1990 (in preparation).

9. C.A. Klein and S. Kittivatcharapong, "Optimal force distribution for the legs of a walking machine," IEEE J. of Robotics and Automation, Vol. 6, No. 1, February 1990, pp. 73-85.

10. C.A. Klein, K.W. Olson, and D.R. Pugh, "Use of Force and Attitude Sensors for Locomotion of a Legged Vehicle over Irregular Terrain," Int. J. of Robotics Research, Vol. 2, No. 2, 1983, pp. 3-17.

11. H. Kobayashi, "On the Articulated Hand," Robotics Research: The Second International Symposium, H. Hanafusa and H. Inoue, Eds., Kyoto, Japan, August 20-23, 1984, pp. 293-299.

12. V. Kumar and K.J. Waldron, "Force Distribution in Closed Kinematic Chains," IEEE Transactions of Robotics and Automation, Vol. 4, No. 6, December 1988, pp. 657-664.

13. V. Kumar and K.J. Waldron, "Sub-Optimal Algorithms for Force Distribution in Multifingered Grippers," IEEE Transactions of Robotics and Automation, Vol. 5, No. 4, August 1989, pp. 491-498.

14. V. Kumar and K.J. Waldron, "Actively Coordinated Vehicle Systems," ASME J. of Mechanisms, Transmissions, and Automation in Design, Vol. 111, No. 2, June 1989, pp. 223-231.

15. Z. Li and S. Sastry, "Task-oriented Optimal Grasping by Multifingered Robot Hands," IEEE Transactions of Robotics and Automation, Vol. 4, No. 1, February 1988, pp. 32-44.

16. R.B. McGhee and D.E. Orin, "A Mathematical Programming Approach to Control of Joint Positions and Torques in Legged Locomotion Systems," Proceedings of ROMANSY-76 Symposium," Warsaw, Poland, September 1976.

17. V.D. Nguyen, "Constructing Force-Closure Grasps," Proceedings of IEEE Int. Conf. on Robotics and Automation, Raleigh, CA, April 1987, pp. 240-245.

18. B. Noble, "Methods for Computing the Moore-Penrose Generalized Inverse, and Related Matters," Proceedings of an Advanced Seminar, sponsored by the Mathematics Research Center, University of Wisconsin, Ed. M.Z. Nashed, Academic Press, 1976.

19. T. Okada, "Computer Control of Multijointed Finger System for Precise Handling," IEEE Transactions on Systems, Man and Cybernetics, Vol. SMC-9, No. 2, February 1979.

20. D.E. Orin, and S.Y. Oh, "Control of Force Distribution in Robotic Mechanisms Containing Closed Kinematic Chains," J. of Dynamic Systems, Measurements, and Control, Vol. 102, June 1981, pp. 134-141.

21. J.K. Salisbury and J.J. Craig, "Analysis of Multifingered Hands: Force Control and Kinematic Issues," Int. J. of Robotics Research, Vol. 1, No. 1, 1982, pp. 4-17.

22. J.K. Salisbury and B. Roth, "Kinematic and Force Analysis of Articulated Mechanical Hands," J. of Mechanisms, Transmissions and Automation in Design, Vol. 105, 1983, pp. 35-41.

23. K.J. Waldron,"Force and Motion Management in Legged Locomotion," IEEE J. on Robotics and Automation, Vol. RA-2, No. 4, 1986.

24. K.J. Waldron, and K.H. Hunt, "Series-Parallel Dualities in Actively Coordinated Mechanisms," 4th Int. Symposium on Robotics Research, Santa Cruz, CA, 1987 .

25. T. Yoshikawa and K. Nagai, "Manipulating and Grasping Forces in Manipulation by Multifingered Grippers", IEEE Int. Conf. on Robotics and Automation," Raleigh, N. Carolina, March 31-April 3, 1987, pp. 1998-2007.

FREQUENCY ANALYSIS FOR
A DISCRETE-TIME ROBOT SYSTEM

YILONG CHEN

General Motors Research Laboratories
Warren, MI 48090

I. INTRODUCTION

The purpose of the robot arm control is to maintain a prescribed motion for the arm along a desired trajectory by applying corrective compensation inputs to the actuators to adjust for any deviation of the arm from the trajectory. In simple terms, the control problem is to find the necessary inputs to actuate the joint motors at every sample time.

Because of their simplicity, PD (or PID) controllers are widely used with various robot arm control methods, like the approximate linearized method [1][2][3]; the computed torque method [4][5]; the hierarchical control method [6][7]; the feedforward compensation method [8]; nonlinear feedback control method [9][10][11]; the adaptive control method [12][13][14][15]; and etc. Our analysis and tests of a Unimation PUMA 560 have shown that the use of a PD feedback, even with a relatively fast sampling rate (5ms) compared to those rates conventionally used in robot dynamic control, leads to serious trade-offs between static accuracy, system stability and damping of high frequency disturbances. Adding an integral of error to the PD to make a PID feedback often makes the overall system less stable, although it improves the static accuracy.

The purpose of this chapter is to propose the design of a new compensator for the robot feedback control which reduces the trade-offs caused by the conventional PD (or PID) controllers.

Our simulations and real tests have shown that the sampling time is critical to the system performance of robots. So it is necessary to use discrete

CONTROL AND DYNAMIC SYSTEMS, VOL. 39

317

time models instead of continuous time models for the design of compensators. The tool that we propose for designing the feedback loop is the frequency response approach based on the z-transform of the discrete time system model cascaded with delay and hold devices.

The frequency response approach enables us to investigate the static accuracy, both the absolute and relative stabilities, insensitivity to model structure inaccuracies and the high frequency noise rejection of closed-loop systems from a knowledge of their open-loop frequency-response characteristics.

An advantage of this approach is that we do not have to calculate the roots of the close loop system to determine the stability. Also the robustness of the overall system can be studied easily. Another advantage of this approach is that frequency-response tests of the hardware are in general simple and can be made accurately by use of readily available sinusoidal signal generators and precise measurement equipment. Often the transfer function of complicated components can be determined experimentally by frequency-response tests. The third advantage of this approach is that gains at high frequency range of a system may be designed so that the undesirable effects of high frequency noise are negligible and the system is less sensitive to model structure inaccuracies.

Although it is more difficult to apply the frequency-response approach to a discrete time system than a continuous time system, we can use the so-called W-transformation to circumvent the difficulty. Under this transformation, the feedback loop design in the frequency domain for the discrete time robot system model is relatively simple and straightforward. The resulting control should reduce the inherent trade-offs of the PD feedback controller. It will lead higher static accuracy, good system stability and less sensitivity to high frequency noise and model structure inaccuracies.

In this chapter, using the frequency-response approach we have designed and realized a so-called Lag-lead compensator for a robot arm. The analytical comparison between a PD (or PID) controller and a Lag-lead compensator shows that the Lag-lead compensator reduces the trade-offs between the static accuracy, system stability and insensibility to disturbances on the frequency domain. The implementation and real tests verify the encouraging results on the time domain. The Lag-lead compensator reduces the tracking errors considerably.

Although a robot is an inherently nonlinear and coupled system, a linear model for the Unimation PUMA 560 serves as a specific example to motivate the discussion in this chapter. Nevertheless, by nonlinear feedback control method [10][11][16], or computed torque method, or feedforward compensation method, we can have a desired linear and decoupled (or approximately linear and decoupled) system for which the above analysis and control principles are valuable. Even in the case that none of the above technique is applied and a Lag-lead compensator is used alone as a servo control, the advantage of reducing the trade-offs still exists. In fact, to have certain static accuracy, the system is more stable and less sensitive to model structure inaccuracy and noise by using the Lag-lead compensator. So the system is more robust and we depend less on the model and the prior knowledge of robot dynamics than using a PID controller. This is also proved by our real tests. In either case, the proposed Lag-lead feedback compensator may replace the conventional PD (or PID) controllers which are so widely used in the various current control strategies for different kinds of robot arms. It results in a small increase of off-line designing and tuning effort and on-line computational load, but the improvements in robot performance are significant.

In Section II, we briefly derive the discrete time system model which includes a computational time delay and a zero order holder for a robot system. The frequency responses of the system with different sampling rates are shown. The conventional PD controller and its limitations are analyzed in the frequency domain in detail. In Section III, we designed and analyzed the Lag-lead compensator by the frequency-response approach. In section IV, we accomplished the realization of the Lag-lead compensator and derived the algorithm for the proposed feedback control law. The system and test results for the implementation are presented and discussed in Section V. The conclusion is given in Section VI.

II. FREQUENCY-RESPONSE OF A DISCRETE TIME ROBOT SYSTEM

In a sampled data system, the choice of sampling rate is often governed by conflicting criteria. For a high performance robot arm under digital computer control, it is often necessary to restrict the time delay of the reconstructed error signal. Besides, the signals may be contaminated by high-frequency disturbances if too low a sampling rate is used. On the other hand, the complexity of a robot control system and the resulting real time computational burden places a maximum on the sampling rate. In practice, the sampling rate for robots often becomes critical. It is not high enough for a high performance robot arm to be considered as a continuous time system. In such cases, we need to deal with a discrete time system model, which we use for feedback control design.

First let us briefly derive the discrete time model of a robot, including a one period computing time delay and a zero order holder (ZOH) for the example shown in Figure 1 with a feedforward and feedback loop. To focus on the comparison of a Lag-lead compensator to a PD controller, a simple linear model is used in this chapter. A more sophisticated discrete nonlinear system model is preferred if it is available [17].

In Figure 1 each axis of a robot arm is described by a 3rd order motor model with a voltage input [14]. For the jth axis, we have

$$\frac{L_j J_j}{K_T^j} \dddot{X}_j + \left[\frac{L_j B_j + R_j J_j}{K_T^j}\right] \ddot{X}_j + \left[\frac{R_j B_j}{K_T^j} + K_e^j\right] \dot{X}_j = U_j \qquad (1)$$

where

U_j = (applied) armature voltage,

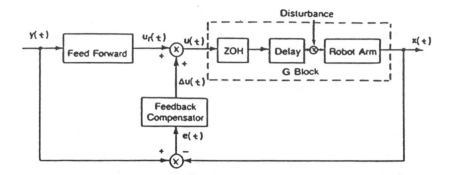

Figure 1. A Control System Of A Robot With a Feedback Loop.

X_j = motor position,
R_j = resistive component of armature circuit,
L_j = inductive component of armature circuit,
K_e^j = voltage constant of motor (back emf),
B_j = viscous dampimg coefficient,
J_j = motor inertia,
K_T^j = torque constant of the motor.

for ease of notation, we drop the subscript j, since we will deal with the axis individually. We may define

$$G_m(s) \triangleq \frac{X(s)}{U(s)} = \frac{b}{s^3 + a_1 s^2 + a_2 s}, \qquad (2)$$

where X(s) and U(s) are the Laplace transformations of x(t) and u(t) respectively, and

$$a_1 = (LB + RJ)/LJ,$$
$$a_2 = (RB + K_e K_T)/LJ, \qquad (3)$$
$$b = K_T/LJ.$$

We can convert the Laplace transfer function (Eq.2) to a z-transfer function by defining

$$X(z) = Z(x(t)) = \sum_{n=0}^{\infty} x(nT)z^{-n} \qquad (4)$$

(where T is the sampling period) and derive the transfer function of the discrete time system for a robot arm [18][19][20].

Considering the zero order holder (ZOH) and denoting $Z[G(s)]$ as the z-transformation converted from $G(s)$ [18], we have

$$G_{zm}(z) \stackrel{\Delta}{=} Z[G_m(s) \text{ and } ZOH]$$

$$= (1 - z^{-1}) Z[G_m(s)/s]$$

$$= (1 - z^{-1}) \left[\frac{b}{a_2} \cdot \frac{Tz}{(z-1)^2} - \frac{ba_1}{a_2^2} \cdot \frac{z}{z-1} + \right.$$

$$\frac{ba_1}{a_2^2} \cdot \frac{z^2 - ze^{-a_1 T/2} \cos(\omega_0 T)}{z^2 - 2ze^{-a_1 T/2} \cos(\omega_0 T) + e^{-a_1 T}} +$$

$$\left. \frac{1}{\omega_0} \left(\frac{b}{2} \cdot \frac{a_1^2}{a_2^2} - \frac{b}{a_2} \right) \cdot \frac{ze^{-a_1 T/2} \sin(\omega_0 T)}{z^2 - 2ze^{-a_1 T/2} \cos(\omega_0 T) + e^{-a_1 T}} \right]$$

where $\omega_0^2 = a_2 - a_1^2/4$

In consideration of a one period computing time delay (see Figure 2) the transfer function of G block in Figure 1 becomes

$$G(z) = \frac{1}{z} \cdot [G_{zm}(z)]$$

$$\tag{5}$$

$$= \frac{\alpha z^2 + \beta z + \gamma}{z(z-1)[z^2 - 2ze^{(-a_1 T/2)} \cos(\omega_0 T) + e^{(-a_1 T)}]}$$

where

$$\alpha = \frac{b}{a_2}T + \frac{ba_1}{a_2^2}e^{-a_1 T/2}\cos(\omega_0 T) - \frac{ba_1}{a_2^2} + \frac{1}{\omega_0}\left[\frac{b}{2}\cdot\frac{a_1^2}{a_2^2} - \frac{b}{a_2}\right] \\ \cdot e^{-a_1 T/2}\sin(\omega_0 T)$$

$$\beta = -\frac{2bT}{a_2}e^{-a_1 T/2}\cos(\omega_0 T) - \frac{ba_1}{a_2^2}e^{-a_1 T} + \frac{ba_1}{a_2^2} - \frac{2}{\omega_0}\left[\frac{b}{2}\frac{a_1^2}{a_2^2} - \frac{b}{a_2}\right] \\ \cdot e^{-a_1 T/2}\sin(\omega_0 T) \tag{9}$$

$$\gamma = \frac{b}{a_2}Te^{-a_1 T} + \frac{ba_1}{a_2^2}e^{-a_1 T} - \frac{ba_1}{a_2^2}e^{-a_1 T/2}\cos(\omega_0 T) \\ + \frac{1}{\omega_0}\left[\frac{b}{2}\frac{a_1^2}{a_2^2} - \frac{b}{a_2}\right]\cdot e^{-a_1 T/2}\sin(\omega_0 T)$$

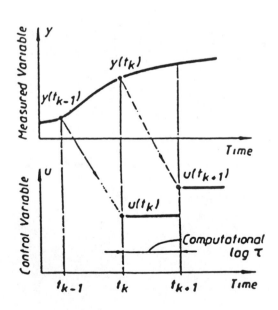

Figure 2. One Period Delay Due To The Computational Lag τ.

Next, before conducting a frequency analysis of the just derived discrete time robot system model, we will introduce and motivate the w - transformation.

First we will discuss frequency analysis methods for continuous and discrete time system. Frequency design methods based on Bode plot are useful and convenient for designing compensators for continuous time systems described by transfer functions. A Bode plot consists of two graphs: one is a plot of the logarithm of the magnitude of a complex valued transfer function; the other is a plot of the phase angle; both are plotted against the frequency in logarithm scale. The usefulness of the method depends on the simplicity of drawing the Bode plots and on rules of thumb for choosing the compensator. The rules of thumb are based on the predictable manner in which Bode plots of transfer functions, generally rational functions in s, are modified by different choices of feedback compensators.

However, frequency curves for discrete-time systems are more difficult to analyze since the pulse-transfer functions are not rational function in s, but in $z = \exp(sT)$. In order to circumvent this difficulty and to use Bode's design techniques, we may use the w-transformation which is defined as

$$w = \frac{2\,(z-1)}{T\,(z+1)}.$$

The relation between the real frequency ω, where $i\omega = s$, and the w-plane frequency ν can be found from $i\nu = w = i(2/T)tan(\omega T/2)$, where $i^2 = -1$. This expression gives a measure of the distortion of the approximation of using w instead of s.

Under the w - transformation, we substitute

$$z = \frac{1 + wT/2}{1 - wT/2}$$

into (5), then we have

$$G(w) \triangleq G[z(w)]$$

$$= K_0 \cdot \frac{(1 - wT/2)^2(\phi_2 w^2 + \phi_1 w + 1)}{(1 + wT/2)w(\theta_2 w^2 + \theta_1 w + 1)}$$

(10)

where

$$K_0 = \frac{1}{T} \cdot \frac{\alpha + \beta + \gamma}{1 - 2e^{-a_1 T/2}cos(\omega_0 T) + e^{-a_1 T}} = \frac{b}{a_2}$$

(11)

$$\phi_1 = \frac{T(\alpha - \gamma)}{\alpha + \beta + \gamma}$$

$$= \frac{a_2}{b} \cdot \left[\frac{bT}{a_2} + (\frac{2ba_1}{a_2{}^2})e^{-a_1 T/2} \cos(\omega_0 T) - \frac{ba_1}{a_2{}^2} - (\frac{bT}{a_2})e^{-a_1 T} \right.$$

$$\left. - (\frac{ba_1}{a_2{}^2})e^{-a_1 T} \right] \div \left[1 - 2e^{-a_1 T/2} \cos(\omega_0 T) + e^{-a_1 T} \right]$$

$$\tag{12}$$

$$\phi_2 = \frac{T^2(\alpha - \beta + \gamma)}{4(\alpha + \beta + \gamma)}$$

$$= \frac{T}{4} \cdot \frac{a_2}{b[1 - 2e^{-a_1 T/2} \cos(\omega_0 T) + e^{-a_1 T}]}$$

$$\cdot \left[\frac{bT}{a_2} - \frac{2ba_1}{a_2{}^2} + \frac{4(-b/a_2 + ba_1{}^2/2a_2{}^2)}{\omega_0} \cdot e^{-a_1 T/2} \sin(\omega_0 T) \right.$$

$$\left. + \frac{2bT}{a_2}e^{-a_1 T/2} \cos(\omega_0 T) + \frac{2ba_1}{a_2{}^2}e^{-a_1 T} + \frac{bT}{a_2}e^{-a_1 T} \right]$$

$$\tag{13}$$

$$\theta_1 = \frac{T - Te^{-a_1 T}}{1 - 2e^{-a_1 T/2} \cos(\omega_0 T) + e^{-a_1 T}}$$

$$\tag{14}$$

$$\theta_2 = \frac{\frac{T^2}{4} + \frac{T^2}{2}e^{-a_1 T/2} \cos(\omega_0 T) + \frac{T^2}{4}e^{-a_1 T}}{1 - 2e^{-a_1 T/2} \cos(\omega_0 T) + e^{-a_1 T}}$$

$$\tag{15}$$

The Bode plot of the uncompensated discrete system G(w) with sampling time T = 0.005s is shown in Figure 3. We can see that the w-plane crossover frequency $\nu_c = 4$. To understand better the limitations of PD (or PID) controllers, it is useful to know how a continuous time system is modified by sampling, computing delay and a zero order holder. These can be seen from comparing the transfer functions $G_m(s)$ (Eq. 2) of the continuous time case and G(w) (Eq. 10) of the discrete time case (and their Bode plots, Figure 4 and Figure 3).

Note that the gain K_0 of G(w) is precisely the same as for $G_m(s)$. After a calculation we can find that, in (Eq. 10), the denominator looks very similar to that of $G_m(s)$ in (Eq. 2), except the term (1+wT/2) which is due to the one period computing time delay. In addition, the denominator will be the same as T approaches zero.

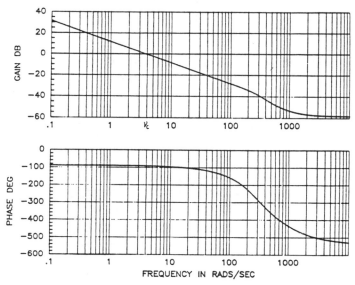

Figure 3. Bode Diagram Of Uncompensated $G(w)$ With $T=0.005s$.

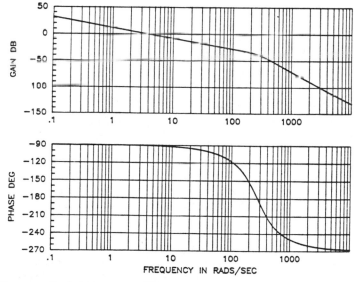

Figure 4. Bode Diagram Of Uncompensated Continuous System $G(s)$.

More important, the comparison shows the creation of right-hand plane zeros of $G(w)$ at $2/T$ and the creation of other two fast zeros when compared to $G_m(s)$. These zeros can be attributed to the sampling-and-hold operations and thus depend on the sampling rate. They become slower and thus more important to the transient response of the system as the sampling rate is decreased. More significantly, in addition to increasing the gain at high frequency, the zero at $2/T$ is in the right hand side of the complex plane, so that it causes an extra phase lag.

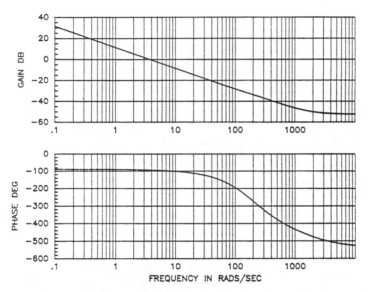

Figure 5. Bode Diagram Of Uncompensated $G(w)$ With T=0.01s.
© 1989 IEEE

The Bode plots of the system with sampling rates 0.01s, 0.02s, and 0.05s are also shown in Figure 5 - 7 respectively. The effect of different sampling rates is clear, especially in the high frequency region.

After observing these important features, now it is interesting to see what happens when we use a conventional PD controller for a robot arm system.

Let us see an example. In our real test shown in Figure 1, after carefully tuning, the PD controller was chosen as $H(s) = g_1 + g_2 s$, where $g_1 = 24, g_2 = 0.225$. From Figure 8, we may see that the continuous time system model with the PD controller has a crossover frequency $\omega_c = 180 \; rad/sec$ and static error coefficient $K = 100$.

Note that this system model has very large positive gain margin ($=+\infty \; db$) and phase margin ($=+90^0$). Also, for the continuous time system model, the gain at high frequency attenuates quickly. When g_1, g_2 were

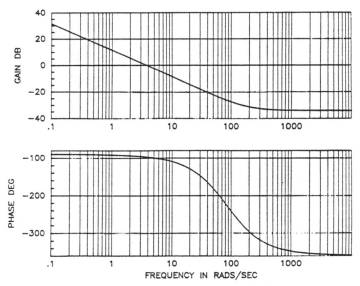

Figure 6. Bode Diagram Of Uncompensated G(w) With T=0.02s.

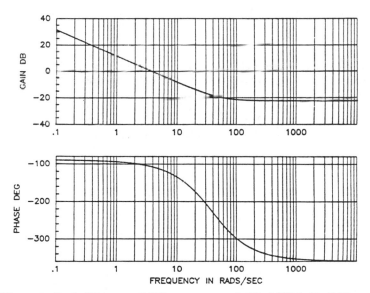

Figure 7. Bode Diagram Of Uncompensated G(w) With T=0.05s.

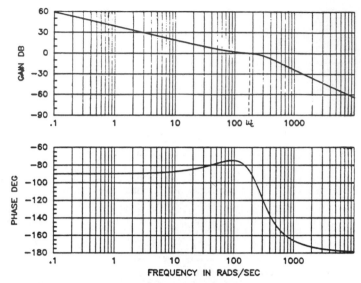

Figure 8. Bode Diagram Of G(s) with PD Controller.

increased, according to the continuous time model, the system would remain stable. However, this conclusion conflicts with our real test results. To understand this fact we need to analyze the more realistic discrete time system model $G(w)$.

The Bode plot of $G(w)$ with the digital implementation of the PD controller is shown in Figure 9. Comparing Figure 9 with Figure 8, we can see that the crossover frequency ω_c and static error coefficient K remain about the same. However, the gain margin ($=+1$db) and phase margin ($=+15^0$) from Figure 9 become critical. It is clear that the system could be unstable when we increase the gain or when the system parameters vary from those of the model.

Figure 9 shows that the PD controller leads to serious trade-offs between the static accuracy, stability, and high frequency noise rejection.

For example, in order to have larger positive gain margin and phase margin for system stability, we must decrease the gain K, which means we have to sacrifice static accuracy. On the other hand, to have better static accuracy and therefore to increase the gain K, the high-frequency noise response of the compensated system with the digital PD controller will be even worse than that we see from Figure 9.

Normally, high frequency noise does not significantly influence the function of analog control devices because of their natural low-pass behavior. However, in the case of digital signal processing, noise is sampled and transmitted and it may be amplified. These trade-offs become more serious as sampling rate decreases (see Figure 10).

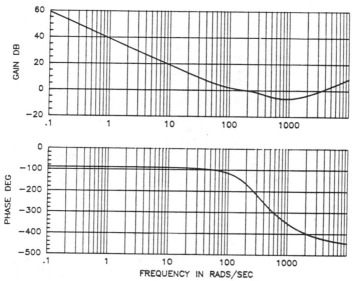

Figure 9. Bode Diagram Of G(w) With T=0.005s.

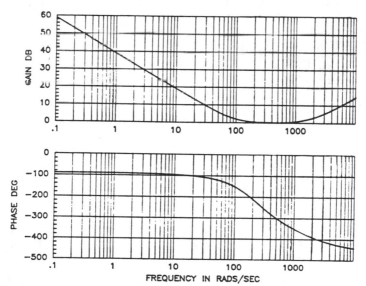

Figure 10. Bode Diagram Of G(w) With PD Controller And T=0.01s.

Adding an integral of error to the PD to make a PID feedback often makes the overall system less stable, although it improves the static accuracy.

We would like to point out that the system behavior is well predicted by the discrete time system model and is consistent with the real robot tests performed on a PUMA 560 arm at the Advanced Engineering Staff of General Motors.

The detailed study shows that even for a relatively fast sampling rate (T=0.005s), the real discrete dynamic system of a robot behaves quite differently from the continuous time dynamic system model. Further, conventional PD (or PID) controllers cannot meet the requirement for high performance because of unacceptable trade-offs between static accuracy, system stability, insensitivity to high frequency noise. Therefore, we must use a discrete time system model and design a more sophisticated feedback compensator.

III. LAG-LEAD COMPENSATOR DESIGN

In dealing with the problem of the compensating control system via frequency-domain techniques, it is useful to understand the general relation between the frequency domain and the time domain. The low-frequency region (the region below the crossover frequency) of the Bode plot indicates the static accuracy of the closed-loop system. The medium frequency region (the region near the crossover frequency) of the plot indicates the absolute and relative stability. The high-frequency region (the region above the crossover frequency) indicates the sensitivity of the system to high frequency noise perturbations and model structure inaccuracies.

The compensation we design reflects our goal of high static accuracy, adequate system stability, insensitivity of model structure inaccuracies and noise rejection. In order to accomplish this goal it is necessary to reshape the system frequency-response curve. The gain in the low frequency region should be large enough to achieve high static accuracy. Near the crossover frequency the slope of the log-magnitude curve in the Bode diagram should be -20db/decade[20], and therefore have a less than 180^0 phase lag. This slope should extend over a reasonably wide frequency band to assure proper phase margin and gain margin. For the high frequency region, the gain should be attenuated as rapidly as possible in order to minimize the effects of noise and model structure inaccuracies.

Based on these requirements, we propose to design a Lag-lead compensator, which has more flexibility than PD (or PID) controllers to reshape the frequency- response curve for our robot system. The suggested Lag-lead compensator in w-plane is:

$$H(w) = K \cdot \frac{(w/L + 1)}{(\beta_1 w/L + 1)} \cdot \frac{(w/d + 1)}{(w/\beta_2 d + 1)}$$

$$\stackrel{\Delta}{=} K \cdot H_L(w) \cdot H_d(w)$$

(16)

where K, L, d, β_1 and β_2 are constant parameters to be selected. The number K is the gain of the compensator and L and d are known as the corner frequencies. The corner frequency of a lag (or lead) is the frequency at which the two asymptotes of the lag (or lead) meet. The Bode diagram of the Lag-lead compensator with sampling time T=0.005s can be seen in Figure 11.

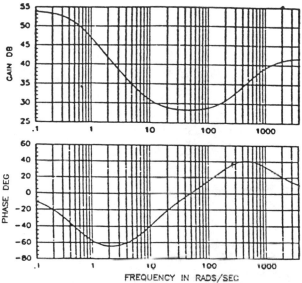

Figure 11. Bode Diagram Of Lag-lead Compensator With T=0.005s

Note that the suggested compensator in (Eq. 16) can also have the following form in w-plane:

$$H(w) = K' \cdot \frac{(w/L + 1)}{w} \cdot \frac{(w/d + 1)}{(w/\beta_2 d + 1)}, \qquad (17)$$

which can be viewed as a special case of (Eq. 16) when β_1 is very large.

The phase lead portion $H_d(w)$ of this compensator in (Eq. 16) alters the frequency-response curve by adding phase lead angle and increasing the phase margin at the crossover frequency. It basically improves the system stability and noise rejection.

The phase lag portion $H_L(w)$ of the Lag-lead network provides attenuation near and above the crossover frequency and thereby allows an increase of gain at the low-frequency range to improve the static accuracy. The Lag-lead network possesses two poles and two zeros. Therefore, such compensation increases the order of the system by two.

By substituting the w-transformation

$$w = \frac{2}{T} \cdot \frac{z-1}{z+1}$$

into (16), we have the following compensator in terms of z, which can be realized in a digital computer:

$$\frac{\Delta U(z)}{E(z)} = H(z) \triangleq H(w(z))$$

$$= \beta_2 K \cdot \frac{[(2 + LT)z + (LT - 2)]}{[(LT + 2\beta_1)z + (LT - 2\beta_1)]} \cdot \frac{[(dT + 2)z + (dT - 2)]}{[(\beta_2 dT + 2)z + (\beta_2 dT - 2)]}$$

(18)

For our case it is desired that: the static error coefficient K_s of the compensated overall system be as high as possible; the phase margin be $30^0 - 60^0$, and the gain margin be 3-6 db [20]. We found that K_s may be as large as 2000 while the phase margin and gain margin still can be made to meet the requirements.

First, let us determine the gain K of this compensator. Since the static error coefficient K_0 of the original system G(w) is $b/a_2 = 4$ and $K_s = K_0 K$, we choose K = 500.

The next step in the design is to choose a new crossover frequency. From the phase angle curve for $G(j\nu)$, in Figure 3, we notice that the phase angle $G(j\nu) = -160^0, at \nu = 100$. It is convenient to choose the new crossover frequency at ν_c to be 100 so that the phase lead angle required at $\nu = 100$ is about $10^0 - 40^0$, which is quite attainable by use of a single Lag-lead network.

The third step is to determine the corner frequency of the phase lag portion of the Lag-lead compensator. Let us choose one corner frequency (which corresponds to the zero of the phase lag potion of the compensator) to be one decade below the new crossover frequency, L = 10. If we choose $\beta_1 = 20$, then another corner frequency L/β_1 (which corresponds to the pole of the phase lag portion of the compensator) becomes 0.5.

Finally the phase lead portion must contribute $10^0 - 40^0$ phase lead. We choose the corner frequency d (which corresponds to the zero of the phase lead portion of the compensator) to be d = 200, and $\beta_2 = 5$, thus the other corner frequency $d\beta_2$ (which corresponds to the pole of the phase lead portion) becomes 1000.

Combining the transfer function of the lag and lead portion of the network, we obtain the transfer function of the Lag-lead compensator from (Eq. 16):

$$H(w) = 500 \cdot \frac{(0.1w + 1)(0.005w + 1)}{(2w + 1)(0.001w + 1)}$$

(19)

The magnitude and phase-angle curves of the Lag-lead compensator just designed are shown in Figure 11. The open-loop transfer function of the overall compensated system is

$$H(w)G(w) = KK_0 \cdot \frac{(0.1w + 1)(0.005w + 1)}{(2w + 1)(0.001w + 1)}$$

(20)

$$\cdot \frac{(1 - wT/2)^2(\phi_2 w^2 + \phi_1 w + 1)}{(1 + wT/2)w(\theta_2 w^2 + \theta_1 w + 1)}$$

where $\theta_1, \theta_2, \phi_1, \phi_2$ are listed in Section II.

The magnitude and phase-angle curves of the compensated overall system (Eq. 20) with sampling time T=0.005s are shown in Figure 12. The phase margin of this compensated system is 40^0, the gain margin is 4 db, and the static error coefficient K_s is 2000. All the requirements are therefore met, and the design has been completed. The comparison of Bode plots between the Lag-lead compensator and the PD controller can be seen in Table I. The improvements are clear.

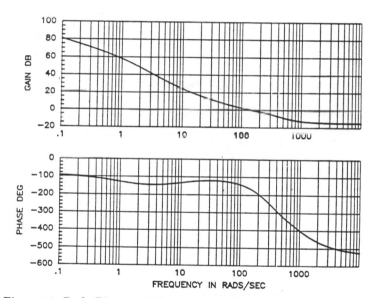

Figure 12. Bode Diagram Of Compensated System By Lag-lead
          Compensator.

We now discuss disturbances to see how they affect the static error of the overall compensated system at this stage. After compensating the gravity, centrifugal and coriolis forces, the disturbance consists mainly of:

	K	Gain Margin	Phase Margin	Noise Attenuation
PD Controller	100	1 db	10^0	Slow
Lag-lead Compensator	2000	4 db	40^0	Fast

Table I. Comparison of PD Controller and Lag-lead Compensator.

(a) Quantization round off error and optical encoder error. They are uniformly distributed white noise and uncorrelated.

(b) Friction, backlash and parameter uncertainty are deterministic (except possibly backlash) and consist mainly of low frequency components. Coulomb friction, particularly, plays an important role in robot arm control.

In real tests, we found that adding a certain amount of voltage to the motor input voltage (about 2 v) for PUMA 560, according to the direction of the movement, reduces static and peak dynamic errors. Therefore, it is reasonable to model Coulomb friction as a 2 unit step function adding to the motor input with a sign opposing the direction of motion. Thus, it is interesting to calculate the static error due to a 2 unit step disturbance $\zeta(t)$. To do this, we may apply the final value theorem because the closed-loop system is stable.

For z-transformation of $\zeta(t)$, we have

$$\zeta(z) = \frac{2z}{(z-1)}.$$

the static error

$$e_s = lim_{n \to \infty} X_\zeta(nT)$$

$$= lim_{z \to 1}(1 - z^{-1}) \cdot \frac{G(z)}{1 + G(z)H(z)} \cdot \zeta(z),$$

If K is large enough, then we have the following approximation:

$$e_s = lim_{n \to \infty} X_\zeta(nT) = \frac{2}{\beta_2 K} \cdot [(LT + 2\beta_1)(\beta_2 dT + 2) +$$

$$(LT + 2\beta_1)(\beta_2 dT - 2) + (LT - 2\beta_1)(\beta_2 dT + 2) +$$

$$(LT - 2\beta_1)(\beta_2 dT - 2)] \div [(2 + LT)(dT + 2) +$$

$$(2 + LT)(dT - 2) + (LT - 2)(dT + 2) + (LT - 2)(dT - 2)]$$

Thus for T $= 0.005$s, $\beta_1 = 20.0$, $\beta_2 = 5.0$, L $= 10.0$, d $= 200$, and K $= 500$, we have the static error $e_s = 2/K = 0.004$ due to the Collomb friction. This theoretical result is very close to the simulation results [23].

IV. REALIZATION ALGORITHM

Different realizations may be obtained by transforming the state-space coordinates. The choice of a suitable realization is very important for numerical conditioning. In particular, the companion forms make the system sensitive to changes in the coefficients, especially with a high sampling rate and with multiple eigenvalues close to the unit circle[21]. To avoid this numerical difficulty, we may represent the compensator as a combination of first-(for distinct real poles) and second-order (for complex poles) systems. In our case, the Lag-lead compensator from (Eq. 18) has simple real poles and zeros and thus may be decomposed as follows:

$$H(z) = \beta_2 K \cdot \frac{[(2 + LT)z + (LT - 2)]}{[(LT + 2\beta_1)z + (LT - 2\beta_1)]}$$

$$\cdot \frac{[(dT + 2)z + (dT - 2)]}{[(\beta_2 dT + 2)z + (\beta_2 dT - 2)]}$$

$$\triangleq \beta_2 K \cdot \frac{\Delta U_1(z)}{E_1(z)} \cdot \frac{\Delta U_2(z)}{E_2(z)}$$

$$\triangleq \beta_2 K \cdot H_1(z) \cdot H_2(z)$$

(21)

where

$$H_1(z) = \frac{\Delta U_1(z)}{E_1(z)} = \frac{[(2 + LT)z + (LT - 2)]}{[(LT + 2\beta_1)z + (LT - 2\beta_1)]}$$

$$= \frac{LT + 2}{LT + 2\beta_1} \cdot \frac{1 + \frac{LT-2}{LT+2}z^{-1}}{1 + \frac{LT-2\beta_1}{LT+2\beta_1}z^{-1}}$$

$$H_2(z) = \frac{\Delta U_2(z)}{E_2(z)} = \frac{[(dT + 2)z + (dT - 2)]}{[(\beta_2 dT + 2)z + (\beta_2 dT - 2)]}$$

$$= \frac{dT + 2}{\beta_2 dT + 2} \cdot \frac{1 + \frac{dT-2}{dT+2}z^{-1}}{1 + \frac{\beta_2 dT-2}{\beta_2 dT+2}z^{-1}}$$

We let

$$LL_1 = \frac{(LT - 2)}{(LT + 2)},$$

$$LL_2 = \frac{(dT - 2)}{(dT + 2)},$$

$$dd_1 = \frac{(LT - 2\beta_1)}{(LT + 2\beta_1)},$$

$$dd_2 = \frac{(\beta_2 dT - 2)}{\beta_2 dT + 2)},$$

$$kk_1 = \frac{(LT + 2)}{(LT + 2\beta_1)},$$

and

$$kk_2 = \frac{(dT + 2)}{(\beta_2 dT + 2)}.$$

The state diagram for this cascade form of the Lag-lead compensator from equation (21) is shown in Figure 13.

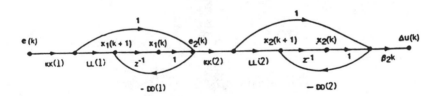

Figure 13. State Diagram For The Cascade Decomposition Of Lag-lead Compensator.

Then we have state equations with states x_1, x_2 and e_2, input e and output Δu as follows:

$$
\begin{aligned}
x_1(n+1) &= -dd_1 x_1(n) + (LL_1 - dd_1)kk_1 \cdot e(n) \\
e_2(n) &= x_1(n) + kk_1 \cdot e(n) \\
x_2(n+1) &= -dd_2 x_2(n) + (LL_2 - dd_2)kk_2 \cdot e_2(n) \\
\Delta u(n) &= \beta_2 K(x_2(n) + kk_2 \cdot e_2(n))
\end{aligned}
\tag{22}
$$

Notice, the input position error e of this Lag-lead compensator can be obtained by the desired position and measuring the real joint position from the optical encoder. The computed output Δu is the corrective voltage (or current if the system model derived with a current input) which we add to and adjust the input voltage (or current) of the robot motor. It is easy to see (Eq. 22) is controllable and observable and therefore is a minimal realization. (Eq. 22) can be easily realized by a digital computer. The computer code skeleton of this implementation for the ith joint of the robot arm is given below:

Procedure Udelta

```
BEGIN
LL(1): = (L*T-2)/(L*T+2)
LL(2): = (d*T-2)/(d*T+2)
DD(1): = (L*T-2*BT1)/(L*T+2*BT1)
DD(2): = (BT2*d*T-2)/(BT2*d*T+2)
KK(1): = (L*T+2)/(L*T+2*BT1)
KK(2): = (d*T+2)/(BT2*d*t+2)
For J: = 1 TO 2 DO
Begin
XNEW(J,I): = -DD(J)*X(J,I)+(LL(J)-DD(J))*KK(J)*E(J,I)
JJ: = J + 1
E(JJ,I): = X(J,I) + KK(J)*E(J,I)
X(J,I): = XNEW (J,I)
End
Udelta = E(3,I)*(K*BT2)
END.
```

For a different set of robot physical configurations, the design procedure and the realization form will be the same. The only changes need to be made are those constant parameters: K, β_1, β_2, d, and L.

Since a compensator is a dynamic system, we need to set the compensator state appropriately when the compensator is switched on. If this is not done, there may be large switching transients. In our cases, compensator states are initially set to 0, which works well in simulations and real tests.

V. IMPLEMENTATION RESULTS

Although the overall compensated system with the Lag-lead compensator looks good on the frequency domain (See Figure 12 and compare to Figure 9), we still need to verify further the system performance on the time domain. Simulation results show that a Lag-lead compensator based on a frequency-response approach indeed leads to higher accuracy while ensuring the stability of the system and insensitivity to disturbances. The overall system is more stable and robust. The detailed simulation results can be found in [23].

Recently some real implementation tests were conducted. The results are also encouraging. To focus on the cmparison between the Lag-lead compensator and the PID controller, we may test each of them alone as a servo control. We can use a PUMA 560 robot arm but bypass the VAL controller. A MicroVax II is used as a target computer to control the arm. A finished Vaxeln Pascal system runs on the target computer by itself, without VAX/VMS or any other operating system present. Vaxeln is a software product for the development of dedicated, realtime systems for VAX processors[24]. It's a faster way to design and implement time-critical applications on micros. Both multitasking and multiprogramming are supported by the Vaxeln Kernel and Vaxeln Pascal programming language.

In the process of developing and building the system for controlling the arm, we may also use the Vaxeln Debugger, thru Ethernet from a VAX 780 computer, to debug the programs in a developed, executing system. The remote debugger can display the states of all processes and jobs in the local-area network and can dynamically change the user's "view" from one process of node to another. Finished Vaxeln systems were loaded down line from VAX 780 into the target computer Macrovax II.

In the standard Unimation system, the VAL software interprets the operating instructions for a robot arm, and the controller transmits these instructions from the computer memory to the arm. From incremental encoders and potentiometers in the robot arm, the controller/computer receives data about arm position. This provides a closed loop control of arm motions. The VAL controller (LSI-11 computer) is interfaced to the joint interface board through a parallel I/O board. The joint interface board then relays joint related information received from the VAL controller to the individual joint microprocessor boards[25].

For our implementation we severed the connection between the joint interface board and the VAL controller by removing the parallel I/O board connection from the VAL controller board. We can then connected the parallel I/O connection to a MicroVax II Q - bus. Thus the MicroVax II is in direct control of the robot arm. we can now communicate with the joint processors by sending the joint microprocessors commands to perform various functions.

Using MicroVax II, we receive point data and build a queue. And the motion planning, inverse kinematics and joint interpolation are then accomplished. By modifying the software, the sampling rates of different executing processes can be varied. We now have two ways to follow a predescribed path:

(1)Input the desired position command from the MicroVax II to the joint microprocessors and use the PUMA PID servo control strategy to get the required currents for the power amplifiers of six joints.

(2)Implemente our own control algorithm, the Lag-lead compensator, and compute the required current in MicroVax II and then send to the joint power amplifiers bypassing the PUMA PID servo control.

We can now compare the conventional PID controller and our Lag-lead compensator by following the same path with the same speed. The only difference between these two tests is the servo control sampling rate. Besides the LSI-11 main CPU, Unimation uses six microprocessors for the servo control of six joints. A sampling rate, 0.875 ms, is used for PUMA PID servo control. In our case, however, in addition of doing queuing, path planning, joint interpolating, timing, error checking and so on, the CPU of MicroVax II has to do the servo control for all six joints also. Due to the large computational load, we can only use no faster than 5 ms sampling rate. Even so, the performance of the Lag-lead compensator is better than the PID controller.

As a example, we let the tool tip of a PUMA 560 follow a path which is a part of a circle with diameter of about 400 mm. The speed for cutting some material is about 80 mm/sec. The running time is 5 seconds. The load is 4 lb. Our system is robust and the parameters of the Lag-lead compensator were fairly easy to be tuned. While the parameters of the PID controller were carefully set by Unimation service people and they are believed to be well tuned. The comparisons of tracking errors for all six joints are shown in Figures 14 - 19. The improvements by the Lag-lead compensator are obvious. They can be also seen in the follwing table. Different paths and payloads were used for tracking, and similar improvements were obtained.

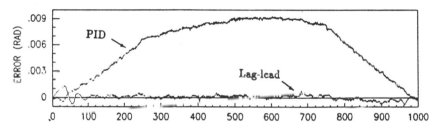

Figure 14. Tracking errors of first joint.

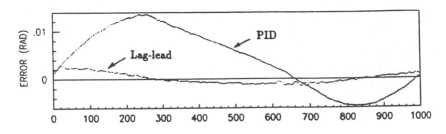

Figure 15. Tracking errors of second joint.

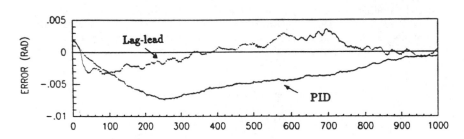

Figure 16. Tracking errors of third joint.

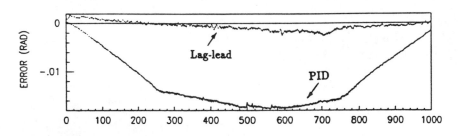

Figure 17. Tracking errors of fourth joint.

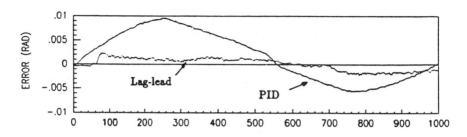

Figure 18. Tracking errors of fifth joint.

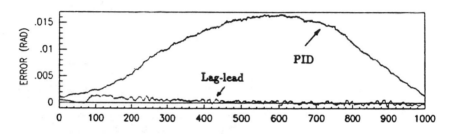

Figure 19. Tracking errors of sixth joint.

JOINT MOTOR	1	2	3	4	5	6
PUMA PID CONTROLLER	0.009	0.013	0.008	0.017	0.009	0.007
LAG-LEAD COMPENSATOR	0.0013	0.003	0.004	0.003	0.003	0.0035
IMPROVING FACTOR	7	4	2	5	3	2

Table II. Maximum Tracking Errors (unit:rad.) By Using PID
Controller and Lag-lead Compensator.

Although the design and tuning for a Lag-lead compensator need a little more effort than for a PID controller, the on-line computational load increased by using the Lag-lead compensator is very small comparing to other parts of computations required by the robot system. In fact, it takes less than 0.2 ms to do the Lag-lead calculation for each axis by using MicroVax II. Moreover, we developed a method which can automatically tune parameters of robots and therefore save time and labour[26][27], the increased tuning effort is not a big problem.

VI. CONCLUSION

In many cases the sampling rate is critical for robot dynamic control. Thus the widely used PD (or PID) controllers have serious trade-offs between the static accuracy, system stability and insensibility to disturbances. When used with the discrete time system models, the classic frequency response analysis and Lag-lead compensators show their important role in improving system performance. Replacing PD (or PID) controllers by Lag-lead compensators results in a small increase of off-line designing and tuning effort and on-line computational load, but the improvements in robot performance are significant.

REFERENCES

(1) R. P. Paul, 1972, "Modelling, Trajectory Calculation and Solving of a Computer-Controlled Arm," Stanford A.I. Lab, Stanford University, AIM 177.

(2) D. L. Pieper, 1968, "The Kinematics of Manipulators Under Computer Control." Ph.D. Dissertation, Stanford University. Memo 72, Stanford, CA. Stanford University Artificial Intelligence Laboratory.

(3) D. E. Whitney, 1969, "Resolved Motion Rate Control of Manipulators and Human Prosthesis. "IEEE Transaction on Man-Machine System, Vol. MMS-10, pp. 47-53.

(4) Y. S. Luh, M. Walker, R. Paul, 1980, "Resolved Acceleration Control for Mechanical Manipulators," IEEE Transaction of Automatic Control, Vol. AC-25, No. 3, June 1980, pp. 468-474.

(5) B. R. Markievicz, 1973, "Analysis of the Computed Torque Drive Method and Comparision with Conventional Position Servo for a Computer - Controlled Manipulator," Jet Propulsion Lab. Tech Memo 33-601.

(6) J. S. Albus, A. J. Barbera, R. N. Nagel, 1981, "Theory and Practice of Hierarchial Control," 23rd IEEE Computer Society International Conference, September, 1981.

(7) C. S. G. Lee, T. N. Mudge, J. L. Turney, 1982, "Hierarchial Control Structure Using Special Purpose Processors for the Control of Robot Arms", Proc. of the 1982 Pattern Recognition and Image Processing Conference, pp. 634-640.

(8) W. Khalil, "Contribution to Automatic Manipulator Control Using a Mathematical Mode of the Mechanisms", thesis, University of Vermont, Montpelier, 1978.

(9) A. K. Bejczy, T. J. Tarn, Y. L. Chen, 1985, "Computer Control of Robot Arms", Proc. of 1st. IEEE International Conference on Robotics and Automation, St. Louis, Missourl, March 1985.

(10) E. Freund, 1982, "Fast Nonlinear Control with Arbitrary Pole-Placement for Industrial robots and Manipu8lators," The International Journal of Robotics Research, Vol. 1, No. 1.

(11) T. J. Tarn, A. K. Bejczy, A. Isidori and Y. L. Chen, 1984, "Nonlinear Feedback in Robot Arm Control," Proc. 23rd IEEE Conference on Decision and Control," Las Vegas, Nevada, Dec. 12-14, 1984.

(12) S. Dubowsky, D. T. DesForges, 1979, "The Application of Model-Referenced Adaptive Control to Robotic Manipulators," J. Dynamic Systems Measurement and Control, V101 (1979), pp. 193-200.

(13) R. M. Goor, 1982, "Continuous Time Adaptive Feedforward Control: Stability and Simulations," General Motors Research Publicatin GMR-4105.

(14) R. M. Goor, 1984. "A New Approach to Minumum Time Robot Control," General Motors Research Publication GMR-4869.

(15) A. J. Koivo, T. H. Guo, 1981, "Control fo Robotic Manipulator with a Adaptive Controller," Proc. 20th IEEE Conference on Decision and Control, San Diego, CA, Dec. 16-18, 1981.

(16) Y. L. Chen, 1984, "Nonlinear Feedback and Computer Control of Robot Arms." D. Sc. Dissertation, Department of System Sciences and Mathematics, Washington University, St. Louis, Missouri, December 1984.

(17) C. P. Newman, V. D. Tourassis, 1985, "Discrete Dynamic Robot Model." IEEE Trans. on Systems, Man, and Cybernetics, Vol. SMC-15, No. 2, March/April 1985.

(18) G. F. Franklin, J. D. Powell, 1980, *Digital Control of Dynamic Systems.* Reading, Mass. Addison - Wesley, 1980.

(19) B. C. Kuo, 1970, *Discrete-Data Control Systems*, Prentice-Hall, Englewood Cliffs, New Jersey.

(20) Katsuhiko Ogata, 1970, *Modern Control Engineering,* Pren-tice-Hall, Englewood Cliffs, New Jersey.

(21) K. J. Astrom, B. Wittenmark, 1984, *Computer-Controlled Systems, Theory and Design*, Prentice-Hall, Inc., Englewood Cliffs, New Jersey.

(22) Y.L. Chen, 1987, "Frequency Response of a Discrete-time Robot system – Limitations of PD Controllers and Improvements by Lag-lead Compensation", Proc. 1987 IEEE International Conference on Robotics and Automation, Raleigh, North Carolina, March 1987.

(23) Y.L. Chen, 1986, "A Lag-lead Compensator for Robot Feedback Control", General Motors Research Publication GMR-5480, August 1986.

(24) Digital Equipment Corporation, 1985, "VAXELN Reference Manual", Maynard, Massachusetts.

(25) Unimation Robotics, 1982, Unimate PUMA Robot 550/560 series Volume I - Equipment Manual 398H1A, May 1982.

(26) Y.L. Chen, 1987, "A Formulation Of Automatic Parameter Tuning For Robot Arms", Proc. 23rd IEEE Conference on Decision and Control, Los Angeles, California, December 1987.

(27) Y.L. Chen, 1987, "Auto-tuning by Learning for Robot Feedback Control", General Motors Research Publication GMR-5766, April 1987.

MINIMUM COST TRAJECTORY PLANNING FOR INDUSTRIAL ROBOTS

Kang. G. Shin and Neil D. McKay[1]
Department of Electrical and Computer Engineering
The University of Michigan
Ann Arbor, Michigan, 48109

I Introduction

During the past several years a great deal of attention has been focused on industrial automation techniques, especially the use of general-purpose robots. Those industries which in the past have constructed special-purpose devices for manufacturing their products are now looking at the possibility of using robots instead. Unlike special-purpose tools, robots' behavior can easily be modified, so that retooling is kept to a minimum. In some cases they can also make feasible the manufacture of a product in lots which would be too small to justify the creation of a special-purpose machine for their manufacture. Mechanical maintenance is also simpler, since there presumably would be only a few types of robots performing many different tasks. (It should be noted, though, that maintenance of special-purpose machines is replaced with maintenance of special-purpose programs.)

Since robots may be controlled in virtually any manner, one legitimately may ask how a robot should best be controlled. The obvious answer to this question is that the robot should produce as large a profit as possible per unit time. The usual assumption is that material costs and fixed costs dominate the cost per item produced, so that it is desirable to produce as many units as possible in a given time.

---

[1] Neil D. McKay is now with Computer Science Dept.,General Motors Research Labs, Warren, MI 48090.

CONTROL AND DYNAMIC SYSTEMS, VOL. 39

There are a variety of algorithms available for manipulator control. These algorithms usually assume that the control structure of the robot has been divided into two levels. The upper level is called *path* or *trajectory planning*, and the lower level is called *path control* or *path tracking*. Path control is the process of making the robots actual position and velocity match some desired values of position and velocity; the desired values are provided to the controller by the trajectory planner. The trajectory planner receives as input some sort of spatial path descriptor from which it calculates a time history of the desired positions and velocities. (The term "Path Planner" is often used in the literature; this is a misnomer, since it does not plan paths but rather supplies timing information to a pre-planned geometric curve. The term "trajectory planner" will therefore be used here.) The path tracker then compensates for any deviations of the actual position and velocity from the desired values.

The reason for dividing the control scheme in this way is that the process of robot control, if considered in its entirety, is very complicated, since the dynamics of all but the simplest robots are highly nonlinear and coupled. Dividing the controller into the two parts makes the whole process simpler. The path tracker is frequently a linear controller (e.g., a PID controller). While the nonlinearities of manipulator dynamics frequently are not taken into account at this level, such trackers can generally keep the manipulator fairly close to the desired trajectory. More sophisticated methods can be used, though, such as resolved motion rate control[34], resolved acceleration control[23], and various adaptive techniques[3, 5, 6, 12, 15, 16, 8, 9].

Unfortunately, the simplicity obtained from the division into trajectory planning and path tracking often comes at the expense of efficiency. The source of the inefficiency is the trajectory planner. In order to use the robot efficiently, the trajectory planner must be aware of the robot's dynamic properties, and the more accurate the dynamic model is, the better the robot's capabilities can be used. However, most of the early trajectory planning algorithms assume very little about the robot's dynamics. The usual assumption is that there are constant or piecewise constant bounds on the robots velocity and acceleration [17, 21, 23]. In fact, these bounds vary with position, payload mass, and even with payload shape. Thus in order to make the constant-upper-bound scheme work, the upper bounds must be chosen to be global greatest lower bounds of the velocity and acceleration limits; in other words, the worst case limits have to be used. Since the moments of inertia seen at the joints of the robot, and hence the accereration limits, may vary by an order of magnitude as the robot moves from one position to another, such bounds can result in considerable inefficiency or under-utilization of the robot.

One of the early minimum-time trajectory planning systems was developed by Luh and Walker [22]. It describes the desired manipulator path in terms of its initial and final points and a set of intermediate points. Each branch of the path has a maximum velocity assigned to it, and each intermediate point is assigned a maximum acceleration and a maximum position error. The time taken to go from the initial to the final point is, then, the sum of the times taken to traverse each branch of the path plus the sum of the times required to make the transitions from one branch to the next. The minimum possible sum of these times can then be found using linear programming. In this case, one must still choose appropriate maximum velocities and accelerations, and this cannot be done properly without either knowing the dynamic properties and actuator characteristics of the robot or having some experimetal results which give maximum velocities and accelerations for given robot configurations. Also, since maximum accelerations and velocities are assumed to be constant over some interval, it is necessary to choose them to be lower bounds of the maximum values over the given interval, i.e., worst case bounds on acceleration and velocity. Since these bounds will in general depend on position and velocity, this could result in under-utilization of the robot's capabilities.

Luh and Lin [21] present a modification of the scheme described above which uses nonlinear programming to generate the minimum-time trajectory. The major difference between the method of Luh and Lin and that of Luh and Walker is in a more careful treatment of the calculation of the times required for the transitions from one path segment to the next. Also, an efficient technique for solving the non-linear programming problem is presented, along with a convergence proof.

Lin, Chang and Luh [17] present a third variation on this trajectory planning method. Instead of using path segments which are straight lines in Cartesian space, they convert points in Cartesian space into the equivalent joint coordinates, and pass a cubic spline through these points. The maximum velocities and accelerations are, then, joint velocities and accelerations, which are easier to compute from the robot cynamics and actuator characteristics. There is, however, still some calculation (or measurement) required in order to determine these quantities. This work, incidentally, also allows for limits on the jerk (time-derivative of acceleration). Placing limits on jerk helps prevent mechanism wear.

Kim and Shin [11, 13] have presented a method which is similar in some respects to the linear programming methods presented previously, but which uses the robot dynamic equations to obtain approximate acceleration bounds at each corner point. They also point out a set of conditions under which the linear programming problem reduces to a set of local optimization problems, one for each corner point. This represents a change in computational complexity of from $O(n^3)$ to $O(n)$.

Another type of trajectory planner has been developed by Bobrow et. al.[2], Shin et. al.[29, 30], and Pfeiffer et. el.[27]. In these minimum-time trajectory planners, it is assumed that the desired path is given a parameterized form. The parametric equations of the path can then be substituted into the manipulator dynamic equations, giving a set of second order differential equations in the (scalar) path parameter. Given these dynamic equations, bounds on individual joint torques can be converted into bounds on parametric accelerations (second time-derivatives of the path parameter); the allowable sets of second derivatives (one set per joint) are intersected, giving a single allowable set. Bounds on velocities (first derivatives of the path parameter) can also be found from these equations, since at some velocities there are no admissible accelerations. Then, using the fact that the minimum-time solution will be bang-bang in the acceleration, it is possible to construct phase plane plots which give the optimal trajectory in terms of the parameter and its derivatives. (The results found in [30] will be presented in Section IV.) Several related techniques may be found in [31, 32, 28, 24].

While most people have taken the separate trajectory planning/trajectory tracking approach, several authors have made attempts at unified approaches to robot control. One early attempt was the near-minimum time control of Kahn and Roth [10], who linerized the dynamic equations, transformed the equations so as to eliminate coupling terms, and generated switching curves for this lenear approximation. The result is a sub-optimal control which seems to give fairly good results if the initial and final states of the robot are fairly close. It does, however, have some problems with overshoot, as one would expect from such an approximation.

Another controller using an approximate dynamic model to generate optimal controls is the near-minimum time-fuel method of Kim and Shin [12]. They use a model which is linear over one sample period, and use coefficients in their linear model which result from averaging the coefficients at the current point on the trajectory and at the final point. The controls which result from this approximation depend upon whether time or fuel is the predominant term in the objective function, and upon the sampling rate. The trajectories for minimum fuel were slower than those for minimum time but had less overshoot. Increasing sampling rate also had the effect of reducing overshoot; thus here two parameters could be varied so as to find a good compromise between manipulator speed and overshoot.

One level up from trajectory planning is spatial planning, the process of finding collision-free paths for the manipulator to follow. Although the spatial planning problem is still largely unsolved, some work has been done, mostly with objects which are assumed to be spheres, cylinders, or convex polyhedra. Lozano-Perez [19] has reduced the spatial plannig problem to two problems: find-space and find-path. The find-space problem amounts to determining "safe" positions for the object being moved, i.e., positions where the object does not overlap any obstacles. Find-path is the process of finding a continuum of safe positions which takes the object from its initial configuration to a desired final configuration. In [19], Lozano-Perez presents algorithms for solving these problems in both two and three dimensions, though the three-dimensional algorithm does not in general give the "best" (minimum-distance) solution. In [18], Lozano-Perez describes the manipulator spatial planning problem in terms of volumes swept out by the links of the manipulator, and introduces polyhedral approximations of these swept volumes in order to use the results which are available for polyhedra.

Luh and Campbell [20] describe spatial planning interms of obstacles and "pseudo-obstacles" which the manipulator must avoid. The psueudo-obstacles arise because all the links, and not just the payload, must avoid the real obstacles. Luh and Campbell determined the shapes of some of these pseudo-obstacles for the Stanford arm. In particular, they considered the problem of keeping the back end of the Stanford arm's single prismatic link from bumping into things, which has the effect of creating a pseudo-obstacle on the opposite side of the arm from the real obstacle which generates it. They also generated polyhedral approximations to these pseudo-obstacles.

More recently, Gilbert and Johnson [7] have developed a technique for determining optimal controls in the presence of obstacles. Their technique starts with some feasible path, given as a cubic spline. They then apply a gradient technique which both moves the points interpolated by the spline and changes the robot's velocity in such a way as to optimize some performance measure. Obstacle avoidance is accomplished by including penalty functions in the performance index.

This chapter assumes that the geometric path planner has generated a parameterized curve in joint space, as described in [2] and [30]. The trajectory planning problem can then be reduced to a problem of small dimension by converting all the dynamic and actuator constraints to constraints on the single parameter which is used to describe the path, and the parameter's time derivatives. Within this framework, a variety of optimization techniques can be applied. The optimization methods described here apply to both minimum time problems and to more general minimum cost control problems. It is also possible to modify the constraints to take uncertain dynamics into account at the trajectory planning stage[32].

The remainder of this chapter is structured as follows: Section II formally states the problems to be solved. Sections IV through VI present three related but distinct trajectory planning algorithms, and gives results of these algorithms as applied to either a simple two degree-of-freedom arm or the first three joints of the Bendix PACS robot arm. The chapter concludes with Section VII.

II Problem Statement

The goal of automation, as previously stated, is to produce goods at a low a cost as possible. In practice, costs may be divided into two groups: fixed and variable. Variable costs depend upon details of the manufacturing process, and include, in the cases where robots are used, that part of the cost of driving a robot which varies with robot motion, and some maintenance costs. Fixed costs include taxes, heating costs, building maintenance, and, in the case of a robot, the portion of the electric power which the robot uses to run its computer controller and other peripheral devices. If one assumes that the fixed costs dominate, then cost per item produced will be proportional to the time taken to produce the item. In other words, minimum cost is equivalent to minimum production time. This important special case will be treated in some detail later. A loose statement of the minimum-cost path planning problem is as follows:

> What control signals will drive a given robot from a given initial configuration to a given final configuration with as small a cost as possible, given constraints on the magnitudes of the control signals and constraints on the intermediate configurations of the robot, i.e., given that the robot must not hit any obstacles?

While the problem of avoiding obstacles in the robot's workspace is not easily formulated as a control theory problem, the problem of moving a mechanical system with minimum cost is. One way to sidestep the collision avoidance problem, then, is to assume that the desired path has been specified a priori, for example as a parameterized curve in the robot's joint space. If this assumption is added, then one obtains a second, slightly different problem statement:

> What controls will drive a given robot along a specified curve in joint space with minimum cost, given constraints on initial and final velocities and on control signal magnitudes?

This form of the problem reduces the complexity of the control problem by introducing a single parameter which describes the robot's position. The time derivative of this parameter and the parameter itself completely determine the robot's position and velocity. The control problem then becomes essentially a two dimensional minimum-cost control problem with some state and input constraints. This chapter will present several methods for solving the minimum-cost trajectory planning problem.

To state the minimum-cost trajectory planning problem more formally, assume that the geometric path is given in the form of a parameterized curve, say

$$q^i = f^i(\lambda), \ \ 0 \le \lambda \le \lambda_{\max} \tag{2.1}$$

where q^i is the position of the ith joint, and initial and final points on the trajectory correspond to the points $\lambda = 0$ and $\lambda = \lambda_{\max}$. Also assume that the bounds on the actuator torques can be expressed in terms of the state of the system, i.e., in terms of the robot's speed and position, so that

$$u \in E(q, \dot{q}) \tag{2.2}$$

where u is a vector of actuator torques/forces, and $E : R^N \times R^N \to R^N$ is a set function. N is the number of joints the robot has. Given the functions f^i, the set function E, the desired initial and final velocities, and the manipulator dynamic equations (2.4.11) and (2.4.12), problem 1 is to find the controls $u(\lambda)$ which minimize the cost function C given by

$$C = \int_0^{\lambda_{\max}} \phi(u(\lambda), q(\lambda), \dot{q}(\lambda)) d\lambda \qquad (2.3)$$

In summary, the robot trajectory planning problem can be stated as:

> Given a geometric path in joint space expressed as a parameterized curve, actuator torque limits expressed in terms of the robot's position and velocity, the dynamic equations of the robot, desired initial and final velocities, and (possibly) externally-imposed constraints on velocity and jerk (the time-derivative of acceleration), how can the joint torques which minimize a particular cost functional be generated?

Solutions to this problem are the subjects of the next four sections.

III Trajectory Planning

The trajectory planning problem is basically an optimal control problem. One possible approach to the solution of this problem is to apply one of the standard tools of optimal control theory, Pontryagin's minimum principle. However, this approach requires solving a two-point boundary value problem for a non linear system of differential equations with non-linear constraints; clearly, this does not lead to a tractable solution. The minimum principle also sheds little or no light on the other auxiliary problems, such as sensitivity to parameter variations. Therefore, we will take a more intuitive approach.

Three trajectory planners will be presented here. The first method will be referred to as the *phase plane method*. It is so called because it makes use of plots of the "pseudo-velocity" $\mu \equiv \dot{\lambda}$ vs. the position parameter λ. (Recall that the robot is to move along a geometric path in which the joint positions q^i are given by a set of parametric functions, i.e., $q^i = f^i(\lambda)$.) Such a plot, in which a velocity is plotted as a function of position, is generally referred to as a "phase plane plot", hence the name. Actually all three trajectory planners described here make use of this idea in one way or another, and from here on, the term "trajectory" will be taken to mean "phase trajectory", or λ-μ plot.

The phase plane method is in general applicable only to minimum time problems, but this is often the case in which we are most interested. Since only minimum-time solutions are to be considered, it is useful to consider how this restriction on the objective function can be used. Obviously, minimizing traversal times is equivalent to maximizing traversal speed. Given this fact, it is easy to see that, at least in the simplest case, the minimum time solution consists of an accelerating and a decelerating part; the robot should accelerate at its maximum rate, then "put on the brakes" at precisely that time which will bring it to a stop at the destination point. Of course, there will in general be some velocity limits as well as acceleration limits. The velocity limits are imposed by the interaction of velocity-dependent force terms in the dynamic equations and the actuator torque limits; the acuators must generate enough torque to overcome these forces and keep the manipulator on the desired path. If the robot is to avoid these velocity limits, then the trajectory must alternately accelerate and decelerate, and the switching points should be timed so that the trajectory just barely misses exceeding the velocity limits. A more precise description of this method appears in the next subsection, including a derivation of the velocity limits and an algorithm for generating the optimal trajectories. Other complications are also discussed.

As an alternative, dynamic programming can be used to solve the trajectory planning problem. Dynamic programming is an impractical method for solving the general path planning problem for an arm with a large number of joints, since there are two state variables per joint, thus requiring a $2n$ dimensional grid. (This is a classic example of the "curse of dimensionality".) However, when the path is given, there are only two state variables; thus only a 2-dimensional grid is required.

To use dynamic programming, the grid is set up so that the position parameter λ is used as the stage variable. Thus a "column" of the grid corresponds to a fixed value of λ while a "row" corresponds to a fixed μ value. One starts at the desired final state (the last column of the grid, with the row corresponding to the desired final μ value) and assigns that state zero cost. All other states with position λ_{max} are given a cost of infinity. Then the usual dynamic programming algorithm can be applied. The algorithm starts at the last column. For each point in the previous column one finds all the accessible points in the current column, determines the minimum cost to go from the previous to the current column, and increments costs accordingly. For each of the previous grid points, the optimal choice of the next grid point is recorded. When the initial state is reached, the optimal trajectory is found by following the pointer chain which starts at the given initial state. In the case at hand, determining which points are accessible from one column to the next is simply a matter of checking to see if the slope of the curve connecting the two points gives a permissible value. (The slope limits can be found from the limits on the actuator torques.) The incremental cost is easily computed for minimum time-energy problems, so a running sum can easily be kept for the total cost.

Dynamic programming has the advantage that it is a well-established and well-understood optimization method. It also gives the control law for *every* speed-position combination, and so makes it possible for the robot to vary its speed if necessary. (This of course assumes that the robot stays on the desired path.) On the other hand, if it is implemented in the most obvious and straighforward manner, it requires a large array for computations, and if the array size is to be known in advance then an upper bound on the velocity is needed. In practice, there may be artificially imposed velocity bounds, but in general it would be necessary to either calculate velocity bounds in advance or create a new (larger) grid and start all over if the trajectory left the grid. The computation times also increase rather quickly as the density of the grid, and hence the accuracy of the solution, increases. Some modifications to the algorithm will be suggested which should considerably increase its speed.

It should also be noted that dynamic programming may still be used even if the robot's actuator torque constraints are not independent of one another; this is not the case with the phase plane method. Making the phase plane algorithm work for non-independent actuators would require that the space of actuator torques be searched for an acceleration bound. Dynamic programming only requires that a function be available which returns a yes-or-no answer to the question "if this acceleration is desired, will the required torques be realizable?". The dynamic programming algorithm itself performs the search of the actuator torque space.

A third algorithm, called the *perturbation trajectory improvement algorithm*, will be presented. This algorithm is in some respects similar to the dynamic programming algorithm, though like the phase plane method it is only applicable to minimum time problems, at least in the form in which it is presented here. This algorithm starts with an initial feasible trajectory, and perturbs the trajectory in such a way that the traversal time for the trajectory decreases, while the trajectory remains feasible. This method has most of the advantages of the dynamic programming method, and can be modified to generate minimum-time trajectories when there are limits on the *jerk*, or the derivative of the acceleration, as well as limits on joint torques.

A   The Robot Dynamic Equations

Before proceeding, we require the general form of the dynamic equations for a robot manipulator. The derivation is omitted, as it may be found in several places[26, 4]. Using tensor notation, the dynamic equations may be written

$$u_i = J_{ij}\ddot{q}^j + C_{ijk}\dot{q}^j\dot{q}^k + R_{ij}\dot{q}^j + g_i \qquad (3.1)$$

where we have used the summation convention, i.e. when a repeated index appears in a term, we sum over that index[33]. We also have $u_i \equiv$ torque applied to joint i, $q^i \equiv$ position of joint i, $J_{ij} \equiv$ manipulator's inertia matrix, $C_{ijk} \equiv$ array representing Coriolis and centrifugal forces, $R_{ij} \equiv$ viscous damping matrix, and $g_i \equiv$ gravitational terms. Note that J, C, R, and g are all functions of the joint position vector q.

B   Parameterization of the Robot Dynamic Equations

Before delving headlong into the trajectory planning problem, the effects of restricting the manipulator's motion to a fixed path will be investigated. In what follows, the manipulator will be restricted to some geometric path

$$q^i = f^i(\lambda), 0 \leq \lambda \leq \lambda_{\max} \tag{3.2}$$

Differentiating this gives an expression for the velocity v^i,

$$v^i = \dot{q}^i = \frac{df^i}{d\lambda}\frac{d\lambda}{dt} = \frac{df^i}{d\lambda}\dot{\lambda} = \frac{df^i}{d\lambda}\mu \tag{3.3}$$

where $\mu \equiv \dot{\lambda}$ is the *pseudo-velocity* of the manipulator. Equation (3.1)becomes

$$
\begin{aligned}
u_i &= J_{ij}(\lambda)\frac{df^j}{d\lambda}\dot{\mu} + J_{ij}(\lambda)\frac{d^2 f^i}{d\lambda^2}\mu^2 \\
&\quad + C_{ijk}(\lambda)\frac{df^j}{d\lambda}\frac{df^k}{d\lambda}\mu^2 + R_{ij}\frac{df^j}{d\lambda}\mu + g_i(\lambda)
\end{aligned}
\tag{3.4}
$$

The equations of motion along the curve (i.e., the geometric path) are then just Eqs. (3.4) plus the equation $\dot{\lambda} = \mu$.

It is of course assumed that the coordinates q^i vary continuously with λ. It is also assumed that the derivatives $\frac{df^i}{d\lambda}$ and $\frac{d^2 f^i}{d\lambda^2}$ exist, and that the derivatives $\frac{df^i}{d\lambda}$ are never all zero simultaneously. This ensures that the path never retraces itself as λ goes from 0 to $\lambda_{\max}$. Such a retrace would force the parameter λ to take a discontinuous jump in order for the point q^i to move forward continuously.

It should be noted that in practice the spatial paths are given in Cartesian coordinates. While it is in general difficult to convert a curve in Cartesian coordinates to that in joint coordinates, it is relatively easy to perform the conversion for individual points. One can then pick a sufficiently large number of points on the Cartesian path, convert to joint coordinates, and use some sort of interpolation technique (e.g. cubic splines) to obtain a similar path in joint space (see [17]for an example).

Introducing some shorhand notation, let

$$
\begin{aligned}
M_i &\equiv J_{ij}\frac{df^j}{d\lambda}, \\
Q_i &\equiv J_{ij}\frac{d^2 f^j}{d\lambda^2} + C_{ijk}\frac{df^j}{d\lambda}\frac{df^k}{d\lambda}, \\
R_i &\equiv R_{ij}\frac{df^j}{d\lambda}, \\
S_i &\equiv g_i
\end{aligned}
$$

Equation (3.4) may then be written

$$u_i \equiv M_i \dot{\mu} + Q_i \mu^2 + R_i \mu + S_i \qquad (3.5)$$

Note that the quantities listed above are functions of λ. For the sake of brevity, the functional dependence is not indicated in what follows.

IV The Phase Plane Method

With the details of curve parameterization out of the way, the phase plane trajectory planning method may now be derived. (This derivation is substantially the same as that found in [30].) As was mentioned earlier in this section, the phase plane algorithm determines a series of alternately accelerating and decelerating phase trajectory segments. The acceleration and deceleration along these segments are the maximum allowable and minimum allowable values of the pseudo-acceleration $\dot{\mu}$. These values will now be derived.

A Derivation of Pseudo-Acceleration Limits

Consider the constraints on the inputs, namely $u_i^{\min}(q, \dot{q}) \le u_i \le u_i^{\max}(q, \dot{q})$. The torque constraints along the parameterized curve can be found by inserting $f^i(\lambda)$ for q^i and $\frac{df^i}{d\lambda}\mu$ for $\dot{q}^i$. This gives constraints of the form $u_i^{\min}(\lambda, \mu) \le u_i \le u_i^{\max}(\lambda, \mu)$. The dynamic equations (3.5) can be viewed as having the form

$$u_i = \Psi_i(\lambda)\dot{\mu} + \Omega_i(\lambda, \mu)$$

where $\Psi_i(\lambda) = M_i(\lambda)$ and $\Psi_i(\lambda, \mu) = Q_i(\lambda)\mu^2 + R_i(\lambda)\mu + S_i(\lambda)$. For a given state, i.e., given λ and μ, this is just a set of parametric equations for a line, where the parameter is $\dot{\mu}$. The admissible controls, then, are those which are on this line in the input space and also are inside the rectangular prism formed by the input magnitude constraints. Thus the rectangular prism puts bounds on $\dot{\mu}$. The reason for converting from bounds on the input torques/forces to bounds on the pseudo-accerleration $\dot{\mu}$ is that all the positions, velocities, and accelerations of the various joints are related to one another through the parameterization of the path. Given the current state (λ, μ), the quantity $\dot{\mu}$, if known, determines the input torques/forces for *all* of the joints of the robot, so that manipulation of this one scalar quantity can replace the manipulation of n scalars (the input torques) and a set of constraints (the path parameterization equations).

To evaluate the bounds on $\dot{\mu}$ explicity, Eq. (3.5) is substituted into the inequalities $u_i^{\min} \le u_i \le u_i^{\max}$ so that

$$u_i^{\min}(\lambda, \mu) \le M_i \dot{\mu} + Q_i \mu^2 + R_i \mu + S_i \le u_i^{\max}(\lambda, \mu) \qquad (4.1)$$

If $M_i = 0$, then these inequalities put no constraints on $\ddot{\mu}$. However, the inertia matrix J_{ij} is positive definite, and by hypothesis the derivatives $\frac{df^i}{d\lambda}$ are not all zero simultaneously at any point on the curve. Therefore,

$$J_{ij} \frac{df^i}{d\lambda} \frac{df^j}{d\lambda} = M_i \frac{df^i}{d\lambda} > 0 \tag{4.2}$$

for all values of λ. But then we must have at least one non-zero M_i, so that there will always be *some* constraint on the pseudo-acceleration. In those cases where $M_i \neq 0$, manipulation of inequalities (4.1) gives

$$\frac{u_i^{\min} - Q_i\mu^2 - R_i\mu - S_i}{|M_i|} \leq sgn(M_i)\ddot{\mu} \leq \frac{u_i^{\max} - Q_i\mu^2 - R_i\mu - Si}{|M_i|} \tag{4.3}$$

or

$$LB_i \leq \ddot{\mu} \leq UB_i \tag{4.4}$$

where

$$LB_i \equiv \frac{u_i^{\min}(M_i > 0) + u_i^{\max}(M_i < 0) - (Q_i\mu^2 + R_i\mu + S_i)}{M_i} \tag{4.5}$$

and

$$UB_i \equiv \frac{u_i^{\max}(M_i > 0) + u_i^{\min}(M_i < 0) - (Q_i\mu^2 + R_i\mu + S_i)}{M_i} \tag{4.6}$$

The expression $(M_i > 0)$ evaluates to one if $M_i > 0$, zero otherwise. Since these constraints must hold for all n joints, $\ddot{\mu}$ must satisfy

$$\max_i LB_i \leq \ddot{\mu} \leq \min_i UB_i, \quad \text{or } GLB(\lambda,\mu) \leq \ddot{\mu} \leq LUB(\lambda,\mu) \tag{4.7}$$

Note that it has *not* been assumed here that $u_i^{\min}$ and $u_i^{\max}$ are constants; they may indeed be arbitrary functions of λ and μ. Later these quantities will be assumed to have specific, relatively simple forms, but these forms should be adequate to describe most of the actuators used in practice.

The difference between the trajectory planing algorithm to be presented and those which are conventionally used can be seen in terms of Eq. (4.7). Assume that the parameter μ is arc length in Cartesian space. Then μ is the speed and $\ddot{\mu}$ the acceleration along the geometric path. Since most trajectory planners put *constant* bounds on the acceleration over some set of position intervals, one would have $GLB(\lambda,\mu) \leq \ddot{\mu}_{\min} \leq \ddot{\mu} \leq \ddot{\mu}_{\max} \leq LUB(\lambda,\mu)$, where $\ddot{\mu}_{\min}$ and $\ddot{\mu}_{\max}$ are constants. The conventional techniques, then, restrict the acceleration more than is really necessary Likewise, constant bounds on the velocity will also be more restrictive than necessary.

B   Formulation of the Optimal Control Problem

With the manipulator dynamic equations and joint torque/force constraints in suitable form, we can address the actual control problem. Problems which require the minimization of cost functions subject to differential equation constraints can be expressed very naturally in the language of optimal control theory. The usual method of solving such a problem is to employ Pontryagin's maximum principle [5]. The maximum principle yields a two-point boundary value problem which is, except in some simple cases, impossible to solve in closed form, and usually is difficult to solve numerically as well. We will therefore not use the maximum principle, but will use some simpler reasoning, taking advantage of the specific form of the cost function of the controlled system.

In the case considered here, minimum cost is equated with minimum time, thus maximizing the operating speed of the robot. The cost function can then be expressed as $T = \int_0^{t_f} 1 \cdot dt$ where the final time t_f is left free. It is assumed here that the desired geometric path of the manipulator has been pre-planned, and is provided to the minimum-time controller in parametric form, as described earlier.

With this parameterization, there are two state variables, i.e., λ and μ, but $(n + 1)$ equations. One way to look at the system is to choose the equation $\dot{\lambda} = \mu$ and one of the remaining equations as state equations, regarding the other equations as constraints on the inputs an on μ. However, the problem has a more appealing symmetry if a single differential equation is obtained from the n equations (4.7) by multiplying the ith equation by $\frac{df^i}{d\lambda}$ and sum over i, giving

$$u_i \frac{df^i}{d\lambda} = M_i \frac{df^i}{d\lambda}\dot{\mu} + Q_i \frac{df^i}{d\lambda}\mu^2 + R_i \frac{df^i}{d\lambda}\mu + S_i \frac{df^i}{d\lambda} \qquad (4.8)$$

or

$$U = M(\lambda)\dot{\mu} + Q(\lambda)\mu^2 + R(\lambda)\mu + S(\lambda) \qquad (4.9)$$

where, expanding the values of M_i, Q_i, R_i, and S_i, we have

$$M(\lambda) \equiv J_{ij}(\lambda)\frac{df^i}{d\lambda}\frac{df^j}{d\lambda} \qquad (4.10)$$

$$Q(\lambda) \equiv J_{ij}(\lambda)\frac{df^i}{d\lambda}\frac{d^2 f^j}{d\lambda^2} + C_{ijk}(\lambda)\frac{df^i}{d\lambda}\frac{df^j}{d\lambda}\frac{df^k}{d\lambda} \qquad (4.11)$$

$$R(\lambda) \equiv R_{ij}\frac{df^i}{d\lambda}\frac{df^j}{d\lambda} \qquad (4.12)$$

$$S(\lambda) \equiv g_i(\lambda)\frac{df^i}{d\lambda} \qquad (4.13)$$

$$U(\lambda) \equiv u_i\frac{df^i}{d\lambda} \qquad (4.14)$$

This formulation has a distinct advantage. Note that the coefficient of $\dot{\mu}$ is quadratic in the vector of derivatives of the constraint functions. Since a smooth curve can always be parmeterized in such a way that the first derivatives are never all equal to zeo simultaneously, and since the inertia matrix is positive definite, the whole equation can be divided by the non-zero, positive coefficient of $\dot{\mu}$, providing a solution for $\dot{\mu}$ in terms of λ and μ. Now there are only two state equations, and the original n dynamic equations can be regarded as constraints on the inputs and on $\dot{\mu}$.

The M term in Eq. (4.9) is a quadratic form reminiscent of the expression for the manipulator's kinetic energy. In fact, if the parametric expressions for the $\dot{q}_i$ are substituted into the formula for kinetic energy, one obtains the expression $K \equiv \frac{1}{2} M \mu^2$.

The Q term represents the components of the Coriolis and centrifugal forces which act along the path plus the fictitious forces generated by the restriction that the robot stay on the parameterized path. The R term represents frictional components, and S gives the gravitational force along the path. U is the projection of the input torque vector onto the velocity vector.

With this formulation, the state equations become

$$\dot{\lambda} = \mu \tag{4.15}$$

$$\dot{\mu} = \frac{1}{M}\left[U - Q\mu^2 - R\mu - S\right] \tag{4.16}$$

The traversal time of the path, T, can be written in terms of λ and μ as

$$T = t_f = \int_0^{t_f} 1 \cdot dt = \int_0^{\lambda_{max}} \frac{dt}{d\lambda}d\lambda = \int_0^{\lambda_{max}} \frac{1}{\mu(\lambda)}d\lambda \tag{4.17}$$

Given these forms for the dynamic equations and the cost function, we have the Minimum Time Path Planning (MTPP) problem as follows.

Problem MTPP:

Find $\mu^*(\lambda)$ and $u_i^*(\lambda, \mu^*(\lambda))$ by minimizing T subject to Eqs. (4.15–4.16), the torque constraints $u_i^{min}(\lambda, \mu^*(\lambda)) \leq u_i^*(\lambda, \mu^*(\lambda)) \leq u_i^{max}(\lambda, \mu^*(\lambda)), 0 \leq \lambda_{max}$, and the boundary conditions $\mu^*(0) = \mu_0$ and $\mu^*(\lambda_{max}) \equiv \mu_f$.

C   Phase Plane Plots

At this point, it is instructive to look at the system's behavior in the phase plane. The equations of the phase plane trajectories can be obtained by dividing Eq. (4.16) by Eq. (4.15). This gives

$$\frac{d\mu}{d\lambda} = \frac{\frac{d\mu}{dt}}{\frac{d\lambda}{dt}} = \frac{\dot{\mu}}{\mu} = \frac{1}{\mu M}\left[U - Q\mu^2 - R\mu - S\right] \tag{4.18}$$

Noting again that the total time T that it takes to go from initial to final states is $T = \int_0^{\lambda_{max}} \frac{1}{\mu} d\lambda$, it is easily seen that minimizing time requires that the pseudovelocity μ be made as large as possible, a result which would be expected intuitively.

The constraints on $\dot{\mu}$ have two effects. One effect is to place limits on the slope of the phase trajectory. The other is to place limits on the value of μ. To obtain the limits on $\frac{d\mu}{d\lambda}$, one simply divides the limits on $\dot{\mu}$ by μ, since $\frac{d\mu}{d\lambda} = \frac{\dot{\mu}}{\mu}$.

To get the constraints on μ, it is necessary to consider the bounds on $\dot{\mu}$. If, for particular values of λ and μ, we have $LUB(\lambda, \mu) < GLB(\lambda, \mu)$ then there are no permissible values of $\dot{\mu}$. Therefore, for each value of λ we can assign a set of values of μ as determined by the inequality $LUB(\lambda, \mu) - GLB(\lambda, \mu) \geq 0$. This inequality holds if and only if $UB_i(\lambda, \mu) - LB_j(\lambda, \mu) \geq 0$ for all i and j. The intersection of the regions determined by these inequalities produces a region of the phase plane outside of which the phase trajectory must not stray. This region will hereafter be referred to as the *admissible region* of the phase plane. Using the equations for the lower and upper bound for all i and j,

$$\frac{u_i^{max}(M_i > 0) + u_i^{min}(M_i < 0) - (Q_i\mu^2 + R_i\mu + S_i)}{M - i} \tag{4.19}$$
$$-\frac{u_j^{min}(M_j > 0) + u_j^{max}(M_j < 0) - (Q_j\mu^2 + R_j\mu + S_j)}{M_j} \geq 0$$

Rearranging this inequality,

$$\left[\frac{Q_i}{M_i} - \frac{Q_j}{M_j}\right]\mu^2 + \left[\frac{R_i}{M_i} - \frac{R_j}{M_j}\right]\mu + \left[\frac{S_i}{M_i} - \frac{S_j}{M_i}\right] \tag{4.20}$$
$$+\left[\frac{u_i^{max}(M_i < 0) - u_i^{min}(M_i > 0)}{|M_i|} - \frac{u_j^{min}(M_j < 0) - u_j^{max}(M_j > 0)}{|M_j|}\right] \geq 0$$

It will prove convenient to "symmetrize" the input torque bounds in the discussion which follows. Each joint has a mean torque u_M^i and a maximum deviation Δ^i given by $u_M^i \equiv \frac{u_i^{max} + u_i^{min}}{2}, \Delta^i \equiv \frac{u_i^{max} - u_i^{min}}{2}$. Equation (4.20) can be rewritten as

$$\left[\frac{Q_i}{M_i} - \frac{Q_j}{M_j}\right]\mu^2 + \left[\frac{R_i}{M_i} - \frac{R_j}{M_j}\right]\mu \tag{4.21}$$
$$+\left[\frac{S_i}{m_i} - \frac{S_j}{M_j}\right] - \left[\frac{u_M^i}{M_i} - \frac{u_M^j}{M_j}\right] + \left[\frac{\Delta^i}{|M_i|} + \frac{\Delta^j}{|M_j|}\right] \geq 0$$

At this point, a specific form for the torque bounds will be assumed. If the maximum and minimum torques for each joint are functions only of the states q^i and $\dot{q}^i$ (i.e., the actuator torques are all independent of one another) and are at most *quadratic* in the velocities $\dot{q}^i$, then this inequality yields a simple quadratic in μ. This allows one to solve for the velocity bounds using the quadratic formula. A particularly simple and useful special case is that encountered when the actuator is a fixed-field D.C. motor with a bounded voltage input. In this case, the torque constraints take the form $u_i^{max} = V_{max}^i + k_m^i \dot{q}^i$ and $u_i^{min} = V_{min}^i + k_m^i \dot{q}^i$ where V_{min}^i and V_{max}^i are proportional to the voltage limits and k_m^i is a constant which depends upon the motor winding resistance, voltage source resistance, and the back E.M.F. generated by the motor. Let $V_{ave} = \frac{V_{max}^i + V_{min}^i}{2}$ and $\Delta^i = \frac{V_{max}^i - V_{min}^i}{2}$. Then, we get

$$u_M^i = V_{ave} + k_m^i \dot{q}^i = V_{ave} + K_m^i \frac{df^i}{d\lambda} \mu \tag{4.22}$$

From here on, the case outlined above will be used for the sake of simplicity. The only changes required for the more general case of quadratic velocity dependence of the torque bounds is a re-definition of the coefficients in some of the equations which follow.

Introducing yet more shorthand notation, let

$$A_{ij} \equiv \left[\frac{Q_i}{M_i} - \frac{Q_j}{m_j} \right] \tag{4.23}$$

$$B_{ij} \equiv \left[\frac{R_i}{M_i} - \frac{R_j}{M_j} \right] - \left[\frac{V_{ave} + k_m^i \frac{df^i}{d\lambda} \mu}{M_i} - \frac{V_{ave} + k_m^j \frac{df^j}{d\lambda} \mu}{M_j} \right] \tag{4.24}$$

$$C_{ij} \equiv \left[\frac{\Delta^i}{|M_i|} + \frac{\Delta^j}{|M_j|} \right] \tag{4.25}$$

$$D_{ij} \equiv \left[\frac{S_j}{M_j} - \frac{S_i}{M_i} \right] \tag{4.26}$$

Noting that (at least in this case) $A_{ij} = -A_{ji}, B_{ij} = B_{ji}, C_{ij} = C_{ji}$, and $D_{ij} = -D_{ji}$, we have the inequalities

$$0 \leq A_{ij}\mu^2 + B_{ij}\mu + C_{ij} + D_{ij} \tag{4.27}$$

$$0 \leq -A_{ij}\mu^2 - B_{ij}\mu + C_{ij} - D_{ij} \tag{4.28}$$

The second inequality is obtained by interchanging i and j and using the symmetry or anti-symmetry of the coefficients. Only the cases where $i \neq j$ need be considered, so there are $\frac{n(n-1)}{2}$ such pairs of equations, where n is the number of degrees of freedom of the robot.

If $A_{ij} = B_{ij} = 0$, we have $C_{ij} - D_{ij} \geq 0$ and $C_{ij} + D_{ij} \geq 0$, which are always true if the robot is "strong" enough so that it can stop and hold its position at all points on the desired path. If $A_{ij} = 0$ and $B_{ij} \neq 0$, then we have a pair of linear inequalities which determine a closed interval for μ. If $A_{ij} \neq 0$, then, without loss of generality, we can assume that $A_{ij} > 0$. Then the left-hand side of the first of the inequalities Eq. (4.27) is a parabola which is concave upward, wheares for the second it is concave downward. When the parabola is concave downward, then the inequality holds when μ is between the two roots of the quadratic. If the parabola is concave upward, then the inequality holds outside of the region between the roots. Thus in one case μ must lie within a closed interval and in the other it must lie outside an open interval, unless of course the open interval is of length zero. In this case, the inequality constraint is always satisfied and the roots of the quadratic will be complex.

Since the admissible values of μ are those which satisfy all of the inequalities, the admissible values must lie in the intersection of all the regions determined by the inequalities. There are $\frac{n(n-1)}{2}$ inequalities which give closed intervals, so the intersection of these regions is also a closed interval. The other $\frac{n(n-1)}{2}$ inequalities, when intersected with this closed interval, each may have the effect of "punching a hole" in the interval. It is thus possible to have, for any particular value of λ, a set of admissible values for μ which consists of as many as $\frac{n(n-1)+2}{2}$ distinct intervals. When the phase portrait of the optimal path is drawn, it may be necessary to have the optimal trajectory dodge the little "islands" which can occur in the admissible region of the phase plane. (Hereafter, these inadmissible regions will be referred to as *islands of inadmissibility*, or just *islands*.) It should be noted, though, that if there is no friction, then $B_{ij} = 0$, which means that in the concave upward case the inequality is satisfied for all values of μ. Thus in this case there will be no islands in the admissible region.

In addition to the constraints on μ described above, we must also have $\mu \geq 0$. This can be shown as follows: if $\mu < 0$, then the trajectory has passed below the line $\mu = 0$. Below this line, the trajectories always move to the left, since $\mu \equiv \frac{d\lambda}{dt} < 0$. Since the optimal trajectory must approach the desired final state through positive values of μ, the trajectory would then have to pass through $\mu = 0$ again, and would pass from $\mu < 0$ to $\mu > 0$ at a point to the left of where it had passed from $\mu > 0$ to $\mu < 0$. Thus in order to get to the desired final state, the trajectory would have to cross itself, forming a loop. But, then, there is no sense in traversing the loop; it would take less time to just use the crossing point as a switching point. Thus we need consider only those points of the phase plane for which $\mu \geq 0$.

Another way of thinking about the system phase portrait is to assign a pair of vectors to each point in the phase plane. One vector represents the slope when the system is accelerating (i.e., $\dot{\mu}$ is maximized) and the other represents the slope for deceleration (i.e., $\dot{\mu}$ is minimized). This pair of vectors looks like a pair of scissors, and as the positon in the phase plane changes, the angles of both the upper and lower jaws of the pair of scissors change. In particular, the angle between the two vectors varies with position. The phase trajectories must, at every point of the phase plane, point in a direction whcih lies between the jaws of the scissors. At particular points of the phase plane, though, the jaws of the scissors close completely, allowing only a single value for the slope. At other points the scissors may try to go past the closed position, allowing no trajectory at all. This phenomenon, and the condition $\mu \geq 0$, determine the admissible region of the phase plane.

D   Determination of Optimal Trajectories

For illustrative purposes, we first present an algorithm for finding the optimal trajectories for which there are no islands in the phase plane which need to be dodged. The only restrictions, then, will be that μ must lie between a pair of values which are easily calculable, given λ. The optimal trajectory can be constructed by the following steps called the *Algorithm for Constructing Optimal Trajectories, No Islands* (ACOTNI).

1. Start at $\lambda = 0, \mu = \mu_0$ and construct a trajectory that has the maximum acceleration value. Continue this curve until it either leaves the admissible region of the phase plane or goes past $\lambda = \lambda_{max}$. Note that "leaves the admissible region" implies that if part of the trajectory happens to coinside with a section of the admissible region's boundary, then the trajectory should be extended along the boundary. It is not sufficient in this case to continue the trajectory only until it touches the edge of the admissible region.

2. Construct a second trajectory that starts at $\lambda = \lambda_{max}, \mu = \mu_f$ and proceeds *backwards*, so that it is a decelerating curve. This curve should be extended until it either leaves the admissible region or extends past $\lambda = 0$.

3. If the two trajectories intersect, then stop. The point at which the trajectories intersect is the (single) switching point, and the optimal trajectory consists of the first (accelerating) curve from $\lambda = 0$ to the switching point, and the second (decelerating) curve from the switching point to $\lambda = \lambda_{max}$.

4. If the two curves under consideration do not intersect, then they must both leave the admissible region. Call the point where the accelerating curve leaves the admissible region λ_1. This is a point on the boundary curve of the admissible region. If the boundary curve is given by $\mu = g(\lambda)$, then search along the curve, starting at λ_1, until a point is found at which the quantity $\phi(\lambda) \equiv \frac{d\mu}{d\lambda} - \frac{dg}{d\lambda}$ changes sign. (Note that since $g(\lambda)$ determines the boundary of the admissible region, there is only one allowable value of $\frac{d\mu}{d\lambda}$. Also note that if $g(\lambda)$ has a discontinuity, $\frac{dg}{d\lambda}$ must be treated as $+\infty$ or $-\infty$ depending upon the direction of the jump). This point is the next switching point. Call it λ_d.

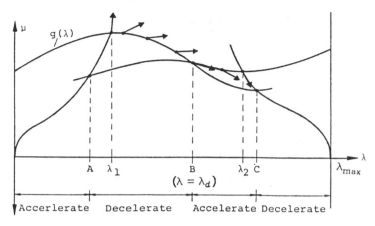

A, B, and C are switching points
B is a point of osculation between
g(λ) and the trajectory

Figure 4.1: A complete optimal trajectory with three switching points

5. Construct a decelerating trajectory backwards from λ_d until it intersects an accelerating trajectory. This gives another switching point (see point A in Figure 4.1).

6. Construct an accelerating trajectory srarting from λ_d. Continue the trajectory until it either intersects the final decelerating trajectory or it leaves the admissible region. If it intersects the decelerating trajectory, then the intersection gives another switching point (see point C in Figure 4.1), and the procedure terminates. If the trajectory leaves the admissible region, then go to 4.

This algorithm yields a sequence of alternately accelerating and decelerating curves which give the optimal trajectory. Before discussing the optimality of the trajectory, one has to show that all steps of the ACOTNI are possible and that the ACOTNI will terminate.

Addressing the first question, 1, 2, 3, 5, and 6 are clearly possible. 4 requires finding a sign change of the function $\phi(\lambda)$. Since $\phi(\lambda)$ must be greater than zero where the accelerating trajectory leaves the admissible region and less than zero where the decelerating trajectory leaves, there must be a sign change. Therefore all steps are possible.

In order to prove that ACOTNI terminates, it must be shown that the search for switching points in step 4 will be performed a finite number of times. To prove this, it is sufficient to prove that the number of isolated zeros of $\phi(\lambda)$, the number of intervals of positive extent over which $\phi(\lambda)$, is zero, and the number of intervals over which $\phi(\lambda)$ does not exist are all finite. To do this, some assumption about the form of the functions $f^i(\lambda)$ must be made. It will be assumed here that the f^i are real valued, piecewise analytic, and composed of a finite number of pieces. (In other words, the f^i are analytic splines.) Under these assumptions, the following theorems prove the convergence and optimality of ACOTNI.

Theorem 1: If the functions f^i are composed of a finite number of analytic, real-valued pieces, then the function $\phi(\lambda)$ has a finite number of intervals over which it is identiclly zero and a finite number of zeros outside those intervals.

Theorem 2: Any trajectory generated by the ACOTNI is optimal in the sense of minimum time control.

Proofs of these theorems may be found in [30].

The whole idea of the algorithm is to generate trajectories which come as close as possible to the edge of the admissible region without actually passing outside it. Thus the trajectories just barely touch the inadmissible region. In practice this would, of course, be highly dangerous, since minute errors in the control inputs or measured system papameters would very likely make the robot stray from the desired path. Theoretically, however, this trajectory is the minimum-time optimum.

We are now in a position to consider the general case, i.e., the case in which friction, copper losses in the drive motor, etc., are sufficient to cause islands in the phase plane. In this case, the algorithm is most easily presented in a slightly different form. Since there may be several boundary curves instead of one, it is not possible to search a single function for zeros, as was done in ACOTNI. Thus instead of looking for zeros as the algorithm progresses, we look for them all at once instead, and then construct the trajectories which "just miss" the boundaries, whether the boundaries be the edges of the admissible region or the edges of islands. The appropriate trajectories can then be found by searching the resulting directed graph, always taking the highest trajectory possible, and backtracking when necessary. More formally, the *Algorithm for Construction of Optimal Trajectories*((ACOT) is:

1. Construct the initial accelerating trajectory. (Same as ACOTNI).

2. Construct the final decelerating trajectory. (Same as ACOTNI).

3. Calculate the function $\phi(\lambda)$ for the edge of the admissible region and for the edges of all the islands. At each of the sign changes of $\phi(\lambda)$, construct a trajectory for which the sign change is a switching point, as in ACOTNI steps 5 and 6. The switching direction (acceleration-to-deceleration or vice-versa) should be chosen so that the trajcectory does not leave the admissible

region. Extend each trajectory until it either leaves the admissible region, or goes past λ_{max}.

4. Find all intersections of the trajectories. These are potential switching points.

5. Starting at $\lambda = 0, \mu = \mu_0$, traverse the grid formed by the various trajectories in such a way that the highest trajectory from the initial to the final points is followed. This is described below in the *grid traversal algorithm* (GTA).

Traversing the grid formed by the trajectories generated in steps 3 and 4 above is a search of a directed graph, where the goal to be searched for is the final decelerating trajectory. If one imagines searching the grid by walking along the trajectories, then one would try to keep making left turns, if possible. If a particular turn leads to a dead end, then it would be necessary to backtrack, and take a right turn instead. The whole procedure can best be expressed recursively, in much the same manner as tree traversal procedures.

The algorithm consists of two procedures, one which searches accelerating curves and one which searches decelerating curves. The algorithm is:

AccSearch:

On the current (accelerating) trajectory, find the last switching point. At this point, the current trajectory meets a decelerating curve. If that curve is the final decelerating trajectory, then the switching point under consideration is a switching point of the final optimal trajectory. Otherwise, call **DecSearch**, starting at the current switching point. If **DecSearch** is successful, then the current point is a switching point of the optimal trajectory. Otherwise, move back along the current accelerating curve to the previous switching point and repeat the process.

DecSearch:

On the current (decelerating) trajectory, find the first switching point. Apply **AccSearch**, starting on this point. If successful, then the current point is a switching point of the optimal trajectory. Otherwise, move forward to the next switching point and repeat the process.

These two algoritms always look first for the curves with the highest velocity, since **AccSearch** always starts at the end of an accelerating curve and **DecSearch** always starts at the beginning of a decelerating curve. Therefore the algorithm finds (if possible) the trajectory with the highest velocity, and hence the smallest traversal time.

The proofs of optimality and convergence of this algorithm are virtually identical to those of ACOTNI, and will not be repeated here. Note that in the convergence proof for ACOTNI the fact that there is only a single boundary

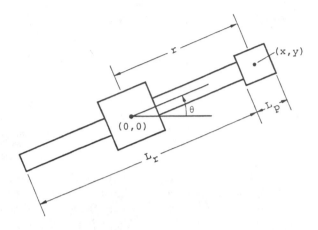

Figure 4.2: The two degree-of-freedom polar robot

curve in the zero-friction case is never used; the same proof therefore applies in the high-friction case.

E   Numerical Examples

To show how the minimum-time algorithm works, a numerical example follows. The robot used in the example is a simple two-degree-of-freedom robot with one revolute and prismatic joint, i.e., a robot which moves in polar coordinates. Despite its simplicity, the example robot is sufficient to show most important aspects of the phase plane trajectory planning method. (More complicated examples are given later). The path chosen is a straight line. Before applying the minimum-time algorithm, we must derive the dynamic equations for the robot. This requires calculation of the inertia matrix, so masses and moments of inertia of the robot must be given.

A drawing of the hypothetical robot is shown in Figure 4.2. The robot consists of a rotating fixture with moment of inertia J_θ through which slides a uniformly dense rod of length L_r and mass M_r. The payload has Mass M_p and moment of inertia J_p, and its center of mass is at the point (x, y) which is L_p units of length from the end of the sliding rod. (The full dynamic equations for this arm are derived in [25].) In the examples presented here, the robot will be moved from the point $(1, 1)$ to the point $(1, -1)$. The equation of the curve can be expressed as $r = \sec \theta$, where θ ranges from $+\frac{\pi}{4}$ to $-\frac{\pi}{4}$. Introducing the parameter λ, one possible parameterization is

$$\theta = \frac{\pi}{4} - \lambda, \quad r = \sec\left(\frac{\pi}{4} - \lambda\right) \tag{4.29}$$

Now introduce the shorthand expressions $M_t \equiv M_r + M_p$, $K \equiv M_r(L_r + 2L_p)$, and $J_t \equiv J_\theta + J_p + M_r(L_p^2 + L_r L_p + \frac{L_r^2}{3})$. Plugging these expressions and the expressions for the derivatives of r and θ into the dynamic equations for the polar manipulator gives (see [25] for a detailed derivation)

$$u_r = -M_t \sec(\frac{\pi}{r} - \lambda)\tan(\frac{\pi}{4} - \lambda)\dot{\mu} - k_r \sec(\frac{\pi}{4} - \lambda)\tan(\frac{\pi}{4} - \lambda)\mu \tag{4.30}$$

$$+ \left[M_t \sec(\frac{\pi}{4} - \lambda)\left(\sec^2(\frac{\pi}{4} - \lambda) + \tan^2(\frac{\pi}{4} - \lambda)\right) + \frac{K}{2} - M_t \sec(\frac{\pi}{4} - \lambda) \right]\mu^2$$

$$u_\theta = -\left[J_t - K \sec(\frac{\pi}{4} - \lambda) + M_t \sec^2(\frac{\pi}{4} - \lambda) \right]\dot{\mu} - K_\theta \mu \tag{4.31}$$

$$+ \mu^2 \left(2M_t \sec(\frac{\pi}{4} - \lambda) - K \right)\sec(\frac{\pi}{4}\lambda)\tan(\frac{\pi}{4} - \lambda)$$

Solving for $\dot{\mu}$, we have

$$\dot{\mu} = \frac{-1}{M_t \sec(\frac{\pi}{4} - \lambda)\tan(\frac{\pi}{4} - \lambda)} \left[u_r + k_r \mu \sec(\frac{\pi}{4} - \lambda)\tan(\frac{\pi}{4} - \lambda) \right. \tag{4.32}$$

$$\left. -\mu^2 \left\{ M_t \sec(\frac{\pi}{4} - \lambda)(\sec^2(\frac{\pi}{4} - \lambda) + \tan^2(\frac{\pi}{4} - \lambda)) + \frac{K}{2} - M_t \sec(\frac{\pi}{4} - \lambda) \right\} \right]$$

and

$$\dot{\mu} = \frac{u_\theta + k_\theta \mu - \mu^2 \left\{ 2M_t \sec(\frac{\pi}{4} - \lambda) - K \right\}\sec(\frac{\pi}{4} - \lambda)\tan(\frac{\pi}{4} - \lambda)}{J_t - K \sec(\frac{\pi}{4} - \lambda) + M_t \sec^2(\frac{\pi}{4} - \lambda)} \tag{4.33}$$

The signs of the coefficients of u_r and u_θ are

$$\mathrm{sgn}(u_r) = \begin{cases} -1 & 0 < \lambda < \frac{\pi}{4} \\ +1 & \frac{\pi}{4} < \lambda < \frac{\pi}{4} \end{cases} \quad \text{and} \quad \mathrm{sgn}(u_\theta) = 1 \tag{4.34}$$

The limits on $\dot{\mu}$ imposed by the θ joint are the same over the whole interval. For simplicity, let $u^i_{max} = -u^i_{max}$ for $i = r, \theta$; then the limits are

$$\dot{\mu} \leq \frac{u^\theta_{max} + \left[\{2M_t \sec(\frac{\pi}{4} - \lambda) - K\}\sec(\frac{\pi}{4} - \lambda)\tan(\frac{\pi}{4} - \lambda)\right]\mu^2 - k_\theta \mu}{J_t - K \sec(\frac{\pi}{4} - \lambda) + M_t \sec^2(\frac{\pi}{4} - \lambda)} \tag{4.35}$$

and

$$\dot{\mu} \geq \frac{-u^\theta_{max} + \left[\{2M_t \sec(\frac{\pi}{4} - \lambda) - K\}\sec(\frac{\pi}{4} - \lambda)\tan(\frac{\pi}{4} - \lambda)\right]\mu^2 - k_\theta \mu}{J_t - K \sec(\frac{\pi}{4} - \lambda) + M_t \sec^2(\frac{\pi}{4} - \lambda)} \tag{4.36}$$

For the r joint, consider the case when $\lambda < \frac{\pi}{4}$. Then we also have

$$
\begin{aligned}
\dot{\mu} \;\leq\; & \frac{1}{M_t \sec(\frac{\pi}{4} - \lambda) \tan(\frac{\pi}{4} - \lambda)} \Big\{ u^r_{max} \\
& + \left[2M_t \sec(\frac{\pi}{4} - \lambda) \tan^2(\frac{\pi}{4} - \lambda) + \frac{K}{2} \right] \mu^2 \\
& - k_r \mu \sec(\frac{\pi}{4} - \lambda) \tan(\frac{\pi}{4} - \lambda) \Big\}
\end{aligned}
\tag{4.37}
$$

and

$$
\begin{aligned}
\dot{\mu} \;\geq\; & \frac{1}{M_t \sec(\frac{\pi}{4} - \lambda) \tan(\frac{\pi}{4} - \lambda)} \Big\{ -u^r_{max} \\
& + \left[2M_t \sec(\frac{\pi}{4} - \lambda) \tan^2(\frac{\pi}{4} - \lambda) + \frac{K}{2} \right] \mu^2 \\
& - k_r \mu \sec(\frac{\pi}{4} - \lambda) \tan(\frac{\pi}{4} - \lambda) \Big\}
\end{aligned}
\tag{4.38}
$$

Finally, we need to determine the differential equations to be solved. These equations are

$$
\begin{aligned}
\dot{\mu} \;=\; & \frac{1}{J_t - K \sec(\frac{\pi}{4} - \lambda) + M_t \sec^4(\frac{\pi}{4} - \lambda)} \Big[-u_\theta \\
& - u_r \sec(\frac{\pi}{4} - \lambda) \tan(\frac{\pi}{4} - \lambda) + 2\mu^2 M_t \sec^4(\frac{\pi}{4} - \lambda) \tan(\frac{\pi}{4} - \lambda) \\
& - \left\{ k_r \tan^2(\frac{\pi}{4} - \lambda) \sec^2(\frac{\pi}{4} - \lambda) + k_\theta \right\} \mu \Big]
\end{aligned}
\tag{4.39}
$$

$$
\dot{\lambda} \;=\; \mu
\tag{4.40}
$$

The numerical values of the various constants which describe the robot are given in [25]. Using this data, the differential equations were solved numerically using the fourth-order Runge-Kutta method, the program being written in C and run under the UNIX[2] operating system on a VAX-11/780[3]. The derivative of the boundary curve $g(\lambda)$ (needed to compute the function $\phi(\lambda)$) was calculated numerically, and the sign changes of $\phi(\lambda)$ found by bisection. The graphs of the resulting trajectories and of the boundary of the admissible region are given in Figure 4.3 for the zero-friction case and in Figures 4.4 and 4.5 for the high-friction case. For $\lambda > \frac{\pi}{4}$ the limits have the signs of u^r_{max} reversed.

Equating upper and lower limits on $\dot{\mu}$ gives the boundary of the admissible region. For $\lambda < \frac{\pi}{4}$, Eq. (4.36) and Eq. (4.40) give

$$
A\mu^2 + B_\mu + C \geq 0
\tag{4.41}
$$

[2] UNIX is a trademark of Bell Laboratories.
[3] VAX is a trademark of Digital Equipment Corporation.

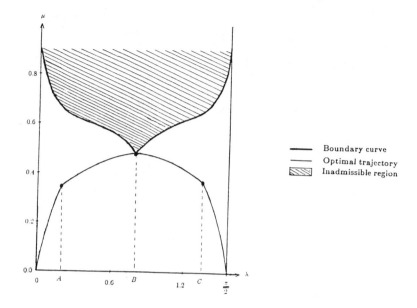

Figure 4.3: Optimal trajectory, zero friction case

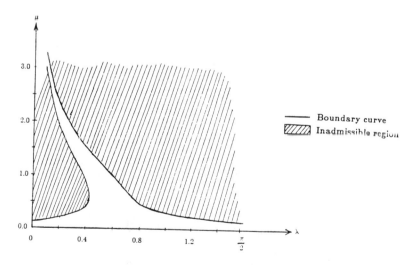

Figure 4.4: Admissible region, high-friction case

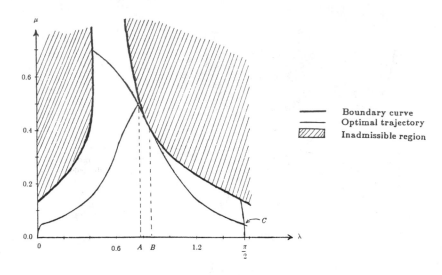

Figure 4.5: Optimal trajectory, high-friction case

where

$$A = -KM_t \sec^4(\frac{\pi}{4} - \lambda) + 2M_t \sec^3(\frac{\pi}{4} - \lambda) \tag{4.42}$$
$$+ \frac{3}{4}KM_1 \sec^2(\frac{\pi}{4} - \lambda) - (2M_tJ_t + \frac{K^2}{2}) + \sec(\frac{\pi}{4} - \lambda) + \frac{KJ_t}{2}$$

$$B = (J_tK_r - M_tk_\theta)\sec(\frac{\pi}{4} - \lambda)\tan(\frac{\pi}{4} - \lambda) \tag{4.43}$$
$$- Kk_r \sec^2(\frac{\pi}{4} - \lambda)\tan(\frac{\pi}{4} - \lambda) + M_tK_r\tan(\frac{\pi}{4} - \lambda)\sec^3(\frac{\pi}{4} - \lambda)$$

$$C = u^r_{\max}(J_t - K\sec(\frac{\pi}{4} - \lambda) + M_t\sec^2(\frac{\pi}{4} - \lambda)) \tag{4.44}$$
$$+ u^\theta_{\max}M_t\tan(\frac{\pi}{4} - \lambda)\sec(\frac{\pi}{4} - \lambda)$$

Likewise, Eq. (4.35) and Eq. (4.38) give

$$- A\mu^2 - B\mu + C \geq 0 \tag{4.45}$$

The same inequalities, with $u^r_{\max}$ negated, work when $\lambda \geq \frac{\pi}{4}$.

Note in particular the shape of the admissible region boundary in Figure 4.4. For values of λ less than about 0.42 there is not a single range of admissible velocities, but two ranges. Thus there is an "island" in the phase plane, though the island is chopped off by the constraint that λ be positive. While the existence of such islands may at first seem to defy intuition, the example shows that they do indeed exist. In this case, the island does not really come into play in the calculation of the optimal trajectory. Nevertheless, the example does demonstrate that there may be situations where the admissible region has a fairly complicated shape. Since most practical manipulators have more than two joints and have more complicated dynamic equations than those of the simple robot used here, it is conceivable that the admissible region of the phase plane for a practical robot arm could have quite a complicated shape.

As a final example, to demonstrate clearly the existence of islands in the phase plane, we include a sketch of the admissible region of the phase plane for a two-dimensional Cartesian robot moving along a circular path. In this case, the dynamic equations are a simple pair of uncoupled, linear differential equations with constant coefficients, i.e., $u_x = m\ddot{x} + k_x\dot{x}, u_y = m\ddot{y} + k_y\dot{y}$ where $m \equiv$ mass of x and y joints, $k_x \equiv$ coefficient of friction of x joint, and $k_y \equiv$ coefficient of friction of y joint.

Moving this manipulator in a unit circle, say in the first quadrant, requires that

$$x = \cos \lambda, \quad y = \sin \lambda, \quad 0 \leq \lambda \leq \frac{\pi}{4} \tag{4.46}$$

Plugging these expressions and their derivatives into the dynamic equations gives

$$u_x = -m\dot{\mu} \sin \lambda - m\mu^2 \cos \lambda - k_x\mu \sin \lambda \tag{4.47}$$

$$u_y = m\dot{\mu} \cos \lambda - m\mu^2 \sin \lambda + k_y\mu \cos \lambda \tag{4.48}$$

Now let the torque bounds be $-T \leq u_x, u_y \leq +T$. Then the bounds on $\dot{\mu}$ are

$$\frac{-T + m\mu^2 \cos \lambda - k_x\mu \sin \lambda}{m \sin \lambda} \leq \dot{\mu} \leq \frac{+T - m\mu^2 \cos \lambda - k_x\mu \sin \lambda}{m \sin \lambda} \tag{4.49}$$

and

$$\frac{-T + m\mu^2 \sin \lambda - k_y\mu \cos \lambda}{m \cos \lambda} \leq \dot{\mu} \leq \frac{+T + m\mu^2 \sin \lambda - k_y\mu \cos \lambda}{m \cos \lambda} \tag{4.50}$$

The admissible region consists of the region where the inequalities given above allow some value of the adcceleration $\dot{\mu}$, as previously described. Simplifying the resulting inequalities gives the admissible region as that area of the phase plane where

$$m\mu^2 + (k_x - k_y)\mu \sin \lambda \cos \lambda + T(\sin \lambda + \cos \lambda) \geq 0 \tag{4.51}$$

and

$$-m\mu^2 - (k_x - k_y)\mu \sin \lambda \cos \lambda + T(\sin \lambda + \cos \lambda) \geq 0 \tag{4.52}$$

Using the values $m = 2$, $k_x = 0$, $k_y = 10$, and $T = \sqrt{2}$ gives the region plotted in Figure 4.6 and clearly shows the island.

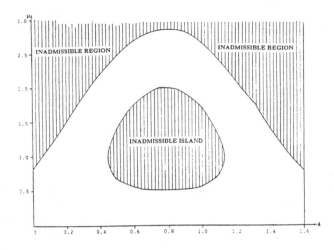

Figure 4.6: Admissible region for cartesian manipulator

V   The Dynamic Programming Method

The algorithms presented in the previous section are adequate for solving the minimum-time trajectory planning problem, provided that the actuator torque limits do not depend on the joint velocity in a complicated way and are independent of one another. However, in situations where driving the robot consumes large amounts of power, the assumption that minimum time is equivalent to minimum cost may not be valid; then the phase plane algorithm gives a solution to the wrong problem. Interdependence of actuator torque constraints is another very real possibility that the phase plane method cannot handle. This interdependence may occur, for example, when a robot uses a common power supply for the servo amplifiers for all joints, or when all joints of a hydraulic robot are driven from a common pump. Finally, it is assumed that the joint torques can be changed instantaneously. This is only approximately true, and indeed it is sometimes desirable to limit the derivatives of the joint torques (or, equivalently, the jerk, or derivative of the acceleration) to prevent excessive mechanism wear.

One means of eliminating these limitations is to use a more general optimization technique. The method proposed here is to use dynamic programming [14] to find the optimal phase plane trajectory. The dynamic programming technique places few restrictions on the cost function that is to be minimized. Putting limits on jerk is also (theoretically) possible[31], and interdependence of torque bounds can be handled fairly painlessly, as will be seen later. One of the major drawbacks of dynamic programming, the "curse of dimensionality", is not an issue in the trajectory planning problems considered here, since the use of parametric functions (Eq.(3.2)) reduces the dimension of the state space from $2n$ for an n-jointed manipulator to two.

A Problem Formulation

As before, the manipulator is assumed to move along a path given by Eq. (3.2), and the dynamics are written in the form of Eq. (3.5). Initally, it is also assumed that the set of realizable torques can be given in terms of the state of the system, i.e., in terms of the robot's position and velocity. Then we have

$$u = (u_1, u_2, \cdots, u_n)^T \in E(q, \dot{q}) \tag{5.1}$$

where $\dot{q}$ is the first derivative of q with respect to time, and u_i is the ith actuator torque/force. E is a function from $R^n \times R^n$ to the space of sets in R^n. In other words, given the position and velocity, E determines a set in the input space. The input torques u_i are *realizable* for position q and velocity $\dot{q}$ if and only if the torque vector u is in the set $E(q, \dot{q})$. Note that independence of the actuator torque limits is *not* assumed.

The cost of C given by Eq. (2.1) may be transformed into

$$C = \int_0^{\lambda_{max}} L(\lambda, \mu, u_i) d\lambda \tag{5.2}$$

The minimum-cost trajectory planning problem then becomes that of finding $\mu = \mu^*(\lambda)$ which minimizes Eq. (5.2) subject to Eq. (3.5) and the inequalities Eq. (5.1).

B Trajectory Planning Using Dynamic Programming

To see how dynamic programming can be applied to this problem, first note that by using the parameterized path Eq. (3.2), the dimensionality of the problem has been reduced; there will be only <u>two</u> state variables λ and μ, *regardless* of how many joints the robot has. The "curse of dimensionality" has therefore been avoided. To apply dynamic programming, one first must divide the phase plane into a discrete grid. Then, the cost of going from one point on the grid to the next must be calculated. Note that since u_i will be determined as a function of λ and μ, Eq. (5.2) can be written strictly in terms of λ and μ; thus the cost computation can be done entirely in phase coordinates. Once costs have been computed, the usual dynamic programming algorithm can be applied, and positions, velocities and torques can be obtained from the resulting optimal trajectory and Eq. (3.2) and Eq. (3.5).

The informal description given above describes the general approach to the MCTP problem. In detail, there are some complications. Therefore, some simplifying but realistic assumptions will be made as we proceed. First, the grid's

λ-divisions are assumed to be small enough so that the functions M_i, Q_i, R_i, S_i, and $\frac{df^i}{d\lambda}$ do not change significantly over a single λ-interval. Then, the coefficients of Eq. (3.5) are effectively constant. So are the coefficients of Eq. (4.18), which we will use as our (single) dynamic equation. (Other assumptions, such as piecewise linearity, are possible. However, these assumptions complicate the analysis considerably). Note that Eq. (4.18) does not explicitly depend on time. Therefore, for the purpose of carrying out the dynamic programming algorithm, we may treat the quantities λ and μ as a stage variable and a single state variable rather than two state variables. Using Eq. (4.18) as our (single) dynamic equation, and noting that M, Q, R, and S are approximately constant over one λ-interval, we need to find a solution to Eq. (4.9) which meets the boundary conditions

$$\mu(\lambda_k) = \mu_0, \quad \mu(\lambda_{k+1}) = \mu_1 \tag{5.3}$$

in the interval $[\lambda_k, \lambda_{k+1}]$. In order to do this, some form for the inputs u_i needs to be chosen. It should be noted that as the DP grid becomes finer, the precise form of the curves joining the points of the grid matters less. As long as the curves are smooth and monotonic, the choice of curves makes a smaller and smaller difference as the grid shrinks. The implication of this is that we may choose virtually any curve that is convenient, and as long as the grid size is small, the results should be a good approximation to the optimal trajectory.

We will use the form

$$u_i = Q_i \mu^2 + R_i \mu + V_i \tag{5.4}$$

for the input, where the V_i are constants that may be chosen to make the solution meet the boundary conditions Eq. (5.3). The form giben by Eq. (5.4) was chosen because it yields particularly simple solutions.

In what follows, we first obtain a solution without torque bound iteration, and then extend the solution to accommodate torque constraints of a much more general type.

C   Case of Non-Interacting Torque Bounds

When the joint torque bounds do not interact, the sets E in Eq. (5.1) are given by

$$E(q, \dot{q}) = \left\{ (u_1, \cdots, u_n)^T \mid u^i_{\min}(q, \dot{q}) \le u_i \le u^i_{\max}(q, \dot{q}) \right\} \tag{5.5}$$

Taking the projection of the input torque vector, as given by Eq. (5.4), onto the velocity vector $\frac{df^i}{d\lambda}$ gives $U = Q\mu^2 + R\mu + V$, where $V = V_i \frac{df^i}{d\lambda}$. Plugging this into the differential Eq. 4.18 gives

$$M \frac{d\mu}{d\lambda} + Q\mu + R + \frac{1}{\mu}(S - Q\mu^2 - R\mu - V) = 0 \tag{5.6}$$

or

$$\frac{d\mu}{d\lambda} = -\frac{1}{\mu}\frac{(S-V)}{M}.$$ (5.7)

Solving this equation, we have

$$\lambda = K - \frac{M}{2(S-V)}\mu^2.$$ (5.8)

Evaluating the constant of integration K and the constant V so that Eq. (5.8) meets the boundary conditions given by Eq. (5.3), one obtains

$$\lambda = \frac{\lambda_k(\mu_1^2 - \mu^2) + \lambda_{k+1}(\mu^2 - \mu_0^2}{\mu_1^2 - \mu_0^2}$$ (5.9)

Solving for μ in terms of λ gives

$$\mu = \sqrt{\frac{(\lambda_{k+1} - \lambda)\mu_0^2 + (\lambda - \lambda_k)\mu_1^2}{\lambda_{k+1} - \lambda_k}}$$ (5.10)

Now that the path is known over one λ-interval, we need to know the inputs u, and the components of the incremental cost.

To evaluate the input torques, we may use Eq. (3.5) and value of μ. Noting that $\dot\mu \equiv \frac{\dot\mu}{\mu} \cdot \mu \equiv \mu\frac{d\mu}{d\lambda}$ and using Eq. (5.7), we obtain

$$\dot\mu = \frac{V - S}{M} - constant.$$ (5.11)

The quantities M and S are given, and using Eq. (5.12) and Eq. (5.11), V can be calculated to be

$$V = S + \frac{M}{2}\cdot\frac{\mu_1^2 - \mu_0^2}{\lambda_{k+1} - \lambda_k}$$ (5.12)

which gives

$$\dot\mu = \frac{\mu_1^2 - \mu_0^2}{2(\lambda_{k+1} - \lambda_k)}$$ (5.13)

Therefore, the equations for u_i become

$$u_i = Q_i\mu^2 + R_i\mu + S_i + M_i \cdot \frac{\mu_1^2 - \mu_0^2}{2(\lambda_{k+1} - \lambda_k)}.$$ (5.14)

Assuming the joint torque limits are independent, determining whether joint i ever demands any unrealized torques requires that we know the maximum values of u_i over the interval $[\lambda_k, \lambda_{k+1}]$. Since λ is a monotonic function of μ over the interval under consideration, we may instead find the maximum and minimum values over the interval $[\mu_{\min}, \mu_{\max}]$, where $\mu_{\min} = \min(\mu_0, \mu_1)$ and

$\mu_{\text{max}} = \max(\mu_0, \mu_1)$. The maxima/minima may occur at one of three μ values, namely μ_0, μ_1, and that value of μ that maximizes or minimizes u_i over the unrestricted range of μ. In the latter case, the value of μ is

$$\mu_m = -\frac{R_i}{2Q_i}. \tag{5.15}$$

If the condition

$$\min(\mu_0, \mu_1) \leq \mu_m \leq \max(\mu_0, \mu_1) \tag{5.16}$$

holds, then the point μ_m needs to be tested. Otherwise the torques must be computed and checked only at the end points of the interval. (If M_i, Q_i, R_i, and S_i were assumed to be piecewise linear in λ, then the formula analogous to Eq. (5.14) would be a quartic rather than a quadratic, and in theory three "midpoints", found by solving a cubic, would have to be tested. But as a practical matter, testing the endpoints of the interval is probably adequate).

Given the formulae for the velocity and the joint torques, the incremental cost can be found using the formula

$$C = \int_{\lambda_k}^{\lambda_{k+1}} L(\lambda, \mu, u_i) d\lambda \tag{5.17}$$

where μ and u_i are given as functions of λ by Eq. (5.10) and Eq. (5.14), respectively. It may be possible to evaluate this integral directly; if not, then the integral may be approximated by any of the standard techniques. Section G shows that the DP algorithm converges when the integral is approximated using the Euler method. Using more sophisticated algorithms should give faster convergence than the Euler mehtod.

With these formulae at hand, it is now possible to state the dynamic programming algorithm in detail. The algorithm, given the dynamics (Eq. (3.5)), the equations of the curve (Eq. (3.2)), the joint torque constraints (Eq. (5.5)), and the incremental cost (Eq. (5.17)), is:

1. Determine the derivatives $\frac{df^i}{d\lambda}$ of the parametric functions $f^i(\lambda)$, and from these quantities and the dynamic equations determine the coefficients of Eq. (3.5).

2. Divide the (λ, μ) phase plane into a rectangular grid with N_λ divisions on the λ-axis and N_μ divisions on the μ-axiz. Associate with each point (λ_m, μ_n) on the grid a cost C_{mn} and a "next row" pointer P_{mn}. Set all costs C_{mn} to infinity, except for the cost of the desired final state, which should be set to zero. Set all the pointers P_{mn} to null, i.e., make them point nowhere. Set the column counter α to N_λ.

3. If the column counter α is zero, then stop.

4. Otherwise, set the current-row counter β to 0.

5. If $\beta = N_\mu$, to to 12.

6. Otherwise, set the next-row counter γ to 0.

7. If $\gamma = N_\mu$ go to 11;

8. For rows β and γ, generate the curve that connects the $(\alpha-1,\beta)$ entry to the (α,γ) entry. For this curve, test, as described in the previous papagraphs, to see if the required joint torques are in the range given by inequalities Eq. (5.5). If they are not, go to 10.

9. Compute the cost of the curve by adding the cost $C_{\alpha\gamma}$ to the incremental cost of joining point $(\alpha - 1,\beta)$ to point (α,γ). If this cost is less than the cost $C_{\alpha-1,\beta}$, then set $C_{\alpha-1,\beta}$ to this cost, and set the pointer $P_{\alpha-1,\beta}$ to point to that grid entry (α,γ) that produced the minimum cost, i.e.set $P_{\alpha-1,\beta}$ to γ.

10. Increment the next-row counter γ and go to 7.

11. Increment the current-row counter β and go to 5.

12. Decrement the column counter α and go to 3.

Finding the optimal trajectory from the grid is then a metter of tracing the pointers P_{mn} from the initial to the final state. If the first pointer is null, then no solution exists; otherwise, the successive grid entries in the pointer chain give the optimal trajectory. Given the optimal trajectory, it is then possible to calculate joint positions, velocities, and torques.

D   Case of Interacting Torque Bounds

It has been asumed in the preceding discussion that the joint torque limits do not interact, i.e., that increasing the torque on one joint does not decrease the available torque at another joint. This assumption manifests itself in the form of the torque constraint inequalities (5.5). This assumption is probably correct in many cases, but in others it certainly is not. Consider, for example, a robot that has a common power supply for the servo amplifiers for all joints. The power source will have some finite limit on the power it can suppply, so that the sum of the power consumed by all the joints must be less than that limit. A similar situation arises when a single pump drives several hydraulic servoes. The pump will have finite limits on both the pressure and the volume flow it can produce. In fact, it may be desirable to have torque limits interact, in the sense that using a single power source to drive the robot may cost less than using several independent power sources. In most cases multiple power sources would not be used at their maximum capacities simultaneously, so that there would be little to be gained by having independent power sources, while the cost of a single large power source would quite possibly be considerably less than that of several small ones.

Because of the possibility of torque limit interaction, it will be assumed here that the inequalities Eq. (5.5) are replaced with the constraint Eq. (5.1), namely $(u_1, u_2, \cdots, u_n)^T \in E(q, \dot{q})$. It is interesting to note that the limits described in the previous papagraph all produce set functions E in Eq. (5.1) which are *convex*. For example, if the sum of the power consumed (or produced) in all the joints is bounded, one obtains the bounds

$$P_{\min} \leq u_i \dot{q}^i \leq P_{\max}. \tag{5.18}$$

For any given velocity, this is just the region between a pair of parallel hyper-planes in the joint space. Likewise, for independent torque bounds, the realizable torques are contained in a hyper-rectangular prism, another convex region. Since the intersection of any number of convex sets a convex set, any combination of these constraints will also yield a convex constraint set. In this light, it is reasonable to make the assumption that the set $E(q, \dot{q})$ is convex. This assumption is important in the analysis that follows.

To see how we may make use of this convexity condition, consider the test for realizability of torques used in the method presented thus far. This test made explicit use of the assumption that the torque bounds do not interact. In order to handle interacting torque bounds using an approach like that in Section C, it must be possible to determine whether *all* torques are realizable over any given λ-interval. If the torques have the form used in Eq. (5.4), then this is in general not possible with any finite number of tests; even in the two-dimensional case, the torques trace out conics in the input space, and there is no general way to determine whether a segment of a conic is entirely contained within a convex set.

Though the question of whether a set of torques is realizable cannot in general be given a definite answer, the realizability question can be answered in some cases. To see how this can be done, consider again the tests for realizability previously described. The maximum and minimum torques for each joint are determined, and these torques are checked. While Eq. (5.4) describes a curve in the joint torque space, the individual torque limits describe a box-shaped volume. The curve describing the joint torques will be entirely contained inside this box. Thus if every point in the box is admissible, then so is every point on the curve. This "reduces" the problem of determining whether every point of a higher-dimensional set is realizale to the problem of determining whether every point of a higher-dimensional set is realizable. However, this higher-dimensional set has a special shape; it is a convex polyhedron, and will be contained in the (convex) set E if and only if all its vertices are in E. Thus by testing a finite number of points, the question of whether a particular set of torques is realizable may sometimes be given a definite "yes" answer.

If this test does not give a definite answer, then the set of inputs in question must be discarded, even though that set may in fact be realizable. however, as the grid size shrinks, the size of the bounding box for the torques also shrinks, so that in the limit the test becomes a test of a single point. Therefore as the grid shrinks, the percentage of valid torques thrown away approaches zero, and the optimal solution will be found.

This method of handling interacting torque bounds requires only one change in the DP algorithm. Step 8, which checks to see if the torques are realizable, must be replaced with a step that generates all corners of the bounding box and tests these points for realizability. If any of the corners does not represent a realizable set of torques, then the test fails. Thus we have

8' For rows β and γ, generate the curve that connects the $(\alpha - 1, \beta)$ entry to the (α, γ) entry. For this curve, generate the maximum and minimum torques at each joint. Check each torque n-tuple formed from the maximum and minimum joint torques. (These are the corners of the bounding box). If any of these n-tuples are not contained in the set E, then go to 10.

E   Algorithm Complexity

The usefulness of the dynamic programming technique depends on its being reasonably efficient in terms of use of computing resources, i.e., it must run reasonably fast and must not use too much memory. Since the trajectory planning is done offline, the algorithm's time requirements are not particularly critical; nevertheless, the time required must not be exorbitant if trajectory planning is to be worthwhile. Likewise, computer memory is relatively inexpensive, but nevertheless puts some limits on the accuracy with which the dynamic programming algorithm can be performed. In this section we present an approximate analysis of the time and memory requirements of the algorithm. Of course, precise numbers will depend rather heavily upon such variables as the computer on which the algorithm is to run, the language in which it is implemented, the compiler used, and the skill of the programmer who writes the code, so the expressions derived here contain a number of implementation-dependent constants.

It is easy to compute the storage requirements for the algorithm. The memory allotted to the program itself is essentially fixed. The size of the grid used for the dynamic programming algorithm varies with the fineness of the grid and the amount of storage required per point on the grid. The grid has N_μ rows and N_λ columns. Each entry must contain a cost C and a pointer P. The size of an entry will then be

$$GS = S_c + S_p \tag{5.19}$$

where GS is the storage requirement for a single point of the grid and S_c and S_p are the amounts of storage required to record the cost and the pointer to the next row, respectively. In the implementaiton presented here, parameterized curves are represented as arrays of points. If one assumes that there is one point per λ-division, then there is an additional $S_d + N_\lambda S_i$, where S_i is the storage required for one interpolation point on the curve and S_d is a certain fixed storage per curve. Multiplying GS by the number of grid entries and adding the amount of storage PS required for the program and the storage required for the curve gives total storage TS as

$$TS = PS + N_\lambda N_\mu (S_c + S_p) + N_\lambda S_i + S_d. \qquad (5.20)$$

For the numerical example presented in this paper, all arithmetic was done in double precision, and integers and pointers are four bytes long. Then for a six-jointed arm the storage required is, ignoring the program storage,

$$TS = 12N_\lambda N_\mu + 80 + 448N_\lambda \qquad (5.21)$$

For example, a 20 × 80 grid requries 28,240 bytes. This can, of course, be reduced considerably by using single rather than double precision; however, even using double precision, the storage required is generally available on small microprocessors.

Calculating the time required to perform the dynamic programming algorithm is somewhat more difficult. There will be $N_\lambda - 1$ steps, where each step requires testing to see if each of the N_μ points in one column can be connected to each of the N_μ points in the next column. Each test must be done, but some of the tests are simpler than others. If the cost at the next grid point is infinite, then there is no point in doing any further calculations. If on the other hand the cost is finite, then input torque bounds must be checked, and if the input torques are admissible, then costs must be calculated and compared. Though actual computation times will vary with the particular problem being solved, the way the time varies with grid size can be roughly determined. To get a bound on this time, assume that *all* the tests and computations must be performed. Then each step of the dynamic programming algorithm requires $K N_\mu^2$ seconds, where K is quantity which depends upon the computer being used and the number of joints the robot has. There are $N_\lambda - 1$ such steps, so the time required is less than $K(N_\lambda - 1)N_\mu^2$. In other words, the execution time is roughly proportioned to the cube of the grid density. In practice, the value of the constant K must be evaluated experimentally. This has been done for the numerical example in Section H, which does indeed show a time dependence proportional to $(N_\lambda - 1)N_\mu^2$.

The dependence of execution time on the number of joints n, i.e., the dependence of the constant K on n, is more difficult to access. K in the equation above depends on both n and the representation used to describe the curve to be traversed. The functions M_i and R_i depend on the matrices J_{ij} and R_{ij} respectively, and the Coriolis term Q_i depends on the three-dimensional array C_{ijk}. In general, then, it might be expected that the evaluation of the function Q_i might take time proportional to the cube of the number of joints. (See, for example, [14]). In any case, the time required for evaluation of the dynamic coefficients is heavily dependent upon the configuration of the robot. Fortunately, in practical cases the number of joints would usually be no more than six, and almost certainly would be less than eight. Since these functions only need to be evaluated once per λ-division of the DP grid, their evaluation will probably be only a minor part of the total time consumed. This being the case, the dependence of execution time on n is not an important factor. (For the numerical example considered here, this is certainly true.)

If the algorithm for handling interacting joints is used, then the dependence of the time on the number of joints increases exponentially with the number of joints, since there are 2^n corners on the bounding box for the input. While this would seem to make the algorithm useless, it should be noted that the size of the bounding box decreases as the grid size shrinks. In practice it may be sufficient to test, for example, the end points of the current segment, rather than all 2^N corners.

F   Algorithm Speedup

Even though solution of the trajectory planning problem by dynamic programming requires only a two-dimensional grid, the algorithm uses large amounts of computer time when the grid gets fine enough to give accurate answers. Part of the reason for this is the exhaustive testing of paths in the grid; when connecting points one column to points in the next, *all* pairs of points are tested. Also, the dynamic programming algorithm generates the optimal trajectories from the *all* points in the grid to the desired goal state. If we can avoid generation of the unused trajectories, considerable speedup should result.

The approach described here involves multiple iterations of the dynamic programming algorithm using a sparse, irregularly spaced grid. Suppose that an approximate solution to the trajectory planning problem is available, say as a result of using dynamic programming with a coarse grid. Then we may plot this approximation in the phase plane, and draw a "swath" around it, indicating the uncertainty of the solution. Then, instead of superimposing a grid on the entire phase plane, we may superimpose a grid on the uncertainty swath. This grid may have a small, fixed number of μ values for each λ value, so that the cost of doing dynamic programming on this grid is relatively low.

If the original trajectory met all the constraints, and for each λ value on the grid one of the grid points is on the original trajectory, then solution will certainly be found when the dynamic programming algorithm is performed. Assuming that the gird includes the points on the upper and lower limits of the swath, the resulting trajectory then must either stay entirely within the swath, or touch the swath's edge. Now a new swath should be drawn, centered on the new optimal trajectory. If the new optimal trajectory touches the edge of the old swath, then the new swath should have the same grid density. If it doesn't touch the edge, then the size of the new swath should be decreased. This process may be repeated until the swath is narrow enough to guarantee that the solution is within desired accuracy limits.

Roughly speaking, the algorithm finds an approximate solution in a reduced search area. If the solution touches the boundary of the search area, then the search area boundary is moved away from the solution. If the solution does not touch the edge of the search area, then a new smaller search area is tried, resulting in a more accurate approximation. By limiting the dynamic programming algorithm's attention to a small area of the phase plane, computation times are kept correspondingly small.

This technique will not be used in any of the examples worked in this section, but a related technique will be described in Section VI.

G   Convergence Properties

The previous section describes the complexity of the DP algorithm. It is obvious from the discussion that the fineness of the DP grid will have a significant impact on the running time of the algorithm. It will also affect the accuracy of the results. This section describes the effect of the grid density on the accuracy of the DP solution in a quantitative manner.

Bellman proved in [1] that discrete approximations to a continuous optimal control problem will converge (in a sense to be defined) as the step size of the DP stage variable decreases. However, the class of systems to which Bellman's proof applies does not cover those considered in this paper. In particular, Bellman assumes that the dynamic equations of the system are not functions of the stage variable, which is the same as λ in this paper. Here we prove a theorem which is an extension of that of Bellman in that it allows the dynamic equation and cost function to be (possible discontinuous) functions of the stage variable. The proof presented here also corrects some minor errors in Bellman's proof.

Like Bellman's proof in [1], we will prove that a sequence of discrete dynamic programming processes with decreasing step sizes will produce, under appropriate conditions, a convergent sequence of return functions. It should be noted that the optimal control policy may not converge even though the return functions do. But since the return function is of primary interest, not the details of the control policy, control policy convergence is not generally important.

From the discussion thus far it is clear that the manipulator dynamics and required constraints take the form:

$$\frac{d\mu}{d\lambda} = G(\lambda, \mu, v) \tag{5.22}$$

where the function G is piecewise continuous in λ and has a finite number N_d of discontinuities, provided the path consists of a finite number of piecewise smooth segments. The control variable v (this is the same as $\dot{\mu}$ in the previous discussions) must meet some set of constraints:

$$\Omega_q(\lambda, \mu, v) \leq 0, q = 1, 2, \cdots, M. \tag{5.23}$$

Also rewrite the objective function $J = -C$ to be maximized as follows.

$$J(v) = \Theta(\mu(\lambda_{\max})) + \int_0^{\lambda_{\max}} F(\lambda, \mu, v) d\lambda \tag{5.24}$$

subject to the initial condition $\mu(0) = \mu_0$. Note that the boundary condition $\mu(\lambda_{\max}) = \mu_f$ can be enforced by taking $\Theta(\mu(\lambda_{\max}))$ to be zero if $\mu = \mu_f$ and $-\infty$ otherwise. The dynamic programming method approximates this continuous problem by discretizing the dynamic equation and objective function using the Euler method, giving:

$$\mu_{k+1} = \mu_k + G(\lambda_k, \mu_k, v_k)\Delta \tag{5.25}$$

$$J(\{v_k\}) = \Theta(\mu_N + \sum_{k=0}^{N-1} F(\lambda_k, \mu_k, v_k)\Delta \tag{5.26}$$

where N is the number of subintervals into which we divide the interval $[0, \lambda_{\max}]$, $\Delta = \lambda_{\max}/N$, $\lambda_k = k\Delta$, $\mu_k = \mu(\lambda_k)$, $v_k = v(\lambda_k)$, and the inputs v_k are constrained by

$$\Omega_q(\lambda_k, \mu_k, v_k) \leq 0, \quad q = 1, 2, \cdots, C. \tag{5.27}$$

Now define $f_n(c)$ for $n = 0, 1, \cdots, N$ by

$$f_n(c) = Sup_{\{v_k\}} \left[\Theta(\mu_N) + \sum_{k=N-n}^{n-1} F(\lambda_k, \mu_k, v_k)\Delta \right], \mu_{N-n} = c \tag{5.28}$$

Then we have

$$f_0(c) = \Theta(c) \tag{5.29}$$

$$f_{n+1}(c) = Sup_v \left[F(\lambda_{N-n-1}, c, v,)\Delta + f_n(c + G(\lambda_{N-n-1}, c, v)\Delta) \right] \tag{5.30}$$

Note that Sup has been used instead of max. This is done to allow the use of nonclosed constraint sets and discontinuous functions. It does not materially change the results of the dynamic programming process in that we may make the return function f_n as close to the optimal value as we please. To see this, consider a single stage of an N-stage process. For each k and $\epsilon > 0$, we may make f_k to be within $\epsilon/2^k$ of its optimal value, thus making f_N be within 2ϵ of the optimum. Since ϵ may be a small as we please, a control strategy can be constructed which will make the return function agree with the optimum value to within any desired tolerance.

Given this form for the dynamic programming problem, we have the following theorem:

Theorem 3: Let the input v satisfy $0 \leq v \leq 1$, and let F and G satisfy the Lipschitz conditions

$$|F(\lambda_1, c_1 v) - F(\lambda_2, c_2, v)| \leq K|c_1 - c_2|^\alpha + B|\lambda_1 - \lambda_2|^\beta$$
$$|F(\lambda, c_1, v) - F(\lambda, c_2, v)| \leq K|c_1 - c_2|^\alpha$$
$$|G(\lambda_1, c_1, v) - G(\lambda_2, c_2, v)| \leq L|c_1 - c2|^\gamma + C|\lambda_1 - \lambda_2|^\delta$$
$$|G(\lambda, c_1, v) - G(\lambda, c_2, v)| \leq L|c_1 - c_2|^\gamma$$

where $\alpha > 0, \gamma \geq 1, \beta \geq 0$, and $\delta \geq 0$, for all admissible λ and v, for all $\mu_{\min} \leq c_1, c_2 \leq \mu_{\max}$, and for all λ_1, and λ_2 such that the interval $[\min(\lambda_1, \lambda_2), \max(\lambda_1, \lambda_2)]$ does not contain any of the N_d points of discontinuity $d_1, d_2, \cdots, d_{N_d}$. Also let F and G satisfy

$$|F(\lambda_1, c_1, v) - F(\lambda_2, c_2, v)| \leq K|c_1 - c_2|^\alpha + B$$

and

$$|G(\lambda_1, c_1, v) - G(\lambda_2, c_2 v)| \leq L|c_1 - c_2|^\gamma + C$$

for all admissible $\lambda_1, \lambda_2, c_1, c_2$, and v. Then the discrete dynamic programming process yields return functions which converge to a limit as the step size Δ goes to zero, provided that μ_k can be guaranteed to stay in the interval $[\mu_{\min}, \mu_{\max}]$.

Proof: The proof is too lengthy to be included here. The interested reader is referred to [25].

Roughly speaking, this theorem shows that when the functions F and G are continuous of sufficiently high order in μ and piecewise continuous in the stage variable λ, then the discrete dynamic programming process converges.

One important point is that in order to apply the theorem, it must be guaranteed that μ_k stays within the range in which the Lipschitz conditions are valid. This can be accomplished in several ways. If, for example, $G(\lambda, \mu_{\max}, v) \leq 0$ and $G(\lambda, |\mu_{\min}, v) \geq 0$ for all λ, then the optimal trajectory can never excape the interval $[\mu_{\min}, \mu_{\max}]$. Another way to assure containment in this interval is to construct an objective function which guarantees that trajectories which stray outside the interval are heavily penalized and therefore never selected. This method works for the examples in this paper; since the examples all penalize time, they all have terms which are inversely proportional to velocity, and so keep the velocity μ greater than some $\epsilon > 0$.

H   Numerical Examples

To demonstrate the use of the dynamic programming algorithm, we present several examples. These examples use the first three joints of a cylindrical electrically-driven manipulator, the Bendix PACS arm. The second and third joints of this robot are similar to the degree-of-freedom robot used to demonstrate the phase plane method, so a direct comparison of the methods is possible in those cases in which the phase plane technique does not apply.

Before presenting the example, an explicit form of the objective function must be chosen. The objective function used here has one component proportional to traversal time, T, and another component proportional to frictional and electrical energy losses. The servo drives of the arm are assumed to consist of a voltage source in series with a resistor and an ideal DC motor. This gives the form

$$C = r_t T + r_e \int_0^T R_{ij} \dot{q}^i \dot{q}^j \, dt + r_e \int_0^T \sum_i I_i^2 R_i^m \, dt \qquad (5.31)$$

$$= r_t \int_0^{\lambda_{\max}} \frac{1}{\mu} d\lambda + r_e \int_0^{\lambda_{\max}} \mu^2 R_{ij} \frac{df^i}{d\lambda} \frac{df^j}{d\lambda} d\lambda \qquad (5.32)$$
$$+ r_e \int_0^T \frac{1}{\mu} \sum_i I_i^2 R_i^m \, d\lambda$$

where r_t and r_e are related to revenue generated per item and the energy costs of the motion respectively, and I_i and R_i^m are the motor currents and resistances for joint i. Since the joint torques u_i are related to the motor currents I_i by the relationships

$$u_i = \frac{k_i^m}{k_i^g} I_i \qquad (5.33)$$

where k_i^m and k_i^g are the motor constant and the motor gearing respectively, the sum in Eq. (5.33) can be written as $\sum_i \frac{(k_i^g)^2 R_i^m}{(k_i^m)^2} u_i^2$. The torques u_i are given in terms of λ, μ and $\dot{\mu}$ by Eq. (3.5), which is quadratic in μ, so the iintegral in Eq. (5.33) can be expressed in terms of integrals of powers of μ.

Since the path is known over one λ-interval, we can determine the components of the incremental cost. To do this, we need the integrals $\int_{\lambda_k}^{\lambda_{k+1}} \frac{1}{\mu} d\lambda$, $\int_{\lambda_k}^{\lambda_{k+1}} \mu \, d\lambda$,

$\int_{\lambda_k}^{\lambda_{k+1}} \mu^2 d\lambda$, and $\int_{\lambda_k}^{\lambda_{k+1}} \mu^3 d\lambda$. In general we have

$$\Phi_k \equiv \int_{\lambda_k}^{\lambda_{k+1}} \mu^n d\lambda = \int_{\lambda_k}^{\lambda_{k+1}} \left[\frac{(\lambda_{k+1} - \lambda)\mu_0^2 + (\lambda - \lambda_k)\mu_1^2}{\lambda_{k+1} - \lambda_k} \right]^{\frac{n}{2}} d\lambda \quad (5.34)$$

$$= \frac{\lambda_{k+1} - \lambda_k}{(1 + \frac{n}{2})(\mu_1^2 - \mu_0^2)}(\mu_1^{n+2} - \mu_0^{n+2})$$

Equations (5.33), (5.35), and (3.2) give the incremental cost as

$$C = 2r_t\Phi_{-1} + r_e R_{ij}\frac{df^i}{d\lambda}\frac{df^j}{d\lambda}\Phi_1 \quad (5.35)$$

$$+ r_e \sum_i \frac{(k_i^g)^2 R_i^m}{(k_i^m)^2}\left[Q_i^2\Phi_3 + 2Q_i R_i\Phi_2 \right.$$

$$+ (R_i^2 + 2Q_i(S_i + M_i\dot\mu))\Phi_1 + 2R_i(S_i + M_i\dot\mu)\Phi_0 + (S_i + M_i\dot\mu)^2\Phi_{-1} \Big]$$

It should be noted that this cost function always has a Φ_{-1} term unless the time penalty is zero and the robot is not influenced by gravity. (In this case, the optimal solution is not to move at all!) Therefore the cost function will prevent the trajectory from going to zero velocity unless forced by boundary conditions. Also, since torques and motor voltages are bounded, the velocity μ is bounded. Thus μ in the DP algorithm stays within some interval $[\mu_{\min}, \mu_{\max}]$, as is required for convergence. Since the maximum and minimum values of the control variable $\dot\mu$ can be computed from Eq.(3.2), we may define a new control variable p by the relationship

$$\dot\mu = \dot\mu_{\min}(\lambda, \mu) + (\dot\mu_{\max}(\lambda, \mu) - \dot\mu_{\min}(\lambda, \mu))p \quad (5.36)$$

so that p ranges from zero to one. It is easily shown that the other conditions of the covergence theorem are met if the parameterized curve is suitably well behaved, so the DP algorithm will converge with this cost function.

The dynamic equations and actuator characteristics for the PACS arm are given in [25]. The DP algorithm was implemented in the C programming language running under UNIX on a VAX11-780. The parameterized curves are represented as sequences of points. All computations are done for a path which is a straight line from $(0.7, 0.7, 0.1)$ to $(0.4, -0.4, 0.4)$, all (Cartesian) coordinates being given in meters.

To verify the correctness of the dynamic programming algorithm, it was first applied to the simple two-degree-of-freedom robot which was used in section 10×10 and a 40×40 DP grid are plotted in Figures 5.1 and 5.2, along with the phase plane plots calculated by the phase plane method. Reassuringly, the trajectories calculated by the DP method seem to converge to the correct minimum-time phase plane plot as the grid gets finer.

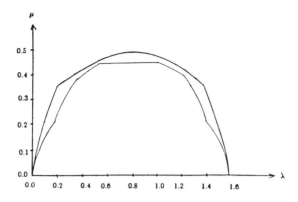

Figure 5.1: Minimum time phase plot for polar manipulator, 10×10 grid

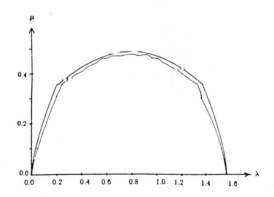

Figure 5.2: Minimum time phase plot for polar manipulator, 40×40 grid

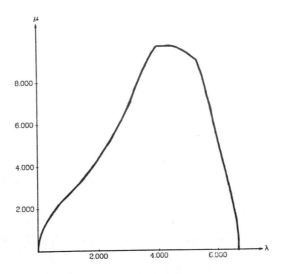

Figure 5.3: Minimum time phase plot for PACS arm

Figure 5.3 shows the phase plane plot for a 40×160 grid with a pure minimum time cost function. The calculated traversal times for 10×10, 20×20, and 40×160 grids are 2.000, 1.972, and 1.905 seconds respectively, compared to 1.782 seconds as calculated by the phase plane method.

Figure 5.4 shows the phase plane plot for a time penalty of 1 unit per second and an energy penalty of ten units per joule. Note that the trajectory is lower than that which is obtained if only time is penalized. The grid size is 20×40.

Figure 5.5 shows the phase plane plot for a minimum time trajectory when, in addition to the torque and voltage constraints, the total power sunk or sourced by the robot is limited to 2 kilowatts.

The time consumed by the algorithm was measured for several different grid sizes, with, however, 400 interpolation points on the curve regardless of grid size. Computation of the dynamic coefficients for 400 points usually took from 0.350 to 1.0 seconds of real time, so the computation time is probably about 0.35 seconds. The computation times for the dynamic coefficients vary as the cube of the number of joints. Thus for a robot with six degrees of freedom instead of three we would expect about eight times as much computation. Taking 100 times the 0.35 seconds, to be very conservative, gives 35 seconds for 400 points. But 400 interpolation points are hardly necessary for a grid with a value of 40 for $N\lambda$; 80 points would certainly be adequate, giving a computation time of about 7 seconds for the dynamic coefficients, not an unreasonable figure if the motion

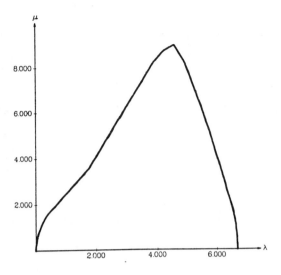

Figure 5.4: Phase plot for PACS arm, minimum time-energy

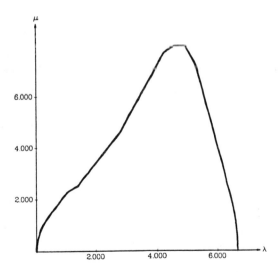

Figure 5.5: Minimum time phase plot for PACS arm with power limit

N_λ	N_μ	minimum	average	predicted
10	10	1.217	1.355	0.6
10	20	3.583	5.260	2.4
20	40	19.917	25.5	19.2
20	80	92.050	110.367	76.8
40	40	29.967	33.856	38.4

Table 5.1: Computation times for different grid sizes (seconds)

is to be repeated a large number of times.

The times given in Table 5.1 are real times for the DP algorithm on a lightly loaded system (average of approximately 2 tasks running concurrently). The times must therefore be regarded as approximations to the actual computation times. Table 5.1 also lists the function $6 \times 10^{-4} N_\lambda N_\mu^2$, which seems to give a good match to the actual running time, as predicted in Section E. It should be noted that the program was run with all debugging features enabled, and that no serious attempt was made to optimize the source code. Indeed, there are redundant computations in several places which could be eliminated.

The effect of grid size on the quality of the results of the dynamic programming algorithm is of practical importance. The grid must be fine enough to give good results but not so fine that the time required to perform the DP algorithm is excessive. Intuitively, varying the number of rows and number of columns in the grid will have different effects on the results. Varying the number of columns (the number of λ-divisions) varies the accuracy of the dynamic model; using a smaller size yields a more accurate approximation to the true dynamic model, since the "pieces" in the piecewise-constant dynamic model will be smaller. Varying the number of rows (the number of μ-divisions) varies the accuracy of the approximation to the true minimum-cost solution; a finer grid size will in general yield a better approximation.

Another important factor is the ratio of the number of rows to the number of columns. If the number of columns is very large and number of rows is very small, then the slope of the curves connecting one point in the DP grid to another must be either zero or very large. Since the torque bounds induce bounds on the slope of the curve, it is possible that the DP algorithm may not even be able to connect a point to its nearest diagonal neighbor. In this case, the alogrithm will no solution at all. Therefore when choosing grid size, one must (i) choose the λ-divisions to be small enough to make the piecewise-constant dynamic model of the robot sufficiently accurate, (ii) choose the μ-divisions to be small enough so that the resulting trajectory is a satisfactory approximation to the true minimum-cost trajectory, and (iii) make sure that the μ-divisions are small enough so that the slope of a curve connecting two adjacent points in the grid is small.

VI   Trajectory Planning by Iterative Improvement

It was pointed out in the previous section that the dynamic programming algorithm, particularly when used with very fine grid, wastes a great deal of computation time by testing admissibility criteria and evaluating cost functions for sections of trajectories which are not part of, and indeed are not even close to, the optimal trajectory. It was then suggested in Section F that reasonable computational accuracy could be obtained from the dynamic programming algorithm by performing the dynamic programming algorithm repeatedly using a grid with a small number of divisions of the state variable, while changing the grid spacing at each iteration. The trajectory planning method proposed here, the *pertubation trajectory improvement algorithm*, or PTIA, is a variation on this theme. In addition, the PTIA may be extended to include more general classes of torque constraints. The inclusion of jerk constraints in the algorithm will be demonstrated.

At the start of PTIA, it is assumed that one phase trajectory can be found which meets all constraints. (In practice, a trajectory with zero velocity usually suffices). This trajectory can then be perturbed to find the one with the shortest traversal time. This perturbation process is made particularly simple by the fact that minimizing time is equivalent to maximizing velocity, we wish to maximize μ, so the phase trajectory should always be pushed upward.

A   The Perturbation Trajectory Improvement Algorithm

In practice, a trajectory planner must deal with a variety of arbitrary parametric curves; two representations for curves which immediately suggest themselves are splines and simple sequences of points. As before, we choose to use the latter representation, i.e., the curve (4.1.1) is represented as an ordered sequence of points (λ_k, q_k); this proves to be the most natural representation for the application of the PTIA.

The trajectory planning process consists of assigning values of the "velocity" μ and "acceleration" $\dot{\mu}$ at each point. For the sake of simplicity, consider only those constraints which can be expressed in terms of position- and velocity-independent bounds on the torque, i.e., ignore jerk constraints for the time being. Then all constraints can ultimately be given as $\lambda-$ and μ-dependent constraints on $\dot{\mu}$, or equivalently constraints on $\frac{d\mu}{d\lambda}$. In terms of the (λ, μ) plot, each point is assigned a set of allowable slopes, In the discrete approximation, this sets limits on the differences between the values of μ at adjacent points. The process of trajectory planing requires that the initial and final points of the curve have zero velocity (or some other fixed velocity) and that the velocities at all the intermediate points be as large as possible, consistent with the constraint that the velocities at neighboring points not differ to much.

One approach to the solution of this problem is to try to push the speed higher at each individual point. The value of μ can be pushed higher at each point insuccession until none of the velocities can be made any larger. (This is just component-wise optimization, where the infinite-dimensional space of functions $\mu(\lambda)$ has been approximated by a finite-dimensional vector of real numbers.) If we call this Algorithm A, then we have

Algorithm A:

1. Set all velocities to values which are realizable (usually all zeroes).

2. Push each intermediate point of the curve as high as possible consistent with the slope constraints.

3. If any of the velocities were changed in step 2, go back to step 2, otherwise exit.

As a practical matter, the search required to find the highest possible velocity in step 2 of Algorithm A may be fairly expensive, especially since it may be repeated many times for a single point. A simpler approach is to just try adding a particular increment to each velocity, and then make the increment smaller on successive passes of the algorithm. This gives

Algorith A':

1. Set all velocities to values which are realizable (usually all zeroes).

2. Set the current increment to some large value.

3. Push each intermediate point of the curve up by an amount equal to the current increment, if this is consistent with the slope constraints.

4. If any of the velocities were changed in step 3, go back to step 3.

5. If the current increment is smaller than the desired tolerance, stop. Otherwise halve the increment and go to 3.

Algorithm A' is really just a combination of gradient and binary search techniques. The direction in which the curve must move (i.e., the gradient direction) is known *a priori*, since increasing the velocity always decreases the traversal time, and the amount of the change is successively halved, as in a binary search, until some desired accuracy is achieved. Algorithm A' is very simple, except possibly for the slope constraint check rquired in step 3. This requires a knowledge of the dynamics and actuator characteristics of the robot. However, this check is a simple "go/no go" check, and can be isolated as a single function call. (Hereafter this function will be called the *constraint function*.) Hence the trajectory planner can be used with other robots by changing a single, though possibly complicated, function.

An important characteristic of the constraint funciton is *locality*. In the case discussed above, the constraints are expressed in terms of λ, μ, and $\frac{d\mu}{d\lambda}$. We need two points to determine the slope $\frac{d\mu}{d\lambda}$, so the constraint depends only upon two points. Therefore when a point of the curve has its μ value changed, it is constrained only by the two adjacent points; the rest of the curve has no influence. This allows much calculation to proceed in parallel. Step 3 of Algorithm A' can be divided into two steps, one which increments the odd numbered points and one which increments the even numbered ones. Since the even numbered points stay the same while the odd numbered ones are being incremented, and vice versa, the points either side of the incremented points remain stationary, so that the constraint checks are valid. (If *all* points were tested simultaneously, then it is possible, for example, to increment two adjacent points; since in each case the constraint check would be made on the assumption that the other point was remaining stationary, it is possible that the new configuration would not meet the required constraints.)

It is easily seen that the process in Algrorithm A' can be extended to more complicated constraints. For example, constraints on the jerk (the derivative of the acceleration) only require a more complicated constraint function. Of course in this case the constraint function needs three points to calculate second derivatives of the speed. Thus the constraints on a single point will be functions of *two* points either side of the point being checked, rather than one point. This affects the degree of parallelism which can be achieved; step 3 would require three passes instead of two. It also may make the algorithm diverge as the number of points on the curve increases, as will be seen later from the numerical examples.

As a simple illustration of how the algorithm works, consider a simple one-dimensional problem. Suppose we wish to move an object of mass m from $x - 0$ to $x = 4$. Further suppose that there is no friciton, and that there are constant bounds on the magnitude of the applied force. There will be only one parametric function f, which may be taken to be the identity function, so that $\lambda = x$. We the have

$$F = ma = m\frac{d^2x}{dt^2} = m\frac{d^2\lambda}{dt^2} = m\frac{d\mu}{dt} = m\frac{d\mu}{d\lambda}\frac{d\lambda}{dt} = m\mu\frac{d\mu}{d\lambda}.$$

If we consider λ-intervals of length 1, then the discrete approximation to the parameterized "curve" will have 5 points. The acceleration $\dot\mu = \mu\frac{d\mu}{d\lambda}$ can be approximated as

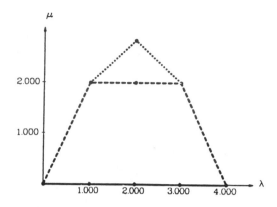

Figure 6.1: Results of Algorithm A.

$$\mu\frac{d\mu}{d\lambda} \approx \mu\frac{\mu_{i+1} - \mu_i}{\lambda_{i+1} - \lambda_i} \approx \frac{\mu_{i+1} + \mu_i}{2}\cdot\frac{\mu_{i+1} - \mu_i}{\lambda_{i+1} - \lambda_i} = \frac{\mu_{i+1}^2 - \mu_i^2}{2(\lambda_{i+1} - \lambda_i)}$$

The torque constraints then become

$$F_{\max} \geq |F| = m|a| =\sim m\left|\frac{\mu_{i+1}^2 - \mu_i^2}{2(\lambda_{i+1} - \lambda_i)}\right|$$

If we use $m = 1, F_{\max} = 2$, and $\lambda_{i+1} - \lambda_i = 1$, this reduces to $|\mu_{i+1}^2 - \mu_i^2| \leq 4$. Now consider what happens if Algorithm A is applied. We may look at the intermediate points of the curve in sequence. First, point 1 can be raised by 2, since the adjacent points have μ values of zero, and $|2^2 - 0^2| = 4$. Raising the middle point, point 2, we are constrained by the fact that $\mu_3 = 0$, which limits μ_2 to 2 also. Likewise, we may change μ_3 to 2. This completes step 2 of Algorithm A. Since some of the μ values changed, we try to increase them again. This time only point 2 can be raised, giving a value of $\mu_2 = 2\sqrt{2}$. On the next pass, no μ values change, so Algorithm A terminates. It is easily verified that the solution obtained from Algoriathm A is indeed the optimal solution to the discretized problem (Figure 6.1 shows the discretized trajectory after passes zero, one and two of Algorithm A.)

Now look at what happens when we use Algorithm A'. Say we start with an increment of 2. Then the result of the first pass of Algorithm A' is the same as the result of the first pass of Algorithm A, namely $\mu_1 = \mu_2 = \mu_3 = 2$. If the increment is cut to 1, then there is no change. Cutting the increment to 1/2, we may raise the middle point to 2.5. Continuing in this fashion, the middle point

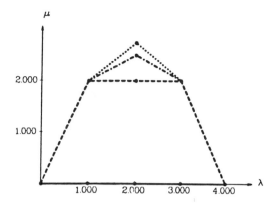

Figure 6.2: Results of Algorithm A'

gets closer and closer to $2\sqrt{2}$, the correct result. (Figure 6.2 shows the trajectory after passes zero, one, three, and four of Algorithm A'.)

B Computational Requirements

The PTIA, unlike dynamic programming, requires relatively little memory; it requires only one floating point number per interpolation point. However, computation of the CPU time requirements is interesting.

Obviously, the computation time must increase at least linearly with the number of interpolation points on the curve. In fact, the time increases as the square of the number of points. To see why this is so, consider what happens when the number of interpolation points is doubled. Since there are twice as many points to check on each pass of the algorithm, the computation time must increase by a factor of two. Recalling that the torque constraints translate into slope constraints, it is clear that the ratio of the amount by which a μ-value may be raised to the distance between λ-values will be approximately constant. Therefore halving the spacing of the interpolation points halves the size of the steps which can be taken in the μ direction, thus doubling the number of steps. This factor of two times the factor of two which results directly from doubling the number of points gives a factor of four increase in computation time. If doubling the number of interpolation points quadruples the computation time, then the time dependence is quadratic in the number of points, i.e., the time dependence is of the form $t = KN^2$, where t is the computation time, N is the number of interpolation points, and K is a constant.

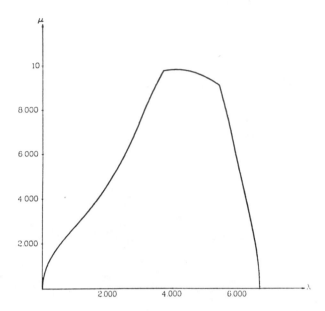

Figure 6.3: Phase plane plot for straight line

C   Numerical Examples

As an example, we again use the Bendix PACS arm. First we consider only constraints on joint torques/forces and motor voltages, without considering constraints on their derivatives. The perturbation trajectory planner was written in the C programming language, and run on a Vax-11/780 under the Unix operating system. The planner was tried with a straight-line path; the traversal time was 1.79 seconds. The phase plane plot is shown in Figure 6.3. Comparing this result to that obtained using the phase plane method shows the traversal time to be virtually identical.

To demonstrate the application of the perturbation technique to problems in which there are constraints on the derivatives of the torques, we consider the same problem with the additional constraint that the time derivatives of the joint torques and forces be less than 100. The time derivatives of the torques are computed using the idenity

$$\frac{du_i}{dt} = \frac{du_i}{d\lambda}\frac{d\lambda}{dt} = \frac{du_i}{d\lambda}\mu$$

The derivative $\frac{du_i}{d\lambda}$ was estimated by calculating the difference between the applied torques on successive intervals and dividing by the average of the lengths

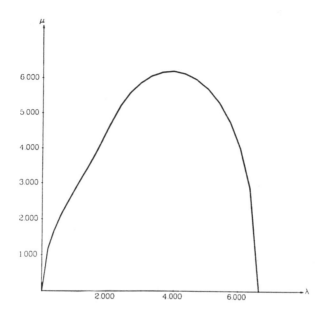

Figure 6.4: Phase plane plot for straight line with jerk constraint, 25 points

of the intervals. For a straight-line path with 25 interpolation points, the traversal time is 2.04 seconds. The μ vs. λ plot is shown in Figure 6.4. For 50 points, the traversal time is 2.26 seconds; the phase plane plot is shown in Figure 6.5.

Note that the trajectory has a "bump" in it; the process has not converged to the proper solution. To understand why this happens, consider the situation shown in Figure 6.6. The solid line shows the current trajectory, and the dashed lines show what happens when either of the two interior points is raised. In either case, a jerk limit is exceeded, even though the jerk constraint would very possibly be met if *both* points were raised simultaneously. Neither point can move before the other does, resulting in a sort of "deadlock". Similar situations can occur with longer sequences of points. If jerk constraints are to be included, then obviously we must prevent this sort of situation from occurring. One means of achieving this goal is to perform the trajectory planning operation several times with some added constraints, relaxing the constraints each time the trajectory is "improved". The constraints used here were simple velocity limits. On each pass, the velocity limit is raised. If the velocity increment is small enough, the top of the phase trajectory remains flat, and the regions of high inflection which cause the anomalies in the phase trajectory never get a chance to appear. With this modification, a velocity increment of 0.025 at each pass gives the phase plane plot in Figure 6.7 for a straight line with 100 points. The calculated traversal time is 2.03 seconds.

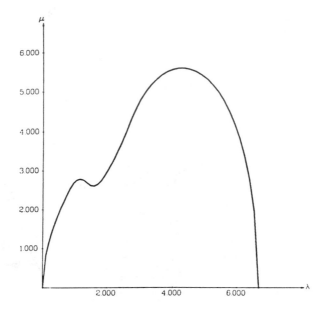

Figure 6.5: Phase plane plot for straight line with jerk constraint, 50 points

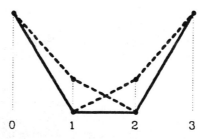

Figure 6.6: Illustration of deadlock caused by jerk constraints

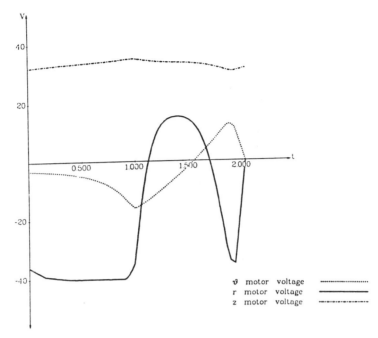

Figure 6.7: Phase plane plot for straight line with jerk constraint, 100 points, velocity increment 0.025

The numerical results obtained using the perturbation trajectory improvement algorithm agree with those obtained by the phase plane method in those cases in which both algorithms can be applied. While the iterative improvement algorithm is not as fast as the phase plane method and not as general as dynamic programming, it is extremely flexible and very easy to program. It is also possible to use this method when there are uncertainties in the robot's dynamics[32].

VII Conclusions

The optimal control problem for robot manipulators can be divided into three parts: geometric path planning, trajectory planning, and tracking. Geometric path planning is the process of creating a curve which the robot is to follow; this curve must avoid obstacles in the workspace. The trajectory planner then assigns timing information to the prescribed curve. Finally, the path tracker attempts to follow the geometric path at the speed which has been determined by the trajectory planner. This chapter has presented three solutions to the optimal trajectory planning problem: the phase plane method, dynamic programming, and the perturbation trajectory improvement algorithm. Each algorithm has strengths and weaknesses.

The phase plane algorithm produces only time-optimal trajectories and it cannot accommodate jerk constraints, but it is very fast and efficient. Dynamic programming, on the other hand, can optimize with respect to very complicated cost functions, but is much more costly when high accuracy is required. Though dynamic programming can produce limited-jerk trajectories, it appears to be difficult to do in practice. The perturbation trajectory improvement algorithm also only produces minimum time trajectories, but it can accommodate jerk constraints very naturally, and the algorithm is very simple. In terms of performance, it is intermediate between the other two methods.

References

[1] R. Bellman. Functional equation in the theory of dynamic programming - vi, a direct convergence proof. *Annals of Mathematics*, 65(2):215–223, March 1957.

[2] J.E. Bobrow, S Dubowsky, and J.S. Gibson. On the optimal control of robotic manipulators with actuator constraints. In *Proceedings of the 1983 American Control Conference*, pages 782–787, June 1983.

[3] M.J. Chung and C.S.G. Lee. An adaptive control strategy for computer-based manipulators. Technical Report RDS-TR-10-82, University of Michigan centr for Robotics and Integrated Manufacturing, Ann Arbor, Michigan, August 1982.

[4] J.J. Craig. *Introduction to Robotics: Mechanics and Control*. Addison-Wesley, Reading, Mass., 1986.

[5] S. Dubowsky. On the adaptive control of robotic manipulators: The discrete-time case. In *Proceedings of the Joint Automatic Control Conference*, pages section TA–2B, June 1981.

[6] S. Dubowsky and D. T. DesForges. The application of model-referenced adaptive control to robotic manipulators. *ASME Journal of Dynamic Systems, Measurement, and Control*, pages 193–200, September 1979.

[7] E.G. Gilbert and D.W. Johnson. Distance functions and their application to robot path planning in the presence of obstacles. Technical Report RSD-TR-7-84, University of Michigan Center for Robotics and Integrated Manufacturing, Ann Arbor, MI, July 1984.

[8] R.M. Goor. Continuous time adaptive feedforward control: stability and simulations. Technical Report GMR-4105, General Motors Research Laboratories, Warren, Michigan, July 1982.

[9] R.M. Goor. A new approach to robot control. In *Proceedings of the 1985 American Control Conference*, pages 387–389, June 1985.

[10] M.E. Kahn and B. Roth. The near-minimum-time control of open-loop articulated kinematic chains. *ASME Journal of Dynamic Systems, Measurement, and Control*, pages 164–172, September 1971.

[11] B.K. Kim and K.G.Shin. An efficient minimum-time robot path planning under realistic constraints. In *Proc. 1984 American Control Conference*, pages 296–303, June 1984.

[12] B.K. Kim and K.G. Shin. An adaptive model following control of industrial manipulators. *IEEE Trans. Aerospace and Electronic Systems*, AES-19(6):805–814, November 1983.

[13] B.K. Kim and K.G. Shin. Minimum-time path planning for robot arms and their dynamics. *IEEE Trans. System, Man, and Cybernetics*, SMC-15(2):213–223, march/april 1985.

[14] D. E. Kirk. *Optimal control theory: an introduction*. Prentice-Hall, Englewood Cliffs, New Jersey, 1971.

[15] A.J. Koivo and T.H. Guo. Control of a robotic manipulator with adaptive controller. In *Proceedings of the 20th IEEE Conference on Decision and Control*, pages 271–276, December 1981.

[16] A.J. Koivo and T.H. Guo. Adaptive linear controller for robotic manipulators. *IEEE Transactions on Automatic Control*, AC-28(2):162–170, February 1983.

[17] C.-S. Lin, P.-R Chang, and J.Y.S. Luh. Formulation and optimization of cubic polynomial joint trajectories for mechanical manipulators. In *Proc. 21 CDC*, pages 330–335, December 1982.

[18] T. Lozano-Perez. Automatic planning of manipulator transfer movements. Technical Report A.I. meno 606, MIT Artificial Intelligence Laboratory, December 1980.

[19] T. Lozano-Perez. Spatial planning: A configuration space approach. Technical Report A.I. memo 605, MIT Artificial Intelligence Laboratory, December 1980.

[20] J.Y.S. Luh and C.E. Campbell. Collison-free path planning for induatrial robots. In *Proceedings of the 21st CDC*, pages 84–88, December 1982.

[21] J.Y.S. Luh and C.S. Lin. Optimum path planning for mechanical manipulators. *ASME Journal of Dynamic Systems, Measurement, and Control*, 102:142–151, June 1981.

[22] J.Y.S. Luh and M.W. Walker. Minimum-time along the path for a mechanical arm. In *Proceedings of the IEEE Conference on Decision and Control*, pages 755–759, December 1977.

[23] J.Y.S. Luh, M.W. Walker, and R.P.C. Paul. Resolved-acceleration control of mechanical manipulators. *IEEE Transactions on Automatic Control*, AC-25(3):468–474, June 1980.

[24] S.P. Marin. Optimal parameterization of curves for robot trajectory design. *IEEE Transactions on Automatic Control*, 33(2):209–214, February 1988.

[25] N.D. McKay. *Minimum-cost Control of Robotic Manipulators with Geometric Path Constraints*. PhD thesis, University of Michigan, Ann Arbor, MI, September 1985.

[26] R.P.C. Paul. *Robot Manipulators: Mathematics, Programming, and Control.* MIT Press, Cambridge, Mass., 1981.

[27] F. Pfeiffer and R. Johanni. A concept for manipulator trajectory planning. *IEEE Journal of Robotics and Automation*, RA-3(2):115–123, April 1987.

[28] G. Sahar and J.M. Hollerbach. Planning of minimum-time trajectories for robot arms. *Int. J. Robotics Research*, 5(3), 1986.

[29] K.G. Shin and N.D. McKay. Minimum-time control of a robotic manipulator with geometric path constraints. In *Proceedings of the 22nd CDC*, pages 1449–1457, December 1983.

[30] K.G. Shin and N.D. McKay. Minimum-time control of a robotic manipulator with geometric path constraints. *IEEE Transactions on Automatic Control*, AC-30(6):531–541, June 1985.

[31] K.G. Shin and N.D. McKay. A dynamic programming approach to trajectory planning of robotic manipulators. *IEEE Transactions on Automatic Control*, AC-31(6):491–500, June 1986.

[32] K.G. Shin and N.D. McKay. Robuast trajectory planning for robotic manipulators under payload uncertainties. *IEEE Transactions on Automatic Control*, AC-32(12):1044–1054, December 1987.

[33] J.L. Synge and A. Schild. *Tensor Calculus.* Dover Publications, New York, 1978.

[34] D. E. Whitney. Resolved motion rate control of manipulators and the human proothcses. *IEEE Transactions on Man-Machine Systems*, MMS 10(2):47–53, June 1969.

TACTILE SENSING TECHNIQUES IN ROBOTIC SYSTEMS

Takeshi TSUJIMURA : Senior Research Engineer
Tetsuro YABUTA : Research Group Leader

NTT Transmission Systems Laboratories
Tokai, Ibaraki-ken, 319-11, Japan

ABSTRACT

This paper proposes an object shape detection system using a force/torque sensor and an insensitive probe. Tactile sensing method has newly been developed, which makes it possible to estimate a contact position on the probe only by measuring contact force. Though the probe itself is not sensitive, the method derives the contact position from force information measured by the sensor. The estimated values do not depend on either the magnitude or the direction of the force. The method is available with any shape and any material of probe. This method has been applied to a manipulator system, and object shape detection experiments have been conducted using two types of probes. Shapes of three dimensional objects have successfully been reproduced. The results confirm that the proposed method enables the manipulator system to detect objects without precise control of the manipulator.

Based on "Object Detection by Tactile Sensing Method Employing Force/torque Information" by Takeshi TSUJIMURA and Tetsuro YABUTA which appeared in IEEE TRANSACTIONS ON ROBOTICS AND AUTOMATION; Vol.5, No.4, pp.444-450; August 1989. ©1989 IEEE.

I. INTRODUCTION

For intelligent robots to replace humans for construction and maintenance work, it is essential that the robots have the ability to recognize environments around them, such as positions of objects of their work and obstacles. "Computer vision" is one of the ways to provide such a capability. Image processing technology, which is essential for robot vision, is still not almighty. For example, image quality is degraded by suspended particles in the air, by poor illumination, and so on. In these situations, some other means are necessary, either supplementing or temporarily replacing the vision system. Also, while the robot is working, it has to gather detailed information for higher accuracy. Recognition of environments is very important especially in performing a task outdoors. A typical example is the noncontact detection method using ultrasonic remote sensor [1] , [2] .

Tactile sensors are another means of detecting objects. All tactile sensors, whether based on conductive silicone rubber, pressure sensitive semiconductors, or piezo-electric elements, detect object position using a single or multiple dispersed contact points [3] - [9] . Salisbury has shown the basic sensor requirements for identifying contact features and a planar sensor prototype without concrete experimental results [10] . Brock et al. have presented a strain-gauge based fingertip sensor which has a spherical sensing surface [11] . No tactile sensor retains a wide sensitive field continuously.

Here, we present a method of estimating contact position from the contact force measured by a force/torque sensor. One distinguishing feature of this method is that, regardless of the shape of the probes used, the position of the contact point can be determined based on the measured value of only one force/torque sensor. As the presented method allows using attached probes of any size and any shape, it is possible to detect the complicated object shape. And, the method is available to measure larger objects accurately without precise probe positioning due to wide detecting surface.

In this paper, we first present the principle of contact position detection employing force input information. Then, we discuss the results of experiments performed to test this principle, in which the shapes of several objects are detected using a manipulator equipped with a force/torque sensor.

II. A CONTACT POSITION DETECTION METHOD EMPLOYING FORCE/TORQUE INFORMATION

When an object hits against a probe and exerts a force on it, the probe transmits the contact force to the attached force/torque sensor. The position information of the contact point is found by calculating the output value of the sensor. Furthermore, the method described here is effective using only a single sensor, regardless of the magnitude or the direction of the external force.

The detection capability of the sensor is based on a three-dimensional system of orthogonal coordinates with three directional force components $\vec{F}$ [$\vec{F} = (F_x, F_y, F_z)^T$] , and three rotational moment components $\vec{M}$ [$\vec{M} = (M_x, M_y, M_z)^T$] , as shown in Fig. 1. The force/torque sensor uses a

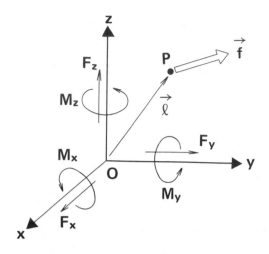

Fig.1   Coodinate system of force/torque.   The sensor measures six force
components Fx, Fy, Fz, Mx, My, Mz at the point O, when force $\vec{f}$ is
applied at point P.

coordinate system with origin O and axes x,y,z. It is assumed that external
force $\vec{f}$ [$\vec{f} = (f_x, f_y, f_z)^T$] is exerted at a point P on vector $\vec{l}$ [$\vec{l} = (l_x, l_y, l_z)^T$]
in space. So long as the space is a satisfactory transmission medium,
external force will be detected by the force/torque sensor for the
directional components $\vec{F}$ and the rotational moment components $\vec{M}$, The
relations among sensor outputs $\vec{F}$ and $\vec{M}$, external force $\vec{f}$, and contact
position vector $\vec{l}$, are clarified in the following equations.

$$\vec{F} = \vec{f} \tag{1}$$
$$\vec{M} = \vec{l} \times \vec{f}. \tag{2}$$

If $\vec{f}$ is eliminated from expressions (1) and (2) , the following expression
is obtained:

$$\vec{M} = \vec{l} \times \vec{F}. \tag{3}$$

By solving for $\vec{l}$, the position of the unknown contact point is given as follows.

$$\vec{l} = (\ \vec{F} \times \vec{M})/|\vec{F}| + k\vec{F}$$
$$= \vec{l^0} + k\vec{F} \qquad\qquad\qquad (4)$$

where $|\vec{F}|$ is the vector $\vec{F}$ norm, $\vec{F} \times \vec{M}$ is the external product of vectors, $\vec{F}$ and $\vec{M}$, and k is a scalar constant.

Since the value of k is arbitrary, the above equation is indefinite; mathematically, each dimensional vector component can be thought of as corresponding to each x,y and z coordinate in a quantitative vector space, where direction of vectors has no relation with that of forces as shown in Fig.2.

i) According to (3), $\vec{l^0}$ and $\vec{M}$ are orthogonal.

i i) According to (4)

$$\vec{l^0} = (\ \vec{F} \times \vec{M})\ /\ |\vec{F}|^2. \qquad\qquad\qquad (5)$$

Therefore, $\vec{l^0}$ and $\vec{F}$, $\vec{l^0}$, $\vec{M}$ are orthogonal. According to i) and i i), $\vec{l^0}$, $\vec{F}$, and $\vec{M}$ are all mutually orthogonal. $\vec{F}$ Equation (4) indicates straight line Lm in Fig.2, which is parallel to $\vec{F}$ and intersects point P^0 on the condition that $\overrightarrow{OP^0} = \vec{l^0}$. The solution to (3) is given by vector $\overrightarrow{OP}$ that shows variable P on the line.

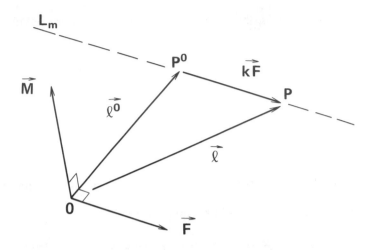

mathematical vector space

Fig.2   Relationship among vectors $\vec{l^0}$, $\vec{l}$, $\vec{F}$, $\vec{M}$ in mathematical vector space. Vectors l0, $\vec{F}$, and $\vec{M}$ meet at right angles to each other, and end point P of vector l always lies on line m.

Physically, the relationship is as follows.   When external force $\vec{f}$ is applied at point P in physical space, sensor outputs $\vec{F}$ and $\vec{M}$ are constant as long as P is somewhere along the line which is parallel to the external force direction and which is represented by line Lm in the mathematical space as shown in Fig.2. Put another way, this indicates that the contact point, where the external force is applied, cannot be uniquely determined based on the outputs of the force/torque sensor alone.   For example, we assume that a force $\vec{f}$ is applied at some point on line Lp which is parallel to $\vec{f}$ in the actual physical space as shown in Fig. 3. Outputs $\vec{F}$ and $\vec{M}$ are constant whether the force is applied on P1 or P2, because the

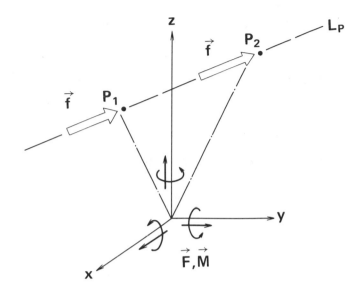

physical space

Fig.3   Schematic of the same forces applied at different points in the physical space. Sensor outputs $\vec{F}$ and $\vec{M}$ are unique as long as force $\vec{f}$ is applied at a point on line Lp which is parallel to vector $\vec{f}$.

two points are both on line Lp. Therefore, only line Lp can be estimated from outputs $\vec{F}$ and $\vec{M}$. For a unique determination, the scalar constant k parameter is needed.

A probe is used for external force to be translated to the sensor. The shape of the probe is manifested as a dimensional curved surface. Parameter k depends on this shape. It is necessary that the shape of the probe be designed so that the applied point of the external force might be uniquely determined. That is, line L intersects the curved surface of the probe at one and only one point.

According to the theory presented here, the location of the force can be calculated, regardless of the magnitude and direction of the exerted force, from outputs of the force/torque sensor. It is represented by the following expression. When the probe equation is represented by $\vec{r}$ and external force $\vec{f}$ is applied on the surface of the probe at a point where $\vec{r} = \vec{r_0}$, the location of the force can be calculated by the equation

$$\vec{r} = (\vec{F} \times \vec{M}) \, / \, |\vec{F}|^2 + k\,(\vec{f})\,\vec{F} \qquad (6)$$

for any external force $\vec{f}$. The proof is as follows:
Since $\vec{M} = \vec{r_0} \times \vec{F}$ according to (2)

$$(\vec{F} \times \vec{M}) \, / \, |\vec{F}|^2 = r_0 - (\vec{r_0} \cdot \vec{F})\,|\vec{F}|^2\,\vec{F} \qquad (7)$$

and therefore, (6) is

$$\vec{r} = \vec{r_0} + (k - (\vec{r_0} \cdot \vec{F}) \, / \, |\vec{F}|^2)\,\vec{F}. \qquad (8)$$

It can be seen that a solution exists along the line which is parallel to $\vec{F}$ and through the point $\vec{r_0}$.

If parameter k can be determined such as

$$k = k(\vec{f}) = (\vec{r_0} \cdot \vec{F}) \, / \, |\vec{F}|^2 \qquad (9)$$

the second term on the right side of (8) becomes zero and (8) is to be independent of $\vec{F}$. Consequently, for example, if force $\vec{f}_1$ is applied on a probe surface $\vec{r}_0$ which is to be estimated, vector $\vec{l}_1^0$ and parameter k 1 are calculated and vector $\vec{r}_0$ is determined as shown in Fig.4. Even if another force $\vec{f}_2$ is applied at the same point, another $\vec{l}_2^0$ and k 2 are calculated and $\vec{r}_0$ is determined to be the same value as the former case.

Actually, equation (9) is useless, because it contains unknown parameter $\vec{r}_0$. In order to find parameter k using a form that does not contain $\vec{r}_0$, it is necessary to define the shape of the probe and to determine parameter k from the geometric conditions. Substituting

$$\vec{l} = \vec{l}^0 + k(\vec{f})\vec{F} \tag{10}$$

into the probe surface equation $G(\vec{l}) = 0$, then

$$G(\vec{l}^0 + k(\vec{f})\vec{F}) = 0. \tag{11}$$

Solving for k yields the solution such that

$$k = K(G, \vec{l}^0(\vec{f}), \vec{F}(\vec{f})) \tag{12}$$

where k is an equation including $G, \vec{l}^0(\vec{f})$, and $\vec{F}(\vec{f})$.

It is the condition for determining k uniquely such that vector $\vec{l}$ and the probe surface $\vec{r}$ meet at one and only one point.

If the probe surface is sufficiently smooth so that friction is not a significant factor, the direction of $\vec{F}$ and the normal direction of the object surface are identical. In this case, if the probe surface has the characteristic that "a normal line extending from any point on the surface has no other intersection with the surface," then an estimated value can be

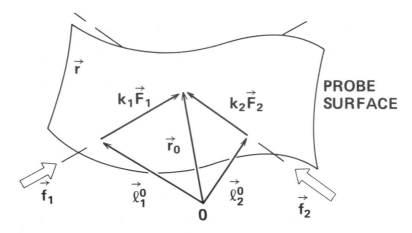

Fig.4   Schematic of different forces applied at the same point. Point $r0$, where force $\vec{f}$ is applied, can be estimated by employing parameter k which is determined from the applied force.

determined uniquely. When friction is a factor, although the direction of $\vec{F}$ and the normal direction of the probe surface are not entirely identical, the theory presented here still holds as far as the probe surface is not too sharply curved.

We have clarified that when a probe is able to transmit force and is constructed with a smooth curved surface, it does not matter where on the probe surface, or from which direction, or how large the external force is exerted, it is possible to estimate the location of the contact point.

For example, parameter k should be defined for two different kinds of probes.

Example 1:

The equation for a semi-spherical surface probe, which is shown in Fig.5, with a radius of R is given by the following expression:

$$|\vec{i}|^2 = R^2. \tag{13}$$

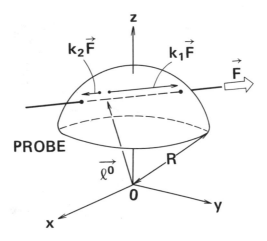

Fig.5 Schematic of a sphere probe. Parameter k is defined by the point
where line $\vec{l}^0 + k\vec{F}$ intersects through the surface of the probe.

Inserting (4) yields

$$| \vec{l}^0 + k\vec{F} |^2 = R^2. \tag{14}$$

Expanding this results in

$$| \vec{l}^0 |^2 + 2k(\vec{l}^0 \cdot \vec{F}) + k^2 |\vec{F}|^2 = R^2 \tag{15}$$

and by solving the equation for k we obtain

$$k = (-(\vec{l}^0 \cdot \vec{F}) \pm ((\vec{l}^0 \cdot \vec{F})^2$$
$$- |\vec{F}|^2 (|\vec{l}^0|^2 - R^2))^{1/2}) / |\vec{F}|^2 \tag{16}$$

$\vec{l}^0$ and $\vec{F}$ are orthogonal, i.e. $(\vec{l}^0 \cdot \vec{F}) = 0$, so

$$k = \pm (R^2 - |\vec{l}^0|^2)^{1/2}) / \vec{F} . \tag{17}$$

Example 2:

In the case of a linear-shaped probe, which is shown in Fig.6, with the

vector directions $\vec{e} = (e_x , e_y , e_z)$, the expression

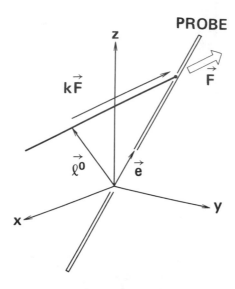

Fig.6   Schematic of a bar probe. Parameter k is defined by the point at
which line $\vec{l}^0 = k\vec{F}$ and the probe meet.

$$\vec{l}^0 + k\vec{F} = p\vec{e}$$
                                                                        (18)

is obtained, where p is a proportional constant. The expression above is a

simultaneous equation of the third degree with both k and p as unknowns,

and is generally impossible to solve.   Moreover, if the injection of

measurement noise and the deformation of the probe cause errors of $\vec{F}$ and

$\vec{e}$, the force line and the probe line practically no longer intersect with each

other. Those are the reasons why the least square method is necessary.

    Solving with the least square method, the following form is obtained

(see the Appendix) :

$$k = ((\vec{I^0} \cdot \vec{F}) \ |\vec{e}|^2 - (\vec{I^0} \cdot \vec{e}) \cdot (\vec{F} \cdot \vec{e})) / (|\vec{F}|^2 |\vec{e}|^2 - (\vec{F} \cdot \vec{e})^2) . \qquad (19)$$

Calculating (4) using parameter k obtained as shown above yields the location of the applied point of an external force. We have shown here that the position of the external force can be calculated based on the outputs of a force/torque sensor alone, regardless of the size and direction of the force.

III. EXPERIMENTAL RESULTS

Based on the principle discussed above, a method of reproducing the shapes of the various objects has been developed and incorporated in a manipulator system. According to the method, objects were felt for by a probe and detected using the contact force data. The experiments confirmed the effectiveness of this new technique.

A. EXPERIMENTAL DESIGN

The object shape detection experiments were carried out using a six-degree-of-freedom articulated manipulator. Figure 7 shows the block diagram of control system of the manipulator and a force/torque sensor. A six-axes force/torque sensor was installed at the end of the manipulator with an attached probe. The manipulator was controlled with an MC68000 microprocessor and RS-232C interface. The manipulator repetition precision was approximately 0.1 mm. The force/torque sensor measures the deformation of a metal block, which is caused by the external force, by wire-strain gauges, and converts the measured values into force and torque components. The measurement range was 600 N for the force, and 12 Nm

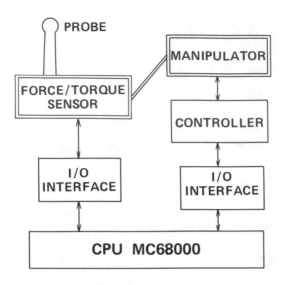

Fig.7 Block diagram of the sensing system.

for the torque within 0.2% accuracy. Two kinds of probes were used. One was a 1 0 cm long aluminum bar aligned with the z-axis.

The second probe was an aluminum semi-sphere with a radius of 1 0 cm, whose center is at the sensor origin point.

The experimental setup is shown in Fig. 8. Positional vector $\overrightarrow{OBj}$, directed at the surface of the object to be measured, agrees with the sum of positional vector $\vec{s}$ directed at the force/torque sensor installed in the end of the manipulator, and vector $\dot{l}$ from the sensor to the object. Vector $\vec{s}$ is measured from the length of the manipulator's links and the rotation-angles of the joints. Vector $\dot{l}$ is obtained by converting a theoretically derived value in the sensor's coordinate system, as was shown in the previous section, into the absolute coordinate system.

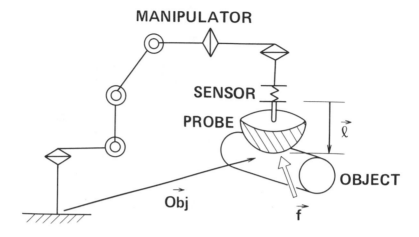

Fig.8 Experimental setup with a bar probe.

The measurement algorithm is shown in Fig. 9. First, a movement path is established around the object, and starting points of measurement are taken on the path. Starting from each measurement point, the sensor moves in minute increments. When the probe and object are brought into contact, detection commences based on output from the sensor. Though the probe contact position varies depending on the shape of the object, the object shape is estimated from the values derived from the contact force based on our theory.

B. EXPERIMENTAL RESULTS

1) Using a bar-shaped probe:

A telephone receiver like that shown in Fig. 1 0 was used as a test object, where the broken line is measuring path of the probe. In the experiment, a

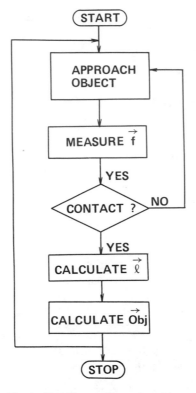

Fig.9   Tactile sensing algorithm.

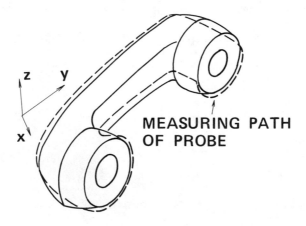

Fig.10   A telephone receiver as object to measure.

bar-shaped probe explored the object and an outline was then detected. To produce a projected image onto the xy plane, the probe was controlled to keep direction of the z-axis. The sensor movement path was established around the object. From this path, measurement was started. Beginning at each measurement starting point, as the counter force supplied by the probe was carefully measured, the probe was moved in 1 mm increments. The counter force was adjusted to be (10 ± 1) N. On satisfying this condition, the method presented earlier was used to estimate the contact point position. Although the object touched different parts of the probe at each contact because of its complex three-dimensional configuration, this method made it possible to detect the object shape precisely.

The measurement results are shown in Fig. 11. The solid line in the figure represents the actual outline of the object. The square points indicate the measured surface of the object. The average and the standard deviation of measurement error were 0.4 mm and 3.9 mm respectively.

2) Using a semi-sphere probe

This detection experiment was carried out using a submarine optical cable joint box[12] to be a test object, as shown in Fig.12. A sensor movement path was laid in the space above the object, and the sensor was lowered 1 mm at a time down the object, starting to take measurement at points along this path.

Again, using the described method, the contact force was detected, and the contact points were estimated. By repeating this operation, the cylindrical shape was reproduced. Since measurements can be taken regardless of the

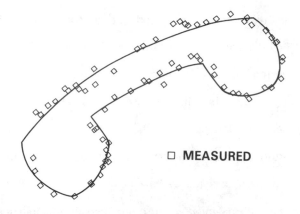

Fig.11 Experimental results of measurement of the receiver.

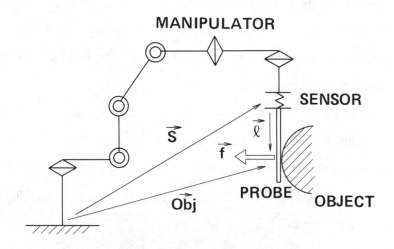

Fig.12 Experimental setup with a sphere probe.

contact position on the probe surface, it is not necessary to precisely control the position of the probe in spite of wide detecting surface.

Measurement results are shown in Fig. 13. Measurement error averaged 2.4 mm and the standard deviation was 1.8 mm. Measurement time took up to 3 0 seconds per point. This was due to the following reasons.

1) Since the volume of movement by the sensor does not vary, but is fixed, when the measurement starting point is separated from the measurement object, a good amount of time is required for the probe to come into contact with the object.

2) In order to maintain uniform measurement conditions, the contact force control values must be set up in advance within a certain effective range, and this is time consuming.

3) Transmission time between the computer and the manipulator must be added.

IV. DISCUSSION

A. EXTERNAL FORCE EFFECTS

As was discussed above in section II, when a probe comes in contact with an external force, the orientation of that force can be calculated from the output of the force/torque sensor. This is possible regardless of the magnitude or the direction of the force and has been confirmed by the following experimental results.

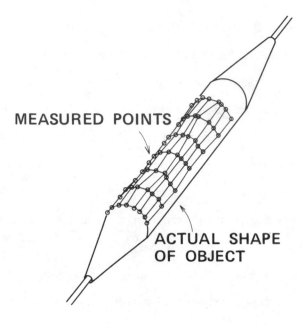

Fig.13 Experimental results of measurement of a submarine-cable joint
       box.

Based on the theory elaborated i n Section II, experiments were carried

out by changing external force exerted on the bar-shaped probe to clarify

how the system would respond to the external force.

1) Changing the point of applied force

First, the point of the external force was determined for a force with a

fixed magnitude and direction of $\vec{f} = (0,10N,0)^T$. The external force was

brought into contact so i t occupied a position of $30 + 10i$ $(i=0,1,\cdots,7)$ mm

on the sensor's z-axis.

The relationship between true values, $\hat{l}$ of the contact position and the values of measured and estimated l_z using the method is shown in Fig. 14. They correspond very closely. The average and the standard deviation of error were 0.9 and 4.4 respectively.

2) Changing the force direction

Fixing the force position at $\vec{l} = (0,0,100mm)^T$, and the magnitude at $|\vec{f}|$ = 10 N, the effect of changing the direction of the external force was investigated. The direction was shifted in 30° degree increments on the xy plane.

The relationship between force directions, $\arg(\vec{f})$ and measured values, Iz is shown in Fig. 15. A close correspondence was found averaging 98.4 mm (i.e., error of 1.6 persent between the estimated positional values and the actual ones. The standard deviation was 4.2 mm (4.2 percent).

3) Changing the force magnitude

Fixing the force position $\vec{l} = (0,0,100mm)^T$, and the y-axis direction at $\vec{f}$ = $(0,fy,0)^T$, the effect of changing the magnitude of the external force was investigated. The force magnitude was adjusted for fy = 5 i (i=0,1,2,⋯ ,8) N.

The relationship between force magnitude, fy and measured values, Iz is shown in Fig. 16. The average of estimated values was 99.3 mm (i.e., error of 0.7 percent) and the standard deviation was 2.5 mm (2.5percent). There is a very close correspondence with the actual value. It was found that when the external force is small, there is a tendency for the estimated value variance to be more pronounced. This is because the measurement error becomes larger with a smaller external force.

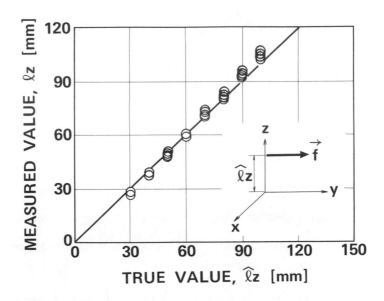

Fig.14   Relationship between true value and measured value of contact
point.

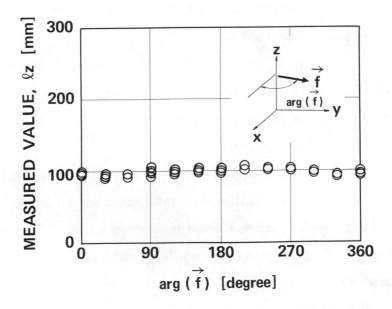

Fig.15 Relationship between force direction and measured value of contact
point.

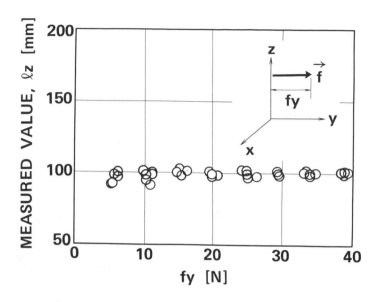

Fig.16 Relationship between force magnitude and measured value of contact point.

B. SENSOR RESOLUTION

Experiments were also conducted based on the theory described in section II using a semi-spherical shaped probe to estimate the point of force exertion. External force was applied at the points on two semicircles where the yz plane and the xz plane respectively intersect the probe surface. The magnitude of the force was controlled at 10 N.

The resulting estimation pattern is shown in Fig. 17. The experimentally estimated values varied from the theoretical ones by an average of 2.7 mm. It is 2.7 percent of the probe diameter. Based on this result and this experimental design, therefore, the resolution of the tactile sensor system is considered accurate to within 2.7 percent.

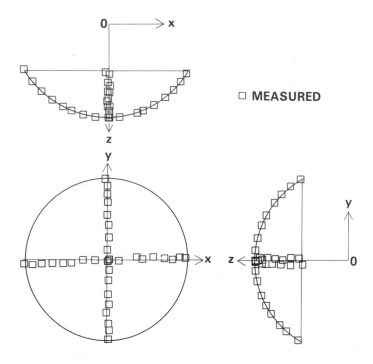

□ MEASURED

Fig.17 Experimental results of estimation of contact points on a sphere probe.

The measurement error may be classified by these three factors; the force measurement error, the length measurement error, and the deformation of the manipulator and the probes. The force measurement error depends on the accuracy of the force/torque sensor. The actual resolutions of measured value of force and moment are 0.1 N and 1 N mm respectively. As force and moment are exerted at 10 N and 1000 N mm in the experiment, those accuracies are to be 1 and 0.1 percent, respectively. The length measurement error is less than 1 mm, that is 1 percent of the probe diameter. The manipulator and the probes deform less than the other errors because of their rigid structure. So, the last factor is minor. Substituting the values into (4), we estimate the total accuracy of the system is less than 3 percent. As a consequence gives a good estimation of the measurement error, which is comfirmed by the experimental results.

V. CONCLUSION

In this paper, first the principle of contact position detection using force information was presented. This method is based on force and moment data measured by a force/torque sensor and is used to estimate the point of the contact force exertion.

Next, shape detection experiments applying this theory were described. Using a manipulator, first with a bar probe and then with a sphere probe, both equipped with a force/torque sensor, the shapes of various objects were detected based on the force information obtained from the sensor. The experimental results show that the proposed method is available to detect the complicated object figures precisely using the attached probes which have wide detecting surface.

Finally, the effect of the contact force itself for estimating the contact point was investigated. Here, it was confirmed that an estimated value can be obtained depending on neither the magnitude nor the direction of the force. The ability of this system to derive an estimated value was found, through the experiments described, to be accurate within 3 percent.

Appendix

Scalar parameters k and p of the following equation:

$$\vec{x} + k\vec{y} = p\vec{z} \tag{A1}$$

can be obtained with the method of least squares, where x, y, and z are al

n $\times$ 1 vectors.

Let Sn be defined as

$$Sn = |\vec{x} + k\vec{y} - p\vec{z}|^2. \tag{A2}$$

Solving the following equations :

$$\partial Sn / \partial k = 0 \tag{A3}$$

$$\partial Sn / \partial p = 0 \tag{A4}$$

yields parameters k and p.

If the i-th component of vectors $\vec{x}, \vec{y}, \vec{z}$ are x_i, y_i, z_i, respectively, (A2) gives

$$Sn = \sum_{i=1}^{n} (x_i + ky_i - pz_i)^2 \tag{A5}$$

Consequently equati=ons (A3) and (A4) give

$$\partial Sn / \partial k = 2 \sum_{i=1}^{n} y_i (x_i + ky_i - pz_i)$$

$$= (2 \sum_{i=1}^{n} y_i^2) k + (-2 \sum_{i=1}^{n} y_i z_i) p + (2 \sum_{i=1}^{n} x_i y_i) \qquad (A6)$$

$$\partial Sn / \partial p = -2 \sum_{i=1}^{n} z_i (x_i + ky_i - pz_i)$$

$$(-2 \sum_{i=1}^{n} y_i z_i) k + (2 \sum_{i=1}^{n} z_i^2) p + (2 \sum_{i=1}^{n} x_i z_i) \quad . \qquad (A7)$$

Then solving equations (A6) and (A7) yields

$$k = (\sum_{i=1}^{n} x_i y_i \sum_{i=1}^{n} z_i^2 - \sum_{i=1}^{n} x_i z_i \sum_{i=1}^{n} y_i z_i) / (\sum_{i=1}^{n} y_i^2 \sum_{i=1}^{n} z_i^2 - (\sum_{i=1}^{n} x_i y_i)^2)$$

$$= (\vec{x} \cdot \vec{y}) |\vec{z}|^2 - (\vec{x} \cdot \vec{z})(\vec{y} \cdot \vec{z})) / (|\vec{y}|^2 |\vec{z}|^2 - (\vec{y} \cdot \vec{z})^2) \qquad (A8)$$

$$p = (\sum_{i=1}^{n} x_i y_i \sum_{i=1}^{n} x_i y_i - \sum_{i=1}^{n} x_i z_i \sum_{i=1}^{n} y_i^2) / (\sum_{i=1}^{n} y_i^2 \sum_{i=1}^{n} z_i^2 - (\sum_{i=1}^{n} y_i z_i)^2)$$

$$= ((\vec{x} \cdot \vec{y})(\vec{y} \cdot \vec{z}) - (\vec{x} \cdot \vec{z}) |\vec{y}|^2) / (|\vec{y}|^2 |\vec{z}|^2 - (\vec{y} \cdot \vec{z})^2). \qquad (A9)$$

ACKNOWLEDGMENT

The authors wish to thank Dr. Sadakuni Shimada, Dr. Akira Sakamoto, and Dr. Koushi Ishihara for their helpful advice and encouragement during this work.

REFERENCES

(1) D.Nitzan et al.:"The Measurement and use of Resistered Reflectance and Range Data in Scene Analysis," Proc. IEEE, Vol.65, No.2, pp.206-220 (1977).

(2) T.Tsujimura,T.Yabuta,T.Morimitsu: "Three-Dimensional Shape Recognition Method Using Ultrasonics for Manipulator Control System," Journal of ROBOTIC SYSTEMS, Vol.3, No.2, pp.205-216 (1986).

(3) E.Carome et al.: "PVF2 Transducers for NDT," Proceedings of the 1979 IEEE Ultrasonics Symposium, pp.346-349 (1979).

(4) J.A.Purbrick: "A Force Transducer Employing Conductive Silicone Rubber," Proc. 1st Int. Conf. Rob. Vis. Sens. Controls (1981).

(5) M.H.Raibert,J.E.Tanner: "A VLSI Tactile Array Sensor," Proc. of 12th ISIR (1982).

(6) P.Dario et al.:"Ferroelectric polymer tactile sensors with anthropomorphic features," Int. Conf. Rob. (1984).

(7) A.R.Grahn, L.Astle: "Robotic Ultrasonic Force Sensor Arrays," Robot 8 Conference(1984).

(8) K.Chun,K.D.Wise: "A high performance silicon tactile imager based on a capacitive cell," IEEE Trans. Vol.ED-32, No.7 (1985).

(9) P.W.Barth et al.: "Flexible circuit and sensor arrays fabricated by monothilic silicon technology," IEEE Trans. Vol.ED-32, No.7 (1985).

(1 0) J.K.Salisbury: "Interpretation of Contact Geometries from Force Measurements," Proc. 1st International Symposium of Robotic Research (1 9 8 3).

(1 1) D.Brock, S.Chiu: "Environment Perception of an Articulated Hand Using Contact Sensors," ASME PED-Vol.15, Robotics and Manufacturing Automation, pp.8 9 - 9 6 (1 9 8 5).

(1 2) S.Furukawa et al.:"Structural design and sea trial results for a submarine optical-fiber cable joint box," IEEE J. of Lightwave Technology, Vol. L T - 2 , No.4 (1 9 8 4).

Sensor Data Fusion in Robotic Systems

J. K. Aggarwal *
Computer and Vision Research Center
The University of Texas
Austin, TX 78712

Y. F. Wang †
Department of Computer Science
University of California
Santa Barbara, CA 93106

Abstract

Many types of sensors may be used to gather information on the surrounding environment. It may be noted that different sensors possess distinct characteristics, are designed based on differing physical principles, operate in a wide range of the electromagnetic spectrum, and are geared toward a variety of applications. A single sensor operating alone provides a limited sensing range, and is inherently unreliable due to operational errors. However, a synergistic operation of many sensors provides a rich body of information on the sensed environment from a wide range of the electromagnetic spectrum. In this paper, we present a brief survey of the techniques/systems for sensor data fusion. We first address sensor data fusion in mobile robotics applications. Since mobile robots operate in an uncertain and constantly changing environment, a steady stream of rich and reliable information on the environment is needed for navigation and path planning. We discuss the feasibility of a few sensors in such applications and present several sensor data fusion techniques for road following and terrain analysis. Next, we review techniques for fusing data from visual, thermal, tactile, and structured-lighting sensors and discuss several general frameworks for sensor data fusion.

⁰* This research was supported in part by a contract from the Army Research Office, DAAL 03-87-K0089

⁰† This research was supported in part by a grant from the National Science Foundation, IRI-8908627

1   Introduction

In this paper, we present a brief survey of the techniques/systems for sensor data fusion. Many types of sensors may be used to gather information on the surrounding environment, for example, visual sensors, visual sensors augmented with structured lighting, thermal (infrared) sensors, proximity sensors, tactile sensors, ultrasonic rangefinders, and laser rangefinders. It should be noted that different sensors possess distinct characteristics, are designed based on differing physical principles, operate in a wide range of the electromagnetic spectrum, and are geared toward a variety of applications. A single sensor operating alone provides a limited sensing range and is inherently unreliable due to operational errors. However, a synergistic operation of many sensors provides a rich body of information on the sensed environment from a wide range of the electromagnetic spectrum.

Research on fusing data from a multitude of sensors has started to gain attention for the following reasons:

- Much effort has been expended to study the design and operation of visual, acoustic, laser ranging, infrared, and tactile sensors. The analysis of visual [10], [15], acoustic [14], [47], laser ranging [13], [47], infrared [19], [60], [61], and tactile [20], [30], [29], [44], [72] sensors/images provides a better understanding of the strengths and limitations of these sensors and facilitates their integration.

- Due to significant advances in sensing, storing, and processing technologies, the acquisition and storage of large amounts of data from multiple sensors are becoming feasible.

- The quest for industrial automation provides the thrust to develop intelligent autonomous robots that replace humans in many assembly and manufacturing tasks. An intelligent robot often operates in an unstructured environment; therefore, it must react to sudden, unexpected events and changes in the environment. To achieve a desirable degree of autonomy and to be able to make intelligent decisions, it is crucial that a robot understand and react to the environment. Sensors provide the basic means to accomplish such interactions.

However, the heterogeneous nature of the sensed data from multiple sensors makes a coherent analysis difficult. Intuitively speaking, sensors that operate in disjoint spectrums (such as visual sensors and infrared sensors) provide a rich set of information with little or no duplication. The heterogeneous nature of the data sets does not allow easy cross verification and checking for correctness. To simply register one set of data with the other can be a challenging task. For sensors sharing similar characteristics, it is likely that the registration problem may be simplified and that different data sets may complement one another more readily.

To illustrate, let us briefly discuss two research projects which implemented the distinct-sensors-maximum-information and similar-sensors-redundant-information concepts. In [60], [61], visual and thermal (infrared) sensors were used together for scene interpretation. Visual and infrared sensors are two distinct physical devices: They sense in different bands in the electromagnetic spectrum, have different spatial resolution, are placed at different locations, and thus observe the same scene from different vantage points. In [60], [61], a visual sensor provides information on surface orientation while a thermal sensor provides information on surface temperature. Together the information from the two sensors enables the computation of the surface heat transfer properties used in region classification (Section 3 presents the details). Such surface thermal properties are difficult, if not impossible, to infer if any single sensor is used. However, it is not a trivial task to register visual and infrared data, nor is there an easy way to verify the correctness of infrared data using intensity data or vice versa.

Wang and Aggarwal [78] present a technique that uses both visual images and structured-light coded images for 3-D modeling. The reasons for selecting these two sensing modules are: (1) they can be made to use the same camera, which simplifies the tasks of calibration and registration; (2) they provide both qualitative and quantitative descriptions of the scene, and sensor data complement each other nicely; (3) a large degree of redundance exists in the sensor data, which facilitates cross checking and verification; and (4) the cost of incorporating a projection device (i.e., a $35mm$ slide projector) in structured lighting is much less than if a new sensor is used. Such an approach uses sensors of similar operating characteristics. Even though the information derived from the second sensor is highly redundant, it is possible to use the new data to verify and correct the old data.

Fusing sensor data involves more than choosing the right sensors and averaging sensor output. As pointed out in [16], the simple averaging of sensor measurements is prone to distortion by individual noisy measurements, and makes overall variance of measurements worse. Further, sensor data may be used in such a way that data from one sensor guide the operation and exploration of another sensor. Such is the case in using a visual sensor with a tactile sensor [5], [6], [7], [70] and in intensity guided range sensing [2]. The key issue in tactile sensing is that blind groping on a surface with a tactile sensor is a poor and inefficient way to understand 3-D structure. Touch needs to be guided, and vision data provide guidance to an active tactile sensor. To improve the efficiency in gathering range data, intensity images may be analyzed first to locate corners, edges, and regions with large variations in the perceived depth. Range data are then acquired for a much smaller set of image points [2] to speed up the data acquisition process.

Our observation is that research on sensor data fusion is still in its embryonic stage. Without a thorough understanding of sensor operations and a careful study of the mechanisms that give rise to different sensor behaviors, an attempt to incorporate a large variety of sensors with vastly different characteristics in robotics systems may be premature. Brady [16] made a similar observation. He observed that different sensors provide different sorts of information (such as depth, orientation, and temperature). Information provided from a sensor may be dense or sparse and may be of a wide range of accuracy and reliability. Sensors may be placed at different vantage points. Hence, techniques for fusing sensor data have to account for all these changing factors.

In this paper, we present a brief survey of the techniques/systems for multi-sensor data fusion. The paper is organized as follows. Section 2 surveys sensor data fusion in mobile robots. Since mobile robots operate in an uncertain and constantly changing environment, a steady stream of rich and reliable information on the environment is needed for navigation and path planning. We address the feasibility of several sensors in such applications and survey several sensor data fusion techniques for road following and terrain analysis. Sections 3 to 6 present techniques for fusing data from a variety of sensors. Section 7 presents several general frameworks for sensor data fusion. Finally, section 8 contains a concluding remark.

2   Sensor Data Fusion in Mobile Robotics

To navigate through its environment, a mobile robot must be able to sense the structure of the environment, to model features presented in the sensor data, to construct a rich description of the environment, to constantly update the world description, and to plan its reaction to unexpected events. Sensor data provide the basis for road following, obstacle avoidance, landmark recognition, hallway detection, cross-country navigation, mapping, and spatial learning.

The complexity of intelligent task planning, the demand of real-time plan execution and navigation, and the need for continuous, smooth motion exceed the capacity of the current computer and sensing technologies. To gain processing speed, guidance, sensing, and planning activities are often executed in parallel on a large number of on-board and off-board computers [18], [34], [48], [75]. It has also been realized that a single sensor (or a single type of sensor) is inadequate to gather data to accomplish all the tasks listed above.

Many universities and industrial laboratories are engaged in research into mobile robotics: Carnegie-Mellon University, Stanford University, the University of Maryland, the FMC Corporation, Hughes, and Martin Marietta to name a few. Research has been conducted on path planning, navigation, terrain mapping, and sensor data fusion. It is beyond the scope of this paper to address all aspects of these research activities. Instead, we discuss, in general terms, how different types of sensors might be used in a mobile platform in indoor, outdoor, and space environments, and the strengths and limitations of different sensors. We also briefly survey sensor data fusion techniques employed in several research projects for road following, terrain analysis, and spatial learning.

An outdoor environment is often not well structured and does not lend itself to human control. Navigating a mobile platform in an outdoor environment must take into consideration road condition, weather, and visibility. A common strategy in navigating an autonomous land vehicle is to do road following, i.e., to maintain the vehicle's course on the road to avoid natural hazards such as ditches, rivers, bushes, and trees. This seemingly trivial task can be deceivingly difficult, since road conditions change with the location, season, weather, and time of day. A road may be covered with snow or fallen leaves, may have shadows cast on it, may not have shoulder, and may not have a clearly defined boundary between the road and shoulder or between the shoulder and vegetation. Simple boundary detection and region growing techniques to locate the road in an image may fail to cope with all the possible variations. Using multiple sensors provides a better tolerance to scene variation. For example, intensity and color cameras have been used to locate road regions in [27], [48], [71], [74], [75].

Visual sensors are also used to detect landmarks or road signs to verify and correct a vehicle's position [79]. Laser ranging devices [23], [24], [27], [48], [71], [74], [75] and ultrasonic sensors [51], [52] have been used for obstacle detection and terrain mapping. A laser rangefinder gathers a dense depth map which is useful in navigation, obstacle avoidance, and terrain mapping. Ultrasonic sensors are relatively inexpensive, and the sparse range readings from sonars are useful in avoiding obstacles.

An indoor environment, on the other hand, is largely man-made and relatively well structured. Lighting and weather conditions may be partially controlled. Man-made objects, such as furniture and machines which are abundant indoors, have clearly defined faces and edges that facilitate computer modeling. The layout of furniture and machines can be more readily changed. More often than not, a description of the environment (such as building blueprints) is available or can be readily constructed to serve as a navigational guide for a mobile robot.

However, in order to negotiate a mobile platform in tight hallways and offices, indoor robots are relatively small. This small physical size limits the amount of on-board computing power as well as the number and size of sensors that can be carried on board. Complicated image analysis and interpretation tasks might have to be performed off-board. The large amount of data flowing back and forth between the on-board processor and the stationary host complicates communication, slows down response, and limits the effective range of a mobile platform. Many indoor robots [14], [18], [26], [28] use simple ultrasonic sensors or proximity sensors for their small size and weight, a less demanding power consumption, and a relatively narrow data bandwidth.

Other applications include space telerobots for a space station or planetary exploration. Sensors are essential for telerobots to interact with the unknown environment. However, the possible hostile and void environments place severe demands on the sensors. As pointed out in [32], active sensors consume excessive power and usually involve mechanical scanning devices that are potentially unreliable. Noncontact sensors have to operate in the wavelengths for which ambient radiation exists in space; that environment precludes the use of ultrasonic sensors. Hence, only visual sensors, and maybe infrared sensors, are viable choices.

From the above discussion, it should be evident that no single sensor is able to perform well under all possible scenarios. Each sensor has its strengths and weaknesses. A synergistic application of different sensors is crucial to the successful operation of a mobile robot. The data bandwidth, reliability, accuracy, power consumption, spatial resolution, and sensing range of each type of sensor has to be carefully studied. Below, we briefly discuss the strengths and limitations of several sensors commonly used in mobile robotics applications.

Vehicle Attitude and Navigation Sensors   These sensors include vehicle speed sensors (shaft encoder, odometer), Doppler sensors, gyrocompasses, dead reckoning sensors and satellite navigation systems. The shaft encoder provides accurate velocity readings at relatively low speeds and over terrain where a vehicle has good traction. The Doppler sensor is accurate at relatively high speeds but is insensitive to wheel slippage. The vehicle position can be tracked by dead reckoning or by a satellite navigation system. As pointed out in [38], dead reckoning provides a continuous position update; however, the update drifts because of the accumulating errors. A satellite navigation system provides accurate absolute position measurements without a drift, but the measurements may not be continuously available. Further, these sensors are used to establish the attitude and velocity of a vehicle, but do not provide information on the surrounding environment.

Visual Sensors   Visual sensors have been widely used in many mobile robot projects. Visual sensors provide data with a high spatial resolution at a fraction of the cost of more expensive laser rangefinders. The major drawback of visual sensors is that they do not report 3-D information directly. Intensity images register a sensor's responses to scene radiation, and a variety of factors may affect this image formation process: the characteristics of the light source; medium attenuation, scattering and diffusion; surface reflectivity and orientation; and imaging geometry. When images are analyzed to infer surface orientation and structure, the interaction of all these factors has to be separated and individual effects isolated. The major difficulty here is that such analyses are not always possible with a single image frame without the accurate modeling of the light source and the sensing configuration and without reliable knowledge of the surface reflectivity of the imaged object. Further, the accurate detection, extraction, and matching of image features are not always automatic in a stereo analysis. The performance of the analysis algorithm and the accuracy of the analysis may degrade significantly due to errors in preprocessing, feature extraction, and matching. To speed up processing, resolution pyramid [71] and focusing [79] techniques might be used to reduce the amount of visual data.

Proximity Sensors   Proximity sensors can be designed in many different ways such as using light emitting diodes (LEDs) or sonars. In [21], a proximity sensor was designed using infrared LEDs. Reflected light intensity is used to determine the distance from the sensed object. Depending on the LED used, such sensors have a limited spatial range (up to 10 ft [31]).

Ultrasonic rangefinders bounce sound waves off objects and use the time-of-flight principle to estimate range. Ultrasonic waves are specular. For example, if the sound wave is directed at a shallow angle to a flat surface (at more than 40° to the normal of the surface), the acoustic wave has a tendency to bounce off mostly concentrated in the direction where the angle of incidence and the angle of reflection are coplanar and equal [47]. Consequently, very little energy is reflected directly back to the detector from this surface, and the estimated range values can be unreliable. Furthermore, the spatial resolution of sonars is usually quite poor. As pointed out in [47], the 30° solid angle of the main lobe of the transducer beam does not allow better than about a 4×4 spatial resolution of a 90° solid angle field. Special acoustic focusing devices, such as a sound absorbing plastic foam tube, are effective in narrowing the directionality to about a 10° solid angle. But this gives at best a 10×10 resolution over a 90° solid angle field—an improvement of about three folds. The major advantages of ultrasonic sensors are their low price, small power consumption, and narrow data bandwidth. Hence, ultrasonic rangefinders are mostly used to provide a rough estimate of range values at sparse locations, such as in avoiding obstacles for mobile robot guidance. The rangfinders are not very suitable for applications demanding high precision and high resolution.

Laser Range Sensors   Rangefinders based on laser technology give much better spatial and range resolutions. Commercial laser ranging systems have used amplitude modulation (such as the ERIM sensor [83] [67]) and triangulation principle (such as the White Scanner of the Technical Arts Corp. [12]) in data acquisition. To construct a dense depth map, scanning is required. Scanning can be a time consuming process. During the scan, objects in the scene have to remain stationary to avoid difficult registration problems. Sophisticated mechanical design is the key to attaining high precision in the sensory data. This is part of the reason why laser ranging devices are much more expensive than CCD-type visual cameras. Furthermore, with moving parts, systems may not be immune to mechanical wear, vibration, and shock in the working environments.

Structured Light Sensors   The structured light (or grid coding) method is a relatively inexpensive way to obtain sparse 3-D surface information [3], [77]. The structured light projection technique replaces one of the stereo cameras with a projection device (a $35mm$ slide projector). Patterns (parallel lines or a grid) are projected to encode the surface for analysis. Since a complete pattern is projected, no time consuming scanning operation is needed and no movable mechanical part is used. However, to discern projected patterns, the spacing between adjacent patterns must be large. Hence, the structured light projection only provides a sparse depth or orientation map. To facilitate the extraction of the perceived pattern from the images, the projection device should operate with a suitable light source to avoid interference from ambient light.

In the following, we survey the sensor data fusion techniques of a few mobile robotics research projects.

Carnegie-Mellon University   The mobile robot projects at CMU include early, small Terregator [33], [34], large Navlab [33], [34], [48], [71], and Ambler [43] design. Both the Navlab and Terregator are equipped with a color camera and an ERIM laser rangefinder. The Ambler is a multi-legged vehicle designed for an exploratory mission on another planet and is still in an experimental stage.

[71] presents the vision and navigation mechanism of the Navlab. The primary function of the color camera is for road following, while the laser rangefinder is used for obstacle detection and detail terrain analysis. In road following, the CMU group abandons the edge tracking analysis, which is believed to be more susceptible to noise and to require an accurate model of the road shape. Instead, an analysis is performed using both color and texture information to classify pixels into road and nonroad classes. In the color analysis, the classifier uses four road and nonroad classes to account for the variation of color due to the change of season, weather, time of day, and place. For each road/nonroad class, the mean m_i and covariance matrix Σ_i of red, green, and blue values are recorded. The confidence that a pixel of color X belongs to a class i is computed using:

$$(X - m_i)^T \Sigma_i^{-1} (X - m_i).$$

Each pixel is classified with the class of the highest probability.

The texture analysis reveals the smoothness of a region. Smoothness is defined as the lack of edge pixels in a small neighborhood. It is generally expected that paved roads appear smoother in an image than nonroad regions of grass, soil, or tree trunks. Hence, if the number of edge pixels in a region is above a certain threshold (35 pixels in an 8×8 neighborhood), that block is classified as nonroad. The classification results from the color (P_i^C) and from the texture (P_i^T) are combined by a weighted average:

$$P_i = (1 - \alpha)P_i^T + \alpha P_i^C.$$

To find the most likely location of the road, a Hough transform is performed to cluster candidate road locations in a parameter space. The parameter space is two dimensional and is spanned by an intercept (P), which is the image column of the road's vanishing point and the orientation (θ) of the road in the image. Each road pixel casts votes for all (P, θ)

combinations to which it belongs, while nonroad points cast negative votes. The (P, θ) pair that receives the most votes is reported as the road.

Color vision alone is not sufficient for navigating a mobile robot. Laser range images provide information on the location of obstacles and terrain type. Obstacles are identified as regions in a range image that do not have vertical surface normals. Terrain analysis attempts to extract edges along range discontinuities, to cluster points of similar surface normal directions into regions, and to expand each region by fitting a planar or quadric surface patch until the fitting error is larger than a preset threshold.

To coordinate various components in the mobile platform, a blackboard architecture is used. A central database, called the local map and managed by a local map builder, is used to store information. Various modules (color vision, range sensor, pilot, etc.) communicate with one another through the blackboard by storing and retrieving information asynchronously. Interesting readers are referred to [71] for detail.

The University of Maryland The mobile robot project at the University of Maryland [8], [25], [79] used a 96:1 scaled wooden terrain board with a painted road network as the simulation platform. A small solid-state CCD black and white camera carried by a robot arm is used in place of a vehicle. Three sensors (spring-mounted variable resistors) are mounted under the camera. As the camera moves, these sensors are dragged over the terrain board. The camera position and orientation are adjusted to maintain a fixed height and tilt from the board.

To speed up the visual processing, the system operates in two different modes: bootstrap and feedforward. In the bootstrap mood, an entire image is processed to locate objects of interest such as roads and landmarks. In the feedforward mood, the system switches to a prediction-verification process in which only a small part of the visual image is processed. The processing is limited to regions that correspond to a predicted object location from the previous object location and vehicle movement.

The whole navigation task is broken into three levels: long-range, intermediate-range, and short-range navigation. Long-range navigation uses given terrain traversability data and landmark visibility data to establish a sequence of regions, from the starting point to the destination, through which the vehicle must move. Intermediate-range navigation selects a corridor of free space for traversal in each intervening region, from the starting point to the destination. Visual sensing is needed to locate such corridors and to establish a vehicle's heading. Short-range navigation is responsible for selecting the actual path to be traversed through the established corridor of free space, for updating the vehicle's position in a local map, and for modifying the contents of the local map based on visual sensor data. Visual sensors are used heavily in both short and intermediate range navigations.

The vision module is responsible for perceiving objects of interest (e.g., roads and land-marks) and representing them in an object-oriented reference frame. The vision module supports low-level image processing, intermediate-level geometry transformations, and high-level knowledge-based recognition. Low-level image processing extracts features (line segments) from the image plane and groups them into significant objects. The 3-D shape of these grouped objects is recovered by the intermediate-level geometry module and then converted from a view-centered coordinate system into an object-centered representation.

In road following, the vision module accepts instruction from the planner to locate the road. A bootstrap stage is initiated to acquire an image frame and to process the whole image frame to locate line features. These line features are grouped into pencils of lines and mapped back into space in the form of parallel lines on a planar surface patch. A knowledge data base stores the pencils of lines and determines whether the grouping and the planar surface descriptions make sense. If so, they are labeled as boundaries, center lines, lane markers, or shoulder boundaries. The vehicle is positioned relative to the lines and information is passed to the navigator. After the vehicle is in motion, small windows in the visual field are selected based on the vehicle heading and previous road boundary locations for processing.

Asada [8] discusses a method for building a 3-D world model using both range and visual sensors. The sensor was designed based on the structured light principle. Planes of light are projected from a rotating mirror, and triangulation principle is used to recover the depth of illuminated points. A range image is transformed into a height map in the vehicle centered coordinate system based on the known height and tilt of the range finder relative to the vehicle. The height map is then segmented into four types of regions: unexplored, occluded, traversable, and obstacle. Unexplored regions are those outside the visual field of the range finder. If range data are not available for a region due to occlusion or inadequate lighting, that region is classified as occluded. Traversable regions are groupings of points close to the ground plane in front of the vehicle.

Obstacle regions are the remaining regions of which the boundary has a high slope, a high curvature, or both. Using information from both visual and range images, these regions are further classified into natural objects and man-made objects. Intuitively speaking, man-made objects have clearly defined edges and surfaces; hence, they possess a relatively uniform slope and curvature in a range image and appear homogeneous in an intensity image. Natural objects such as trees and bushes have fine structures and show a large variation in intensity, orientation, and curvature measurements. Finally, the height maps created at different instants are integrated into a global map by matching the traversable regions between adjacent height maps.

Other Outdoor Mobile Robotics Projects   The mobile robot project at Martin Marietta [27], [74], [75] used both color video and range imagery for navigation. Similar to [71], color video images are used to find road boundaries, and range images are used to locate obstacles within the road's boundary. The mobile robot project carried out by the Hughes Artificial Intelligence Center [23], [24] used the same vehicle as that at the Martin Marietta facility, and hence, used the same sets of sensors. The autonomous vehicle developed at the FMC Corp. [51], [52] used both ultrasonic range sensors for obstacle avoidance and a color camera for road following.

Indoor Mobile Robotics Projects   In [58], stereo vision and sonars were used together to construct a 2-D world description in an indoor environment. A 2-D cellular representation called the Occupancy Grid is used to integrate data from these two sensing modules. Each cell in the Occupancy Grid contains a probabilistic estimate of whether it is empty or occupied by an object in the environment. These estimates are obtained from sensor models that describe the uncertainty in the range data.

For ultrasonic sensor data, cells in a volume in front of the sensor are classified into two classes: probably empty and somewhere occupied. The probability function of whether a cell is empty or occupied is defined by the range distance and the beam width. Basically, the range profile creates a wedge with the sensor at the apex. The boundary of the wedge is determined by the size of the main transmission lobe, and the range profile forms the base. The interior of the wedge (the cells along the ultrasonic wave path) is inferred to be empty. An area around the base is inferred to be occupied. The size and shape of the occupied area are determined by the ultrasonic sensor model. The space outside the wedge is considered to be unknown.

In the stereo analysis, the stereo cameras are separated by a horizontal base line and positioned at about the same height as the ultrasonic range finder. The stereo system processes a narrow band of scanlines at the height of the ultrasonic sensor to produce a 2-D world model commensurate with the sonar. Along these scanlines, near-vertical edges are extracted and matched to localize surface boundaries and markings in space. Similar to the ultrasonic range profile, a wedge shape depth profile is generated. A stereo error model is used to decide the probability of empty (inside the wedge), occupied (around the base), and unknown (outside the wedge).

Depth maps from the two sensors are integrated using a Bayesian model, taking into consideration the probabilistic empty/occupied measures and the uncertainty in such measures derived from the sensor models. The authors point out that the totally different nature of the raw data from the two sensors precludes any simple analytic or geometric approach to their integration. The occupancy Grid provides a means to make the data commensurate and comparable.

In [46], four sensing modalities—vision, acoustics, tactile, and odometry—were used to guide an indoor robot. Stereo vision is used to compute the 3-D location of vertical lines within the field of view of the stereo cameras. The acoustic systems use many ultrasonic rangefinders for obstacle avoidance. As the last line of defense, a tactile sensing system, composed of bumpers with internal tape switches, alerts the robot of physical contacts which fail to be detected by the ultrasonic and vision sensors. Other indoor robotics applications have also used ultrasonic sensors [14], [22], vision sensors [56], [81], and proximity sensors [21].

3   Thermal & Visual Sensors

Traditionally, thermal images have been used in many military applications for target detection and in civilian applications for home energy audits, fault detection in components, and medical diagnoses. Most existing techniques for analyzing thermal images rely on statistical pattern recognition techniques with features that are computed directly from image pixel values [19] [49] [68]. The techniques are simple and straightforward; however, the performance is sensitive to scene radiance, occlusion, and viewing direction. It is desirable to establish a correct mathematical model of thermal sensing.

In [60], [61], a new technique was proposed to integrate information from both thermal (infrared) and visual images. A model for thermal sensing in outdoor environments under bright sunlight was developed, and extensive calibration and experiments were conducted to verify the correctness of the model.

Contrary to wide belief, a thermal camera does not sense and report temperature directly. Instead, a thermal camera senses the irradiation of the environment in the infrared spectrum (for example from $8\mu m$ to $12\mu m$). The registered irradiation is then used to compute the temperature of the imaged radiating source using Planck's equation:

$$\int_{\lambda_1}^{\lambda_2} \frac{C_1}{\lambda^5(e^{C_2/\lambda T} - 1)} d\lambda = K_a L_t + K_b,$$

where L_t is the thermal camera voltage output directly proportional to the net radiation absorbed by the detector. K_a and K_b are calibrated by imaging at least two objects of known thermal properties (i.e., the emissivity and temperature).

As pointed out in [60], [61], temperature alone is not a good indicator for the region classification (under the sun, everything feels hot.) Instead, the authors use the ratio between the heat fluxes conducted from an imaged surface to the interior of the object to the total absorbed heat fluxes (W_{cd}/W_{abs}) for classification. Concrete walls, brick walls, pavements, cars, and vegetation all have quite distinct W_{cd}/W_{abs} ratios (Tables II, III, and IV in [61]).

The authors observe that under thermal equilibrium, the total heat fluxes absorbed by a surface (W_{abs}) equal those lost to the environment. Heat fluxes might leave a surface in three possible ways: through radiation, convection, and conduction, or

$$W_{abs} = W_{cd} + W_{cv} + W_{rad}.$$

Radiation and convection heat transfer depend on the difference between the surface temperature and the temperature of the environment, as well as on the emissivity and the convection heat transfer coefficient of the imaged surface, which can be estimated empirically. The temperature of an imaged surface is reported by the thermal sensor and used to compute W_{cv} and W_{rad}. Hence, if W_{abs} can be estimated, W_{cd} and W_{cd}/W_{abs} can be computed from the above equation to classify imaged surfaces.

The total absorbed heat fluxes W_{abs} are determined by the solar absorptivity of the surface (α_s) and by the orientation of the surface toward the sun (θ_i). Information on α_s and θ_i is not readily available from a thermal image. However, many shape-from techniques have been developed in computer vision research to compute the surface orientation from visual images [4], [10], [15]. In [60], [61], the authors use a simple shape-from-shading technique to estimate θ_i. Assuming that the viewed surface is opaque and Lambertian, a simple reflection model is constructed that relates the perceived brightness to the incident angle θ_i and to the reflectivity of the surface ρ ($= 1 - \alpha_s$):

$$L_v = K_\rho cos\theta_i + C_v = \rho K_v cos\theta_i + C_v,$$

where L_v is the digitized value of the intensity of the visual image, and K_v and C_v are the overall gain and offset of the visual imaging system which are given through calibration. For each image region, θ_i of one pixel is computed from either stereopsis or other shape-from techniques. This angle is used to compute the solar absorptivity of the whole surface. Once the solar absorptivity of the surface is known, the orientation of the remaining pixels can be estimated. Nandhakumar and Aggarwal [60], [61] report experimental results using the W_{cd}/W_{abs} ratio to classify outdoor scenes with cars, pavement, buildings, and vegetation.

4 Tactile & Visual Sensors

Tactile sensing is a relatively new sensing mechanism which obtains surface measurements by making physical contact. Tactile feedback is important in object grasping and manipulation. This feedback can be used to adjust grasping forces to prevent slippage. Tactile sensors can also be used to locate the corners and edges of an object, which are usually feasible grasping locations.

4.1   Design of Tactile Sensors

Contact force induces a change in material properties that affects sensor readings. For example, the resistance of piezoresistive material, such as the thick-film polymer used in [72], varies as a function of the mechanical load. The thick-film polymer has been used to construct a force sensing resistor (FSR) in a voltage divider circuit [72]. The output voltage (V_{out}) of the divider circuitry is related to the applied force (F) by a simple formula:

$$V_{out} = \frac{V_{EE}\ R_{FSR}}{R_{FSR} + R_{fixed}} = \frac{V_{EE}\ kF^{-1}}{kF^{-1} + R_{fixed}},$$

where V_{EE} is the applied voltage to the voltage divider circuitry, and R_{FSR} and R_{fixed} are the resistance (in ohms) of the force sensing resistor and the fixed, by-pass resistor, respectively. For typical piezoresistive material, k is approximately $8.98 \times 10^6\ \Omega \cdot lb$.

A prototype sensor was constructed using force sensing resistors in VLSI fabrication. The sensing area is $0.5in \times 0.5in$ with a 16×16 (or $0.031in$) spatial resolution. The overall package measures $0.8in \times 0.8in \times 0.25in$, which is small enough to be located on a fingertip. The author reports that the sensor is very sensitive at low loads yet is still useful when very large loads are encountered.

In [44], a high resolution tactile sensor was designed. The sensor uses two layers of conductive material to form a cross-bar architecture with pressure sensors built at junctions. One of the layers is made of anisotropically conductive silicone rubber (ACS), which conducts only along one axis in the plane of sheet. The other layer is a printed circuit board etched to conduct perpendicular to the ACS layer. A separator is inserted in between these two layers to pull the conducting layers apart when pressure is released. When pressure is applied, the conductive rubber presses through the separator. The area of contact and the contact resistance vary with the applied pressure. The sensor array can be scanned by applying a voltage to one column at a time. The output voltage of a divider circuitry is used to estimate the contact resistance and the contact force.

In [82], the authors report an experimental tactile sensor design which uses optical fibers. Optical fibers are a type of dielectric waveguides. These waveguides channel light energy by "trapping" it between layers of dielectric materials. In an ideal straight fiber, light propagates in a finite number of modes (ray paths) with virtually no energy loss. If force is applied to bend a fiber, such modes are lost in the bent section. A light ray trapped in the bent fiber must radiate part of its energy. How much energy is lost depends on fiber parameters as well as on the radius of curvature and the spatial frequency of the bend. Such energy loss is detected by a sensing circuitry placed at the end of the optic fiber.

In [82], fibers of graded index, multimodes, and profiles which are approximately parabolic were used because they are much more sensitive to microbending. Experiments were conducted to study the relationship between bending and energy lost. The authors report that for deflections up to $0.102mm$ ($0.004in$), the sensor output exhibits unrepeatable, nonlinear behavior. For larger deflections, it is observed that light attenuation is proportional to the bending in a linear manner.

Other principles in designing tactile sensors include VLSI fabrication [65], capacitive sensing [29], and strain-stress tensor analysis [30]. A tactile sensor using a pressure sensitive VLSI design similar to that in [44] was reported in [65]. In [65], sensors are placed under windows in the overglass (an insulating layer of SiO_2 normally placed on integrated circuits for protection). Simple computing elements are placed in the VLSI chip, where no windows are present, to allow parallel filtering and convolution with programmable masks. In [29], the voltage across variable-distance capacitor plates are related to the applied force, which alters the spacing between plates. Fearing and Hollerbarch [30] conducted an analysis using solid mechanics models and contact theory for tactile sensing.

4.2   Sensor Data Integration

Allen [5] and Allen and Bajsy [6] report a method for combining stereo vision with touch sensing. Visual sensing is used to construct low and medium level descriptions to guide further tactile exploration. The authors point out that stereo vision cannot handle many candidate match points, is unable to match horizontally oriented zero-crossings, and is subject to image quantization error. Hence, the best one can expect from stereopsis is a sparse set of matched points on the contours of regions isolated from visual sensing. It is suggested that touch sensing can be used to collect data on objects where no stereo analysis is performed or where views are occluded in visual images. The key issue is that blind groping on a surface with a tactile sensor is a poor and inefficient way to understand a 3-D structure. Touch needs to be guided and visual data provide guidance to an active tactile sensor.

The same authors [5] [6] analyzed visual images by stereo matching using zero-crossings as features. Epi-polar constraint is used to limit the search for matching zero-crossing pairs. To eliminate multiple matches, the sign and orientation of matched zero crossings need to be consistent. Furthermore, local neighborhoods around matched zero crossings need to show a good pattern correlation. The experimental objects are common kitchen items such as mugs, plates, bowls, pitchers, and utensils homogeneous in color with no discernible textures. The lack of surface detail on the test object limits the number of features that can be used in the stereo matching process. Hence, stereo vision provides only sparse depth measurements of points along contours.

To build a more accurate surface description, a multi-level Coon's patch description is constructed. To construct a Coon's patch, positions, tangents in each of the parametric directions, and the twist vectors (cross derivatives) of the surface patch at knot points need to be specified. A stereo depth map provides information to construct a level 0 description using a 2 by 2 rectangular knot set located on the surface boundary. To refine the description, more knot points are needed. Positions, tangents, and twist vectors at new knot points are obtained by a tactile sensor. The tactile sensor traces in the direction of the midpoints of the level 0 boundary curves using the surface contacts from the tactile sensor to control the robot arm motion. The level 1 surface is then formed by adding tactile traces onto the Level 0 surface constructed from vision. The level 2 surface is formed by adding tactile traces to each of the four patches defined by the level 1 surface. Hence, a hierarchical surface description is constructed using this successive refinement process.

Stansfield [70] proposes another method for combining vision and touch. A vision-guide-touch strategy similar to those in [5], [6] was studied. A stereo analysis on edge pixels along contours and simple region segmentation based on thresholding and grouping are first performed. A tactile sensor is used to further exploit the properties of the edges and regions detected in the visual images.

The tactile sensor detects several simple haptic primitives. The author suggests seven haptic primitives for touch sensing which include elasticity, compliance, roughness, the normal-contact position of a surface, and three types of contact—point, edge, and extended. For example, the compliance of a surface can be measured by first bringing the tactile sensor into contact with the surface, next applying successively greater forces, and finally recording the position of the sensed surface under different external pressures. The greater the deformation, the higher the compliance measurement. Elasticity can be measured by pushing the surface at the contact point, releasing the pressure, and observing whether the surface bounces back to its original shape. The normal-contact position is established by groping the surface from many different directions and measuring the contact area. Because the sensor has a planar shape, when the maximum area is attained, the sensor is approximately tangent to the surface, and the surface normal direction is estimated. Contacts are classified into extended, point, and edge contacts, depending on their shape and size. Simple routines (called specialists) are implemented for these haptic primitives, which form the lowest level of the tactile perceptual system.

Several knowledge based modules are designed which employ the specialists to determine surface shape, part (small shape), contour, edge, and corner. For example, the edge expert does simple active edge following. It is triggered by an edge contact reported by the low-level specialists. Once triggered, the edge expert is primarily data driven by keeping contact with the object edge and by determining whether it has encountered more edge contacts. The contour expert is triggered by the detection of an edge contact. It invokes the edge expert to follow the edge until an edge termination condition is met. The contour expert may then signal a closed contour or invoke a corner expert for corner detection.

An object is divided into components. Each component is further divided into a set of features (surface patch, contour, and part). Visual sensing provides a rough segmentation and localization of features. The active stage explores each of the visible components from several different views (top, front, left, and right). Various haptic specialists and knowledge based modules are used to measure the material properties and the shape of the contact surface. Information is stored in a set of feature frames generated for the object. Sensor fusion occurs at the beginning of the exploration, when visual data are used to determine how active touch exploration is to proceed, and at the intermediate levels, when visual information is used for the motor guidance of the arm/sensor system during its active feature extraction.

5 Thermal & Tactile Sensors

Russel [66] introduces a method to provide tactile feedback using a sensed material's thermal property. This scheme emulates human touch sense. Basically, the blood supply heats our fingers above room temperature. Contact with an object causes heat to be conducted from the finger to the object at a rate depending on the material property of the object. The skin's resulting temperature drop provides information on the material's thermal properties.

The sensor was constructed with a source of heat (a power transistor), a layer of elastic material of known thermal conductivity to couple the heat source to the touched object (silicone rubber), and a temperature transducer to measure the contact-point temperature. The author reports successful qualitative discrimination between aluminum, wax, cork, and polystyrene foam. As expected, aluminum feels cold because of a large temperature drop through heat conduction. Paraffin wax, cork, and plastic foam feel progressively warmer due to a small heat conduction.

6 Structured Light & Visual Sensing

An algorithm [76] [78] was introduced for inferring the surface structure of 3-D objects from multiple viewing directions using the integration of structured light and visual sensing. Construction of the surface description of a 3-D object is decomposed into two stages: (1) The surface orientation and partial structure are first inferred from a set of single views using structured light sensing, and (2) the surface structures inferred from different viewpoints are integrated to complete the 3-D description using both the structured light and visual sensing.

In the first stage, an active structured light technique is used to recover visible surface orientation and partial structure. The structured light technique projects a spatially modulated pattern (e.g., a rectangular grid pattern) to encode object surfaces for analysis. When a regular grid pattern is projected onto an object surface, the perceived pattern shows many forms of distortion. Intuitively speaking, the way the pattern is distorted is affected by the imaged surface structure. A constraint satisfaction algorithm is developed which uses the orientation of the perceived pattern to infer the visible surface orientation using the following equations:

$$A sin \Psi_o + B sin \Theta_o cos \Psi_o = 0$$

$$C cos \Theta_o cos \Psi_o + D sin \Theta_o cos \Psi_o + E sin \Psi_o = 0,$$

where Θ_o and Ψ_o are the pan and elevation angles of the surface normal vector. A, B, C, D, and E are functions of (ρ_1, ρ_2), which are the image orientations of the two stripes in the projected grid pattern. The visible surface structure is recovered by integrating the orientation map inferred from the structured light.

To integrate surface structure from multiple viewpoints, two sources of information are used, namely the occluding contours extracted from the visual images and the surface structure inferred from the structured light. Occluding contours or silhouettes can be readily obtained by a simple thresholding operation on intensity images. They present an estimate of the projective range of the surface structure. The bounding volume description of the object is constructed by intersecting the occluding contours from multiple viewing directions in space. The bounding volume description presents a first approximation to the true object surface structure.

As mentioned before, a structured light technique is used to infer detailed surface structure from multiple viewpoints, and the inferred structure presents strong surface constraints. This structure is used to further refine the 3-D surface description. To position such a partial structure with respect to the bounding volume, its position along the line of sight have to be determined. It can be shown that unique positions exist to place a structure without violating the geometrical constraints imposed by the occluding contours and the bounding volume. Intuitively speaking, a partial surface structure cannot be positioned outside the bounding volume. Otherwise, its projection along a certain viewing direction will be outside the region delineated by the silhouette of the object in that particular direction. Further, when positioning a partial surface structure against the bounding volume, extremal points exit on the structure that must make contact with the bounding volume. Such contact relations uniquely determine the position of a partial structure with respect to the bounding volume.

7   General Sensor Data Fusion Schemes

Grimson and Lozano-Perez [36] present a technique for locating and recognizing partially occluded objects using data gathered from a variety of sensing modules including sonar, laser, tactile, and visual sensors. It is assumed that the object to be recognized can be modeled as a set of planar faces. Only sparse surface position and orientation measurements are available, and these measurements can be from many different sensors.

The essence of the recognition process is to construct a tree of interpretations. Each level of the tree corresponds to a measured planar patch from sensor data. Each node has the same number of fan-outs as the number of faces in the model. Each leaf node is an interpretation which consists of pairing every sensed patch with a model surface. If there are m sensed patches and n faces in the model, then a tree of interpretations has m levels, with n^m nodes at the leaf level. To efficiently find the correct interpretation from the n^m possibilities, it is important to prune infeasible interpretations as soon as they are identified.

Pruning is achieved by testing the consistency of various inner product measurements which are coordinate independent. For example, suppose two sensed patches P_i and P_j are matched with two model faces P_k and P_l; then, the inner product of the two sensed surface normals, i.e., $n_i \cdot n_j$, should be similar to the inner product of the two model surface normals, i.e., $n_k \cdot n_l$. If the two products do not agree, then no interpretation that assigns those patches to these model faces needs to be considered. This corresponds to pruning the entire subtree below the considered node.

If occlusion is allowed, extraneous points and surface patches that may not correspond to any model points and faces might be presented in the sensor data. One more branch that represents the possibility of discarding the sensed patch, called the null face, is included. Further, when traversing the tree of interpretations, the length of the current longest match is kept. At any node in the tree, a null face interpretation is allowed only if the sum of the number of non-null faces in the partial match and the number of remaining data points is at least as large as that of the longest match. In this way, a path which contains many null (but still consistent) interpretations will not be futilely explored.

Ayache and Faugeras [9] present a method for fusing multiple 3-D descriptions from stereo data taken at different positions. Sensor data are gathered by a pair of stereo cameras positioned on a mobile platform. Stereo data are first analyzed to build a 3-D local map. Local maps constructed at different locations are registered to provide a better estimate of the movement of the mobile platform between various viewpoint positions. Finally, the estimated movements of the mobile robot are used to further reduce the positional uncertainty of the geometric primitives found in local maps. The extended Kalman filtering is suggested for building, registering, and updating visual maps.

As pointed out in [9], the tasks of building, registering, and updating visual maps can be put into a common abstract framework that optimizes a certain parameter based on a certain observation. For example, in a stereo reconstruction process, the observation is the location of a feature in the pair of stereo images, and the parameter to be estimated is the 3-D position of the observed feature. In fusing local 3-D maps from different locations, the observation is the 3-D location of a feature from different viewpoints, and the parameters are the translation and rotation between viewpoints. Finally, to update the description of the geometry and the uncertainty of a particular feature visible from a certain viewpoint F_i (a parameter), the feature position from another viewpoint, say F_j, and the translation and rotation between F_i and F_j are used as the observations.

In all these tasks, the observation ($\mathbf{x}$) is corrupted by noise, which is modeled as an additive zero mean Gaussian noise. Given a number of observations $\mathbf{x}_i$, the task is to best estimate the parameter vector $\mathbf{a}$, given that $\mathbf{x}_i$ and $\mathbf{a}$ are related by

$$\mathbf{f}_i(\mathbf{x}_i, \mathbf{a}) = \mathbf{0}.$$

The authors suggest Kalman filtering, which is an optimal linear recursive estimator, for the tasks. The Kalman filter propagates a conditional probability density function $f_{\mathbf{a}|\mathbf{x}_1,\mathbf{x}_2,\cdots\mathbf{x}_i}$ by incorporating observations $\mathbf{x}_i$ recursively without recomputing a global optimization at each iteration. Under the assumptions that the system is linear, the noise in the observations is white (noise values are not correlated) and Gaussian, then it can be proven that all reasonable choices for an "optimal" estimate (mean, mode, median, etc.) coincide. And the Kalman filter can be shown to be the best filter of any conceivable form. Hence, the Kalman filter provides an optimal way to fuse multiple sensor data.

By linearizing the relation $\mathbf{f}_i(\mathbf{x}_i, \mathbf{a}) = \mathbf{0}$, we have

$$\mathbf{y}_i = \mathbf{M}_i\mathbf{a} + \mathbf{u}_i,$$

where $\mathbf{y}_i$ involves observation $\mathbf{x}_i$ and $\mathbf{u}_i$ represents a noise term. Starting with an initial estimate, $\hat{\mathbf{a}}_0$ of $\mathbf{a}$, and its associated covariance matrix, $\mathbf{S}_0 = E((\hat{\mathbf{a}}_0 - \mathbf{a})(\hat{\mathbf{a}}_0 - \mathbf{a})^t)$, Kalman filtering provides a mechanism to deduce an estimate $\hat{\mathbf{a}}_n$ of $\mathbf{a}$ and its covariance matrix $\mathbf{S}_n = E((\hat{\mathbf{a}}_n - \mathbf{a})(\hat{\mathbf{a}}_n - \mathbf{a})^t)$ by taking into account n observations recursively in a statistically optimal way. Details can be found in [9].

Henderson $et\ al.$ [41] propose a sensor data integration system. A multisensor kernel system is configured by defining the low-level and high-level representations and the logical sensor specification. At the low level processing, the system coordinates the operations of several sensors such as turning a sensor on and off, aiming and focusing a camera, etc. To integrate data from various sensors, it is assumed that the raw data from a sensor are in the form of two pieces of information, namely, a feature (point, line, surface, etc.) and a location of that feature. A 3-D structure called the spatial proximity graph is used as the low-level data organizational tool. The nodes of the spatial proximity graph correspond to the positions in space of the features extracted from raw sensory data. Nodes are linked by an edge if they are within some prespecified distance in space.

The high-level object model is assumed to be either a feature model or a structure model. Feature models involve mapping sensed data into a single number or into a vector that represents the data. For example, many industrial vision systems model objects in terms of the surface area, the number of corners, the number of holes, the size of the holes, etc. Structural models provide a description of the components of an object and the relations between the components.

To bridge the gap between the sensor data represented in a proximity graph and 3-D objects represented in either a feature or a structure model, the concept of logical sensor is introduced. Physical sensors are defined by parameters associated with each individual sensor (e.g., spatial resolution and the aspect ratio of a CCD camera). A logical sensor, in contrast, is defined in terms of physical devices and algorithms on the data. For example, an edge image sensor is defined by a physical sensor which acquires images and by an algorithm which extracts edges from the image. The notion of a logical sensor thus allows a flexible hardware/software mix in terms of sensors and processing algorithms on sensed data. Logical sensors can be designed to construct a spatial proximity graph from raw sensor data or to map low level sensor data into high level models.

Henderson *et al.* [42] further develop the multi-sensor integration idea and describe a system to model the structure and behavior of multisensor systems. This multisensor knowledge system is used to describe both the parameters and characteristics of the individual components of a multisensor system and to deduce the global properties of the complete system. The system supports the description of various components in a multisensor system, such as sensors, analysis algorithms, processes, actuators, and interconnection schemes. The multisensor knowledge system is also capable of simulating and monitoring the behavior of such a system to evaluate the system's performance.

8   Concluding Remarks

In summary, we have presented a brief survey of several techniques for sensor data fusion. It is evident from the discussion that the sensor data fusion capacity will become an indispensable ingredient in many intelligent robotics applications. Such a capacity is crucial in industrial automation.

Many issues are still unresolved or only partially answered in sensor data fusion:

- *Sensor models:* How can the behavior of a sensor be better understood through rigid mathematical modeling? Modeling should be general enough to describe both a class of sensors and a particular instance. For example, a generic range sensor might be modeled as a device which provides depth information. A particular range sensor has a different set of parameters such as operating principle (time of flight, triangulation, amplitude modulation, etc.), spatial resolution, spatial range, and price. The reliability and accuracy of a sensor under a variety of operating conditions also need to be studied.

- *Sensor selection strategies:* What is a suitable strategy to match sensors with the task at hand? What are the sensor parameters needed to make such a decision? What task dependent descriptions need to be specified? If task descriptions change dynamically or sensor failure occurs, how can new sensors be incorporated to suit the changing requirements or to replace failed ones?

- *Data structures:* What is a suitable data structure for sensor data fusion? Different sensors provide different information (position, orientation, temperature, pressure, etc.) on different geometric entities (point, line, surface, etc.). It is a nontrivial task to select a coherent and efficient storage structure and treatment method for the heterogeneous set of data.

- *Data fusion strategies:* As mentioned before, the simple averaging of sensor measurements is prone to distortion by individual noisy measurements, and makes the overall variance of measurements worse. Sensor data have different spatial resolution and accuracy, and measure different properties (e.g., thermal or visual) of an object. A synergistic analysis may be difficult indeed.

- *System issues:* How can sensing activities be best organized? Sensors may operate independently and in parallel. It is also possible to operate one sensor using the guidance of another sensor. Because of the wide difference in the operation of sensors, sensing activities are usually asynchronous. How can communication between sensors be achieved?

We believe more research is needed to provide answers to the above questions and to develop feasible sensor data fusion techniques/systems. It should be mentioned that it is neither our intent nor is it possible to include all research work here in the area of sensor data fusion. Interested readers are directed to the list of references at the end of the paper for additional information.

References

[1] A. S. Acampora and J. H. Winters, "Three-Dimensional Ultrasonic Vision for Robotic Applications", *IEEE Transactions on PAMI*, vol. 11, no. 3, 1989, pp. 291-303.

[2] J. K. Aggarwal and M. J. Magee, "Determining Motion Parameters Using Intensity Guided Range Sensing", *Pattern Recognition,* vol. 19, no. 2, 1986, pp. 169-180.

[3] J. K. Aggarwal and Y. F. Wang, "Inference of Object Surface Structure from Structured Lighting—an Overview", in *Machine Vision Algorithms, Architectures, and Systems,* edited by Herbert Freeman, Academic Press, San Diego, CA, 1988, pp. 193-220.

[4] J. K. Aggarwal and C. H. Chien, "3-D Structures from 2-D Images", in *Advances in Machine Vision*, Springer-Verlag, 1989, pp. 64-121.

[5] P. Allen, "Integrating Vision and Touch for Object Recognition Tasks", *International Journal of Robotics,* vol. 7, no. 6, 1988, pp. 15-33.

[6] P. Allen and R. Bajcsy, "Two Sensor Are Better Than One: Example of Integration of Vision and Touch", *Proceedings of the 3rd International Symposium on Robotics Research*, Gouvieux, France, 1986, pp. 59-64.

[7] P. Allen and P. Michelman, "Acquisition and Interpretation of 3-D Sensor Data from Touch", *Proceedings of the Workshop on Interpretation of 3D Scene*, Austin, TX, Nov. 27-29, 1989, pp. 33-40.

[8] M. Asada, "Building a 3-D World Model for a Mobile Robot from Sensory Data", *Proceedings of the IEEE International Conference on Robotics and Automation*, Philadelphia, PA, Apr. 24-29, vol. 2, 1988, pp. 918-923.

[9] N. Ayache and O. D. Faugeras, "Building, Registering and Fusing Noisy Visual Maps", *International Journal of Robotics Research*, vol. 7, no. 6, Dec. 1988, pp. 45-65.

[10] H. G. Barrow and J. M. Tenenbaum, "Computational Vision", *Proceedings of IEEE*, vol. 69, 1981, pp 572-595.

[11] S. Begej, "Planar and Finger-Shaped Optical Tactile Sensors for Robotic Applications", *IEEE Journal of Robotics and Automation*, vol 4. no. 5, 1988, pp. 472-484.

[12] P. J. Besl, "Active Optical Range Imaging Sensors".

[13] P. Besl and R. Jain, "Range Image Understanding", *Proceedings of CVPR*, San Francisco, CA, 1985, pp. 430-449.

[14] J. Borenstein and Y. Koren, "Obstacle Avoidance with Ultrasonic Sensors", *IEEE Journal of Robotics and Automation*, vol 4. no.2, 1988, pp. 213-218.

[15] M. Brady, "Computational Approaches to Image Understanding", *ACM Computing Surveys*, vol. 14, no. 1, 1982, pp. 3-71.

[16] M. Brady, Foreword, Special Issue on Sensor Data Fusion, *International Journal of Robotics Research*, vol. 7, no. 6, 1988, pp. 2-4.

[17] R. A. Brooks, "A Hardware Retargetable Distributed Layered Architecture for Mobile Robot Control", *Proceedings of the IEEE International Conference on Robotics and Automation*, Raleigh, NC, Mar. 31-Apr. 3, vol. 3, 1987, pp. 106-110.

[18] B. L. Burks, G. de Suassure, C. R. Weisbin, J. P. Jones and W. R. Hamel, "Autonomous Navigation, Exploration, and Recognition Using the HERMIES-IIB Robot", *IEEE Expert*, Winter, 1987, pp. 18-27.

[19] M. Burton and C. Benning, "Comparison of Imaging Infrared Detection Algorithms", *Proceedings of SPIE*, vol. 302, 1981, pp. 26-32.

[20] A. Cameron, R. Daniel and H. Durrant-Whyte, "Touch and Motion", *Proceedings of the IEEE International Conference on Robotics and Automation*, Philadelphia, PA, Apr. 24-29, vol. 2, 1988, pp. 1062-1067.

[21] E. Cheung and V. Lumelsky, "Motion Planning for Robot Arm Manipulators with Proximity Sensing", *Proceedings of the IEEE International Conference on Robotics and Automation*, Philadelphia, PA, Apr. 24-29, vol. 2, 1988, pp. 740-745.

[22] J. L. Crowley, "World Modeling and Position Estimation for a Mobile Robot Using Ultrasonic Ranging", *Proceedings of the IEEE International Conference on Robotics and Automation*, vol. 2, 1989, pp. 674-680.

[23] M. Daily, J. Harris, D. Keirsey, K. Olin, D. Payton, K. Reiser, J. Rosenblatt, D. Tseng and V. Wong, "Autonomous Cross-Country Navigation with the ALV".

[24] M. J. Daily, J. G. Harris and K. Reiser, "An Operational Perception System for Cross-Country Navigation", *Proceedings of the IEEE CVPR Conference*, Ann Arbor, MI, June 5-9, 1988, pp. 794-820.

[25] S. J. Dickinson and L. S. Davis, "An Expert Vision System for Autonomous Land Vehicle Road Following", *Proceedings of the IEEE CVPR Conference*, Ann Arbor, MI, June 5-9, 1988, pp. 826-831.

[26] M. Drumheller, "Mobile Robot Localization Using Sonar", *IEEE Transactions on PAMI*, vol. 9, no. 2, 1987, pp. 325-332.

[27] R. T. Dunlay, "Obstacle Avoidance Perception Processing for the Autonomous Land Vehicle", *Proceedings of the IEEE International Conference on Robotics and Automation*, Philadelphia, PA, Apr. 24-29, vol. 2, 1988, pp. 912-917.

[28] A. Elfes, "Sonar-Based Real-World Mapping and Navigation", *IEEE Journal of Robotics and Automation*, vol. 3, no. 3, 1987, pp. 249-265.

[29] R. S. Fearing, "Some Experiments with Tactile Sensing during Grasping", *Proceedings of the IEEE International Conference on Robotics and Automation*, Raleigh, NC, Mar. 31-Apr. 3, vol. 3, 1987, pp. 1637-1643.

[30] R. S. Fearing and J. M. Hollerbach, "Basic Solid Mechanics for Tactile Sensing", *International Journal of Robotics Research,* vol. 4, no. 3, 1984, pp. 40-54.

[31] A. M. Flynn, "Combining Sonar and Infrared Sensors for Mobile Robot Navigation", *International Journal of Robotics Research,* vol. 7, no. 6, 1988, pp. 5-14.

[32] D. B. Gennery, T. Litwin, B. Wilcox and B. Bon, "Sensing and Perception Research for Space Telerobotics at JPL", *Proceedings of the IEEE International Conference on Robotics and Automation*, Raleigh, NC, Mar. 31-Apr. 3, vol. 1, 1987, pp. 311-317.

[33] Y. Goto and A. Stentz, "The CMU System for Mobile Robot Navigation", *Proceedings of the IEEE International Conference on Robotics and Automation*, Raleigh, NC, Mar. 31-Apr. 3, vol. 1, 1987, pp. 99-105.

[34] Y. Goto and A. Stentz, "Mobile Robot Navigation: The CMU System", IEEE Expert, Winter, 1987, pp. 44-54.

[35] W. E. L. Grimson, and T. Lozano-Perez, "Model-based Recognition from Sparse Range or Tactile Data", International Journal of Robotics, vol. 3, no. 3, 1984, pp. 3-35.

[36] W. E. L. Grimson, and T. Lozano-Perez, "Search and Sensing Strategies for Recognition and Localization of Two- and Three-Dimensional Objects", Proceedings of the 3rd International Symposium on Robotics Research, Gouvieux, France, 1986, pp. 59-64.

[37] L. D. Harmon, "Automated Tactile Sensing", International Journal of Robotics, vol. 1, no. 2, 1982, pp. 3-32.

[38] S. Y. Harmon, "The Ground Surveillance Robot (GSR): An Autonomous Vehicle Designed to Transit Unknown Terrain", IEEE Journal of Robotics and Automation, vol. 3, no. 3, 1987, pp. 266-279.

[39] G. Hager and M. Mintz, "Task-Directed Multi-Sensor Fusion", Proceedings of the IEEE International Conference on Robotics and Automation, vol. 2, 1989, pp. 662-667.

[40] T. Henderson and E. Shilcrat, "Logical Sensor Systems", Journal of Robotics Systems, vol 1, no. 2, 1984, pp. 169-193.

[41] T. C. Henderson, W. Fi and C. Hansen, "MKS: A Multisensor Kernel System", IEEE Transactions on Systems, Man, and Cybernetics, vol. 14, no. 5, 1984, pp. 784-791.

[42] T. Henderson, E. Wetz, C. Hansen and A. Mitiche, "Multisensor Knowledge systems: Interpreting 3D Structure", International Journal of Robotic Research, vol. 7, no. 6, Dec. 1988, pp. 114-137.

[43] M. Hebert, and C. Caillas, E. Krotkov, I. S. Kewon, T. Kanade, "Terrain Mapping for a Roving Planetary Explorer", Proceedings of the IEEE International Conference on Robotics and Automation, vol. 2, 1989, pp. 997-1002.

[44] W. D. Hillis, "A High Resolution Imaging Touch Sensor", International Journal of Robotics, vol. 1, no. 2, 1982, pp. 33-44.

[45] S. A. Hutchinson, R. L. Cromwell and A. C. Kak, "Planning Sensing Strategies in a Robot Work Cell with Multi-Sensor Capabilities", Proceedings of the IEEE International Conference on Robotics and Automation, Philadelphia, PA, Apr. 24-29, vol. 2, 1988, pp. 1068- 1075.

[46] D. J. Kriegman, E. Triendl and T. O. Binford, "A Mobile Robot: Sensing, Planning and Locomotion", Proceedings of the IEEE International Conference on Robotics and Automation, Raleigh, NC, Mar. 31-Apr. 3, vol. 1, 1987, pp. 402-408.

[47] R. A. Jarvis, "A Perspective on Range Finding Techniques for Computer Vision", IEEE Transactions on PAMI, vol. PAMI-5, No. 2, Mar. 1983, pp. 122-139.

[48] T. Kanade, C. Thorpe, S. Shafer and M. Hebert, "Carnegie Mellon Navilab Vision System".

[49] J. H. Kim, D. W. Payton and K. E. Olin, "An Expert System for Object Recognition in Natural Scenes," *Proceedings of the 1st Conference on AI Applications,*, Denver, CO, 1984, pp. 170-175.

[50] E. Krotkov and R. Kories, "Adaptive Control of Cooperating Sensors: Focus and Stereo Ranging with an Agile Camera System", *Proceedings of the IEEE International Conference on Robotics and Automation*, Philadelphia, PA, Apr. 24-29, vol. 1, 1988, pp. 548-553.

[51] D. Kuan and U. K. Sharma, "Model-Based Geometric Reasoning for Autonomous Road Following", *Proceedings of the IEEE International Conference on Robotics and Automation*, Raleigh, NC, Mar. 31-Apr. 3, vol. 1, 1987, pp. 416-423.

[52] D. Kuan, G. Phipps, A-C Hsueh, "Autonomous Robotic Vehicle Road Following", *IEEE Transaction on PAMI*, vol. 10, no. 5, 1988, pp. 648-658.

[53] J. J. Le Moigne and A. M. Waxman, "Structured Light Pattern for Robot Mobility", *IEEE Journal of Robotics and Automation*, vol. 4, no. 5, 1988, pp. 541-548.

[54] R. C. Luo and M-H Lin, "Robot Multi-Sensor Fusion and Integration: Optimum Estimation of Fused Sensor Data", *Proceedings of the IEEE International Conference on Robotics and Automation*, Philadelphia, PA, Apr. 24-29, vol. 2, 1988, pp. 1076-1081.

[55] R. C. Luo, M-H Lin and R. S. Scherp, "Dynamic Multi-Sensor Data Fusion System for Intelligent Robots", *IEEE Journal of Robotics and Automation*, vol. 4, no. 4, 1988, pp. 386-396.

[56] R. L. Madarasz, L. C. Heiny, R. F. Cromp and N. M. Mazur, "The Design of an Autonomous Vehicle for the Disabled", *IEEE Journal of Robotics and Automation*, vol. 2, no. 3, 1986, pp. 117-125.

[57] B. Maqueira, C. I. Umeagukwu and J. Jarzynski, "Application of Ultrasonic Sensors to Robotic Seam Tracking", *IEEE Journal of Robotics and Automation*, vol. 5, no. 3, 1989, pp. 337-344.

[58] L. Matthies and A. Elfes, "Integration of Sonar and Stereo Range Data Using a Grid-Based Representation", *Proceedings of the IEEE International Conference on Robotics and Automation*, Philadelphia, PA, Apr. 24-29, vol. 2, 1988, pp. 727-733.

[59] W. T. Miller III, "Sensor-Based Control of Robotic Manipulators Using a General Learning Algorithm", *IEEE Journal of Robotics and Automation*, vol. 3, no. 2, 1987, pp. 157-165.

[60] N. Nandhakumar and J.K. Aggarwal, "Multisensor Integration—Experiments in Integrating Thermal and Visual Sensors", *Proceedings of the First International Conference on Computer Vision*, London, England, June 1987, pp. 83 - 92.

[61] N. Nandhakumar and J. K. Aggarwal, "Integrated Analysis of Thermal and Visual Images for Scene Interpretation", *IEEE Transactions on PAMI*, vol. 10, no. 4, 1988, pp. 469-481.

[62] D. Nitzan, "Three-Dimensional Vision Structure for Robot Applications", *IEEE Transactions on PAMI*, vol. 10, no. 3, 1988, pp. 291-309.

[63] C.H. Oh, N. Nandhakumar and J.K. Aggarwal, "Integrated Modelling of Thermal and Visual Image Generation", *Proceedings of the IEEE Computer Vision and Pattern Recognition Conference*, 1989, pp. 356-362.

[64] L. J. Pinson, "Robot Vision: An Evaluation of Imaging Sensors", *Journal of Robotics Systems*, vol 1, no. 3, 1984, pp. 263-314.

[65] M. H. Raibert and J. E. Tanner, "Design and Implementation of a VLSI Tactile Sensing Computer", *International Journal of Robotics*, vol. 1, no. 3, 1982, pp. 3-18.

[66] R. A. Russell, "A Thermal Sensor Array to Provide Tactile Feedback for Robots", *International Journal of Robotics*, vol. 1, no. 3, 1985, pp. 35-40.

[67] R. E. Sampson, "3D Range Sensor via Phase Shift Detection", *IEEE Computer Magazine*, Aug. 1987, pp. 23-24.

[68] L. Sevigny, G. Hvedstrup-Jensen, M. Bohner, E. Ostevold and S. Grinaker, "Discrimination and Classification of Vehicles in Natural Scenes from Thermal Imagery", *Computer Vision, Graphics and image Processing*, vol. 25, 1983, pp. 229-243.

[69] V. Srinivasan and R. Lumia, "A Pseudo-Interferometric Laser Range Finder for Robot Applications", *IEEE Journal of Robotics and Automation*, vol. 5, no. 1, 1989, pp. 98-105.

[70] S. A. Stansfield, "A Robotic Perceptual System Utilizing Passive Vision and Active Touch", *International Journal of Robotics*, vol. 7, no. 6, 1988, pp. 138-161.

[71] C. Thorpe, M. H. Hebert, T. Kanade and S. A. Shafer, "Vision and Navigation for the Carnegie-Mellon Navlab", *IEEE Transactions on PAMI*, vol. 10, no. 3, 1988, pp. 362-373.

[72] B. Tise, "A Compact High Resolution Piezoresistive Digital Tactile Sensor", *Proceedings of the IEEE International Conference on Robotics and Automation*, Philadelphia, PA, Apr. 24-29, vol. 2, 1988, pp. 760-764.

[73] T. Tsujimura and T. Yabuta, "Object Detection by Tactile Sensing Method Employing Force/Torque Information", *IEEE Journal of Robotics and Automation*, vol. 5, no. 4, 1989, pp. 444-450.

[74] M. A. Turk, D. G. Morgenthaler, K. D. Gremban, M. Marra, "Video Road-Following for the Autonomous Land Vehicle", *Proceedings of the IEEE International Conference on Robotics and Automation*, Raleigh, NC, Mar. 31-Apr. 3, vol. 1, 1987, pp. 273-280.

[75] M. A. Turk, D. G. Morgenthaler, K. D. Gremban and M. Marra, "VITS—A Vision System for Autonomous Land Vehicle Navigation", *IEEE Transactions on PAMI*, vol. 10, no. 3, 1988, pp. 342-361.

[76] Y. F. Wang and J. K. Aggarwal, "On Modeling 3-D Objects Using Multiple Sensory Data", *Proceedings of The IEEE International Conference on Robotics and Automation*, Raleigh, North Carolina, May 31 - Apr. 3, vol. 2, 1987, pp. 1098-1103.

[77] Y. F. Wang and J. K. Aggarwal, "Geometric Modeling Using Active Sensing—an Overview", *IEEE Control Systems Magazine*, vol. 3, no. 2, 1988, pp. 7-13.

[78] Y. F. Wang and J. K. Aggarwal, "Integration of Active and Passive Sensing Techniques for Representing Three-Dimensional Objects", *IEEE Journal of Robotics and Automation*, vol. 5, no. 4, 1989, pp. 460-471.

[79] A. M. Waxman, J. J. LeMoigne, L. S. Davis, B. Srinivasan, T. R. Kushner, E. Liang and T. Siddalingaiah, "A Visual Navigation System for Autonomous Land Vehicle", *IEEE Journal of Robotics and Automation*, vol. 3, no. 2, 1987, pp. 124-141.

[80] L. E. Weiss, A. C. Sanderson and C. P. Newman, "Dynamic Sensor-Based Control of Robots with Visual Feedback", *IEEE Journal of Robotics and Automation*, vol. 3, no. 5, 1987, pp. 404-417.

[81] W. M. Wells, III, "Visual Estimation of 3-D Line Segments from Motion—A Mobile Robot Vision System", *IEEE Journal of Robotics and Automation*, vol. 5, no. 6, 1989, pp. 820-825.

[82] J. G. Winger and K-M Lee, "Experimental Investigation of a Tactile Sensor Based on Bending Losses in Fiber Optics", *Proceedings of the IEEE International Conference on Robotics and Automation*, Philadelphia, PA, Apr. 24-29, vol. 2, 1988, pp. 754-759.

[83] D. M. Zuk and M. L. Dell'eva, "Three-dimensional Vision System for the Adaptive Suspension Vehicle", *Final Report No. 170400-3-F*, ERIM, DARPA 4468, Defense Supply Service, Washington, D. C.

INDEX